INFORMATION AND THE INFORMATION PROFESSIONS

Information and the Information Professions

VOLUME I OF THE SELECTED WORKS OF

Marcia J. Bates

PROFESSOR EMERITA, DEPARTMENT OF INFORMATION STUDIES
GRADUATE SCHOOL OF EDUCATION AND INFORMATION STUDIES
UNIVERSITY OF CALIFORNIA, LOS ANGELES

Ketchikan Press
BERKELEY

Text and figures © 2016 by Marcia J. Bates

Edited by Marcia J. Bates

Published by Ketchikan Press, Berkeley, CA

Drawing on page *xvi* © 1971 Charles E. Martin/The New Yorker Collection/ The Cartoon Bank, with permission.

SERIES DESIGN & PRODUCTION
Chris Hall

ISBN 978-0-9817584-1-1

PUBLISHER CATALOGING-IN-PUBLICATION DATA
Names: Bates, Marcia J.
Title: Information and the information professions: volume I of the selected works of Marcia J. Bates / edited by Marcia J. Bates.
Description: Berkeley, CA : Ketchikan Press, 2016. | Series: Selected works of Marcia J. Bates | Includes index.
Identifiers: ISBN 978-0-9817584-1-1
Subjects: LCSH: Information science. | Library science. | Information scientists.
Classification: LCC Z665.B316 2016 v. I | DDC 020 -- dc23

Text set in Calluna 9.5/12.5; heads and captions set in Calluna Sans

To my mentors

Carl Baumann
POMONA COLLEGE

William J. Paisley
STANFORD UNIVERSITY

Contents

Acknowledgments

My heartfelt thanks go to the talented people who worked with me in preparing these volumes: Chris Hall for his book design, illustrations, and typesetting, Nicholas Carroll and Chris Hall for production management, Leonard Rosenbaum for indexing, and Doris Lechner for proofreading.

I also want to thank my parents, especially my mother, for their hard work in putting me through college. My mother made it possible for me to have the education she could only dream of.

—*Marcia J. Bates*

Introduction to the series

The iconic image associated with Silicon Valley's new world of information technology is that of two guys in a garage tinkering with and launching new inventions that revolutionize the world. The founders of Hewlett-Packard, the founders of Apple, and, more recently, the founders of Google all fit the mold. Investors pour millions, nay, billions of dollars into these projects, with subsequent phenomenally great reward, as the products are taken up and used throughout the world. Many of these products involve information storage, transfer, and use. The average person's sense of how information can be found and utilized, sent to friends, or support study has changed drastically.

But there was a prehistory to this story. People have needed and used information since prehistoric times. They even needed it—a lot—just before the Internet exploded in people's consciousness in the 1990's. What did they do then?

It turns out that the several forms of limited information technologies available then—when investment in information retrieval research was a pittance—forced those of us who thought about these questions to become very knowledgeable about the inner workings of information seeking, the expressing of information needs, the methods of describing and indexing information, as well as the acts of searching, retrieving and utilizing information. In the field of information science, we had to understand the process at a granular level, in order to take advantage of the then-available technologies, which were usually vastly more limited than today's capabilities. We were forced to understand the process deeply, in the way that pilots used to develop a bone-deep feel for flying, before the advent of the automatic pilot and the disengagement by pilots from the core processes of keeping a plane in the air.

Many of the directions we took are now difficult to pursue, because the giant search corporations like Google and Yahoo have labeled all this intellectual territory as proprietary. In a *New Yorker* article ("When G.M. Was Google," *New Yorker,* Dec. 1, 2014), Nicholas Lemann describes a technical problem that Google's engineers "solved." But that problem was well known for decades in the information sciences, and had been addressed in a number of ways in both manual and automated systems. Google might have solved it simply by *doing a search* on the research literature of the subject.

The purpose of these volumes is to republish selected significant publications of mine, representing my work on the topics of information, the information professions, information seeking and searching, information organization, and user-centered design of information systems. The papers demonstrate the work we in my field did to understand information behavior and information system development and use, back when we had to burrow deeply into the human relationship with information in a way that is seldom done these days. There is a great deal to learn from that earlier work, and a great deal of knowledge that is being overlooked in these days of rapture with the new technologies. When the intoxication fades a bit, and people return to wanting to improve the actual ways people find and use information, then perhaps some of the approaches explored in these papers will regain prominence.

One of the "problems" of information science is that it continues to have associations with the discipline that has dealt with information storage and retrieval the longest: library science. It makes sense: There were decades of transition from all-paper information storage to various sorts of electronic storage and retrieval. The transition involved many stages of development, using various combinations of paper and electronic information processing. The computers we worked with started out with minuscule amounts of storage and processing power, and gradually managed to handle vastly more of our storage-intensive resources.

This association of information science with library science is not a problem because of the latter field's expertise and long experience with information. Rather, it is a problem only because of the astounding degree of negative stereotyping conventionally associated with the field. Could there be a greater contrast than that between the two guys in a garage, and the hermetic librarian hunched over his desk, protecting his collection from defilement by users, or the virginal spinster with the pencil stuck in the bun at the back of her head, presiding over the reference desk?

I have been astonished countless times in my life at the stubborn persistence of these stereotypes, despite all facts to the contrary. Every Christmas we are treated to a showing on television of the film *It's a Wonderful Life,* wherein, during the dream sequence, the wife of Jimmy Stewart's George Bailey, who would now have never married without him, becomes a librarian, and is shown locking the library door at night, as she comes home from her lonely work. An analysis of the social role of this librarian stereotype has yet to be written. It is a charged and deeply meaningful archetype in the popular consciousness, and deserves full analysis. In the meantime, however, this stereotype has largely doomed any chance that the new information technology world learns from what librarians know about information and people. I recall recently being in a symposium and observing a graduate student from another field, who had become interested in what we do, hastily denying—indeed, almost panicking at the thought—that her new interest had anything to do with library science.

Information science, when it launched in the 1960's, was the exciting new field that would take the old librarianship into the new world of automation. And it did do that, with great success. But information science had another "flaw" that doomed it in the thinking of many computer scientists. Not only were there more women in the field than in computer science, there was also a great deal of attention to the so-called soft side of technology, its use and integration with people's lives and work, the psychology and sociology of information.

I sat in the audience at a conference while Gerard Salton, the father of modern information retrieval, explained that it was not possible to study the human factors in information system use, because these social aspects were too squishy and unmanageable (not his exact words). It is hard to maintain a balance in a discipline between people with a social and psychological perspective and those with an engineering, technical perspective when applying their expertise to a problem. As a rule, they have fundamentally different cognitive styles, which do not necessarily mix well. Yet both these perspectives are needed where information systems are concerned. Information science has retained that balance, but consequently sometimes puts off computer scientists, many (not all) of whom literally do not see the social side of information, are not interested in it, and are happy to purge that perspective from their university departments.

For all these reasons, the extensive knowledge that has built up within information science is frequently ignored in today's research and scholarship, and in the popular consciousness. The publication of these articles of mine is my bid to draw attention to some of what has been learned in information science about information-related behavior and good information system design. In what follows, the technologies may indeed have been superseded, but pay attention to the thinking around the technologies—what we learned about people and information, what we learned about dissecting an information query and searching on it, what we learned about how people process information when approaching an information system, and what we learned about how to design an information system to support information seeking well. Much of what we learned has *still* not been implemented in information systems, despite the billions of dollars poured into the information technology world.

Finally, these papers are also of interest as a historical record of that time of transition in information research between the 1960's and the early twenty-first century. Coincidentally, the arc of my professional career happened to parallel the arc of development of modern information retrieval, and some of that history is represented in these volumes. I hope you find material of interest and value herein.

The papers reproduced here represent, in my rough estimate, about half my published work, the half I felt would be of most use today. I have always had wide interests within information science, and many of the papers could arguably fall in more than one category. I have grouped them as best I can in the categories below, arranged by volume. One popular article, on "berrypicking," appears in both Volumes II and III. Otherwise, all articles are unique.

Articles and chapters are carefully reproduced exactly as they originally appeared in print, with two exceptions: errors are corrected, and citations in reference lists that were originally listed as "in press" have been supplied with their subsequent publication information. Consequently, a few citations will be more recent than the original publication date of the article in which they appear.

> Volume I: *Information and the Information Professions*
> *Information* (4 articles)
> *The Information Professions* (14 articles)
> *Appendix A: Author curriculum vitae*
> *Appendix B: Content Lists and Index for all volumes*

> Volume II: *Information Searching Theory and Practice*
> *Search Vocabulary Use* (5 articles)
> *Searching Behavior* (10 articles)
> *Appendix: Content Lists and Index for all volumes*

Volume III: *Information Users and Information System Design*
Information Seeking and Interaction (8 articles)
User-centered Design of Information Systems (8 articles)
Appendix: Content Lists and Index for all volumes

"Just think! Every book that's ever been published in the United States
is right here in the Library of Congress."
"Even 'The Poky Little Puppy'?"

Preface to Volume I

"What is the nature of information?" is the core question in information science, and is foundational to the character of all the information professions. I have written on these subjects at various points in my career, and the key works are brought together in this volume.

SECTION ONE: INFORMATION

The question of the nature of information is central to the field of information science, just as the nature of communication is core to communication research. In both cases, the fields have generated countless definitions and debates. In graduate school, I encountered the definition of information as "the pattern or organization of matter and energy." I had an immediate instinctual feeling that this definition was the right one as a foundation for information science. The topic has interested me since I was in graduate school, and I continued to think about and analyze the subject on and off throughout my career. However, I could never quite bring it together in a strong theoretical context until late in my career. Even then, I felt that we were at the dawn of an age where the appropriate philosophical apparatus behind the concept was only just taking form. (When it does take form, it will have, I believe, David Searle's conceptualization in the 2010 book, *Making the Social World: The Structure of Human Civilization* [New York: Oxford University Press], at its heart.) I was running out of years in my career before that larger paradigm fell into place. Finally, I felt I had to take a shot at the topic before I ran out of life.

That meant, however, that I could only develop the model partway. I started with the most fundamental sense of information—as being in contrast to entropy—which I then endeavored to build upon, based on a concept of information pattern detection as something that developed evolutionarily.

I knew I was at risk in proposing any concept of information that had any objective existence in any way. Over my long career I have seen social science scholars swing back and forth between objective and subjective senses of science, often with extreme prejudice toward the opposite side of that argument. At the time of my writing, we were in an extreme subjective phase. This conflict seems absurd to me. We have learned much from both perspectives, and unless we want to disappear into a totally solipsistic perspective in which absolutely only our own subjective experience is real and valid, the very fact that we have conceptualized and acted on these contrary perspectives so many times in the past surely suggests that they both exist meaningfully and usefully in some form. It was my intent, ultimately, to build a subjective framework of information atop the objective sense. I did not get as far as I wanted, however, with the subjective aspects.

In this section, the two major papers I wrote on information are reproduced ("Information and knowledge: An evolutionary framework for information science," 2005, and "Fundamental forms of information," 2006.) In addition, I wrote an entry on the topic "Information" for the *Encyclopedia of Library and Information Sciences*, providing a review of several of the senses in which the term has been used and debated in the field. Finally, in an early (1987) manifestation of my interest in the importance of information as a variable in the field, I published "Information: The last variable," arguing that the information itself is too often the last variable researchers think to address when they do social research in information science.

SECTION TWO: THE INFORMATION PROFESSIONS

Materials in the "Information Professions" section tackle the character of the information professions in several ways. The 1999 article, "The invisible substrate of information science," constitutes a description of both the explicit and the implicit understandings of the field. This paper came together as I collected submissions for a two-issue celebration of 50 years of the American Society for Information Science's flagship *Journal* as guest editor. The article won the *"Journal of the American Society for Information Science* Best Paper of the Year" Award in 2000.

Subsequently, as the interest in information-related activities has exploded in recent years in our society, new information professions seem to appear every few months. I wanted to reconcile some of these new disciplines with the ones we have long understood to constitute the information disciplines, as well as to provide some boundaries for the extent of this increasingly confusing collection of fields. The 2015 article, "The information professions: Knowledge, memory, heritage" builds on the 1999 "Substrate" paper, and adds to it a model of the full range of information disciplines. We could call these two papers a *definition by intension* of this area of research and practice, that is, definitions describing the key attributes or features of research, theory, and practice in the fields.

The next three papers constitute a sort of *definition by extension*. These describe the particular topics that are researched in the fields, that is, they define the field by listing what topics appear in it. Another paper in the 1999 anniversary issue, "A tour of information science through the pages of *JASIS*," groups samples of related article titles from the issues of the journal over the fifty years to give the reader a feel for the issues and topics that occupied researchers who submitted articles to the journal. The 2010 "Introduction to the *Encyclopedia of Library and Information Sciences*" describes the process of selecting topics for the encyclopedia, drawing on a 50-person editorial board as well as my ideas as Editor-in-Chief and those of the Co-Editor, Mary Niles Maack. The "Topical table of contents" from the encyclopedia groups the topics covered conceptually, and thereby displays the underlying structure of the encyclopedia, far better than the alphabetical contents list can. The encyclopedia covered ten information-related disciplines, and was not limited to library and information science.

The remaining papers in this section all address the nature of the information professions, especially library and information science, from a variety of other perspectives that can also shed light on the character of the field(s). My paper, "Many paths to theory: The creative process in the information sciences," addresses the challenge of tapping into one's creativity around the research questions of our field. It is published in a 2016 book directed to researchers and doctoral students: Diane H. Sonnenwald, Ed., *Theory Development in the Information Sciences* (Austin, TX: University of Texas Press). Each article describes some aspect of theory development that interests one of the authors, who are major researchers in the field. In my paper, after discussing means of releasing one's creativity, I describe a number of information science topics that I feel are well worth studying, and which I did not have time to research during my career.

"An introduction to metatheories, theories, and models" (2005) reviews no fewer than 13 different theoretical and/or methodological approaches that are being used currently in the research of library and information science. Exposition on the various approaches, along with references to example articles in each case, is intended to help students sort out their interests and understandings about the various paradigms operating in the field.

"The role of the Ph.D. in a professional field" and "A doctorate in library/ information science" both address doctoral level work in the field. The first, a speech, addresses the particular role that the Ph.D. degree has in a professional, as distinct from an academic, field. The second, appearing in a major professional journal directed at working librarians, was intended to explain the nature of doctoral study and help librarians make a decision on whether to go on for the doctorate.

"Future directions for the library profession and its education," published in 1996, is the final report of meetings held under the sponsorship of the California Library Association regarding recommended future directions for master's-level education for librarians in the state of California. This report came out at a unique moment in the history of library education for the state, as the heads of all three programs then existing in the state—at the University of California at Los Angeles (UCLA), UC Berkeley, and San Jose State University—as well as several other prominent figures in the field in the state, including Kenneth Starr, the State Librarian, agreed on an approach to library education for the state. I was Chair of the Department of Information Studies at UCLA at the time.

"A criterion citation rate for information scientists" is a bibliometric paper addressing an aspect of the analysis of citation rates in the field. Citation analysis is a method by which academic disciplines are studied and evaluated. After its publication, I received a letter, now lost, from the great historian of science, Derek de Solla Price, encouraging me to continue work on this topic. The topic was not central to my interests, however, and I did not pursue it.

The 2004 article, "Information science at the University of California at Berkeley in the 1960's: A memoir of student days," not only provides a personal perspective on the then-developing field of information science at one of the major programs where information science was pursued during those years, it also reviews the content matter of the field as information science was conceived of at the time.

The final two items in this section address the personal experience of being a woman in the years when I pursued my career in the field. The 2005 "Acceptance speech for the American Society for Information Science and Technology Award of Merit" not only expresses my thanks for the award, but also describes what it was like to be a woman professor over the decades of the late twentieth century. "A note from Marcia J. Bates" describes the unfailing support I received from Pauline Atherton Cochrane, as a mentor and senior member of the field.

ONE

Information

Information and knowledge: An evolutionary framework for information science

ABSTRACT

Background Many definitions of information, knowledge, and data have been suggested throughout the history of information science. In this article, the objective is to provide definitions that are usable for the physical, biological, and social meanings of the terms, covering the various senses important to our field.

Argument Information 1 is defined as the pattern of organization of matter and energy. Information 2 is defined as some pattern of organization of matter and energy that has been given meaning by a living being. Knowledge is defined as information given meaning and integrated with other contents of understanding.

Elaboration The approach is rooted in an evolutionary framework; that is, modes of information perception, processing, transmission, and storage are seen to have developed as a part of the general evolution of members of the animal kingdom. Brains are expensive for animals to support; consequently, efficient storage, including, particularly, storage at emergent levels—for example, storing the concept of chair, rather than specific memories of all chairs ever seen—is powerful and effective for animals.

First published as Bates, M. J. (2005). Information and knowledge: An evolutionary framework for information science. *Information Research, 10*(4), paper 239. http://www.informationr.net/ir/10-4/paper239.html

Conclusion Thus, rather than being reductionist, the approach taken demonstrates the fundamentally emergent nature of most of what higher animals and human beings, in particular, experience as information.

Part 1: Development of basic concepts and framework

Introduction

The intent of this essay is to define and conceptualize information for the purposes and uses of information science/studies. The emphasis is three-fold: 1) the information definition is presented and contextualized within an evolutionary framework, 2) *knowledge* and *data* are defined and discussed in relation to the information concept and 3) various theoretical issues surrounding this understanding of information are developed in detail. A companion essay (Bates, 2006) presents the same definition, then develops and justifies a series of fundamental forms of information.

Background

A conceptualization of information is obviously central to a discipline named information science. Many major and minor efforts have been made over the years to develop the term and to provide a framework for theory development and further general development of information science (Belkin, 1978; Belkin & Robertson, 1976; Brier, 1998; Brookes, 1975, 1980; Buckland, 1991; Budd, 2001; Day, 2001; Derr, 1985; Dervin, 1977, 1983, 1999; Dretske, 1981; Floridi, 2002; Fox, 1983; Goonatilake, 1991; Hjørland, 2002b; Losee, 1990, 1997; MacKay, 1969; Meadow & Yuan, 1997; Pratt, 1977; Raber, 2003; Shannon & Weaver, 1975; Thompson, 1968; among others). Major collective efforts, with many contributors, have been published in Machlup and Mansfield (1983) and Vakkari and Cronin (1992). There have been many reviews of the literature on the concept also (Aspray, 1985; Fischer, 1993; Wellisch, 1972; Wersig & Neveling, 1975; among others), of which Cornelius (2002) and Capurro and Hjørland (2003) appear to be the most recent.

A proper review of these many definitions and discussions of information would be book-length and would make impossible an adequate discussion of the definition proposed in this essay. A brief review of common classes of definitions will be presented and discussed below, while the reader is referred to other sources for a detailed discussion of other views.

One source of the variety of approaches may be the many different disciplinary origins of writers on information. Engineering, the natural sciences, a wide array of the social sciences and the humanities have all contributed to the discussion. Ultimately, however, our discipline must surely find a way of thinking about information that is distinctively suitable for our own theoretical and practical uses (Bates, 1987, 1999).

Within information science, it is desirable to develop an understanding of information that is applicable for the various senses in which researchers and practitioners need to talk about information. Buckland (1991) has said that most definitions in the field represent approaches toward information that take information 1) as knowledge, 2) as process, or 3) as thing. Derr (1985), Dretske (1981) and Fox (1983) have followed the knowledge approach, by taking information to be various types of propositional statements. Boulding (1961), Brookes (1980) and Dervin (1977) have taken a different tack within the knowledge approach, by writing of information as constituting additions or changes in a mental map. Finally, after the original submission of this essay, but before the final version was sent in, a paper by A.D. Madden appeared, titled "Evolution and information," where information is defined as "a stimulus which expands or amends the World View of the informed" (Madden, 2004, p. 9). This approach constitutes another of the definitions of the additions-or-changes-in-the-mental-map variety. Madden's paper is a reflection on why and how animals need to perceive and communicate information to each other.

Continuing with Buckland's categories of information definitions, Pratt (1977) has taken the purest process approach by defining information as the moment of being informed, of "in-formation" in the mind. Buckland himself (1991) has emphasized information's *thing-ness,* always of importance in a world with many information objects lying around.

Claude Shannon, the progenitor of the original burst of interest in the concept just after World War II, took the amount of information to be a function of one's freedom of choice in selecting a message (Shannon & Weaver, 1975). Cornelius (2002), Day (2001), Hayles (1999) and Rayward (1992), on the other hand, view information as a social construct and endeavor to extract out the social meanings and history surrounding the term. In order not to exhaust the space constraints of this essay, the reader is directed to the cited resources for more detail on the various conceptualizations of information.

Though a historical approach is not being taken in this essay, it is appropriate to provide some historical context to the discussion that follows, for reasons that will become apparent. If we examine the above range of types of definitions of information, we see that they can be fit within the broader pendulum swings of the sciences and humanities over the past fifty years. In our field and many others in the 1950s forward, an extreme scientism—including the most extreme form, logical positivism—dominated. Social science disciplines gained legitimacy by showing how their subject matters could be studied using adaptations of classic, natural scientific methods.

In more recent years, there has been a reaction to this approach, with a concomitant swing towards the use of what are essentially humanities methods in the social sciences. Now the fashion is to deride the very scientific techniques so recently valorized and to insist that only highly qualitative and subjectivist methods produce credible results. Hermeneutic interpretation, detailed participant observation and historical analysis, among others, are now the methods of choice. Nowadays, it is seldom remembered, however, that the logical positivist approach was itself a reaction to what were deemed ineffective subjectivist research philosophies that preceded it.

In like fashion, attitudes toward information itself have swung between highly objectivist and subjectivist interpretations. When I began teaching in the 1970s, most students in my classes would insist on the objective nature of information; now that position is reversed and most feel that information has no meaning except as it has impact on an experiencing being. Their reactions paralleled the formal literature—and perhaps the spirit of the times as well. I believe that we are missing the most important lesson that should be coming out of these historical swings—the recognition that each of these positions has something to teach us and that the long-term goal should be to develop an approach that allows each perspective to give over to us what it has to teach. Perhaps we can find a way to think about information that effectively allows for both subjective and objective perspectives.

Objectives of this essay

Currently, there are a number of metatheoretical positions, that is, fundamental philosophical and theoretical frames, or "orienting strategies" (Wagner & Berger, 1985), competing for the attention of the discipline of information studies. Ellis (1992), Frohmann (1994) and Hjørland (2002a), among others, have contributed to the elucidation of these positions. They include research metatheories that are historical, statistical and bibliometric, engineering, and design-based, as well as ones arising out of sociology, psychology, and the physical sciences.

These various metatheoretical positions have contributed enormously to information studies, but there is one very productive metatheory in the social and biological sciences that has been largely ignored in information studies, and which will form the basis for the understanding of information presented herein. Over the last several decades, researchers in biology, anthropology, and psychology have come together to study human behavior in a new way. That new framework is rooted in an understanding of evolution and its impact on the cognitive, linguistic, and social structures of human beings. There are several variations on this approach, going by names such as human behavioral ecology (Winterhalder & Smith, 2000), and evolutionary culture theory (Durham, 1991). Probably the most common name used is evolutionary psychology (Barkow et al., 1992).

These researchers have been working to integrate their disciplines' understandings of human beings in ways that promote a richer comprehension than any one perspective provides. This "conceptual integration" (Cosmides et al., 1992) in the human sciences is intended to relate and connect the knowledge among these disciplines, not reduce one field to another. No claims are being made that psychology is *merely* biology. On the contrary, this metatheory not only recognizes but also explicitly draws on the unique learning contributed by each of the biological and social sciences.

For our purposes, it is not necessary to subscribe to the full theoretical program of evolutionary psychology (or any of its variants) to reflect on and use the ideas to be presented in this paper. However, putting these ideas in the evolutionary context enriches one's understanding and provides a possible foundation for building a scientifically-based conception of information that also harmonizes with several of the more social-science-based theories of information in our field. The ultimate goal is to find a basis for an integrative understanding of information for information science.

A definition of information

Where concepts or ideas are drawn from others' work, they are so cited; otherwise the following conceptualization is my own.

THE BASIC DEFINITION

An early think-piece by Edwin Parker in a conference on information needs provided the following definition of information, which is the one used in this essay:

> *Information* is the pattern of organization of matter and energy. (Parker, 1974, p. 10)

The definition is not original with Parker; this approach to information was endemic in the 1970s. However, in the 1970s and early 1980s, when

I became interested in this definition, I was not able to find any formal description or development of it—nor any since—just occasional references to the idea. I believe that we in information science have not developed the potential of this productive definition for the field. Now that the biological and human sciences are developing a deeper and richer understanding of information processing and adaptation, there is a renewed value in considering this definition.

THE DEFINITION OF PATTERN

Pattern is understood to be used in the above definition in two of the senses commonly found in dictionaries. Here are two of those in the *Random House Unabridged Dictionary* (1993, p. 1423). First: "a natural or chance marking, configuration, or design: *patterns of frost on the window.*" Secondly, "a combination of qualities, acts, tendencies, etc., forming a consistent or characteristic arrangement: *the behavior patterns of teenagers.*" The first definition implies a first-order pattern. The design we see on the frosted window is not part of any larger whole; it is simply a grouping or spotting of frost on the window that is not totally chaotic—so it does contain information—but nor is it a system. (To be very precise, ice crystals form in a regular pattern, but since frost must form on something, the irregularity of frost patterns reflects the irregularities of the surface it forms upon ["Frost." Encyclopædia Britannica online].)

The second definition implies a second-order kind of pattern, one in which a variety of features are knitted together in an overall system or integral design. A pattern of behavior implies repeated similar cycles of activity, that is, some coherence above and beyond the bunching and scattering of the frost on the window.

These patterns may be characterized as *emergent,* meaning that the sum of the elements constitutes something new, a whole with its own distinct qualities. When we look at a chair, we see a (first-order) pattern of light and dark, solid colors and edges. But we also are capable of recognizing the chair as a chair, an emergent pattern that we can recognize quickly as a whole, because it possesses certain features we have learned to recognize through experience and with the help of our inherent cognitive abilities.

This is not to imply that a sharp distinction between first- and second-order patterns is being proposed. Patterns form, dissolve, fragment, etc. in many ways continually. Rather, both definitions are provided in order to claim both senses for *pattern* in the above definition. A *pattern of organization,* as used here, does not have to imply coherence, though it often does; just something other than pure entropy.

A question immediately arises upon hearing of such a definition of information. *Whose* pattern of organization is *the* pattern? Is there one objective pattern, or numerous individual subjective patterns, that is being referred to?

To proceed further with elaborating the definition, it is necessary to state two fundamental assumptions that ground the ideas that follow. The first assumption is that there is a real universe out there, that we are not solipsists, that is, what we see as being outside ourselves is not a movie we are running in our minds, but an actual universe independent of ourselves. To posit an independent universe, however, does not necessitate the further assumption that we humans have a complete, clear, or "true" understanding of that independent universe, or that it is ultimately understandable in only one way, just that it exists in some form. The second assumption is that the universe is not total entropy, or total undifferentiated chaos or disorder. To put the latter in the affirmative, the assumption is that there are differences, distinctions, differentiations existing in that universe. For example, there are some elements or compounds in one place and other elements or compounds in a different place, thus making some parts of the universe different from other parts. In our daily lives, of course and in science and other learning, we believe that there are innumerable such differentiations.

But having posited that there is some structure in the universe independent of the experience of living creatures does not necessarily entail the assumption that there is one true structure to that universe and every variant understanding is false. Animals differ tremendously in their perceptions and in their cognitive equipment. The color vision of cats is not as good as that of human beings, but cats have better night vision. Consequently, the world looks significantly different to them from what it looks to human beings—but it is not the case that we see the world *truly* and they see the world *falsely*. Animals react to and interpret their worlds dramatically differently from one another, depending on their inborn equipment and life experiences.

When information is defined here as the pattern of organization of matter and energy, there are patterns of organization that exist in the universe whether or not life exists anywhere in it. There is one shape and structure of a rock here and a different shape and structure of another rock there, whether or not any animals ever see the rocks. At the same time, once life comes along, it is useful for those living things to perceive and interact with their environments. How each living thing experiences its environment will have enormous variations and some similarities.

My pattern of organization is not your pattern of organization, but, at the same time, we both live in the same world and may be responding to virtually the same things. The point here is that there are many patterns of organization of matter and energy; something going on in the universe independent of experiencing beings, as well as all the various perceived and experienced patterns of organization that animals develop out of their interactions with the world. All of these patterns of organization can be looked upon from an observer's standpoint as information; whether they are independent of sensing animals or are the tangible neural-pattern results of processing in an individual animal's nervous system. *In this particular sense,* both of what are usually called objective and subjective senses of pattern of organization are included in the definition as used here.

The evolutionary context

All these differentiations, clusterings and bunchings of matter and energy present in the universe can be the subject of animals' perceptions and interactions with the world, whether with animals, plants, rocks, i.e., matter, or with waves of sound or waves of the earth in an earthquake, i.e., energy. *Animals are able to detect and process those differentiations in the universe and experience them as patterns of organization.* (Plants can respond to light and other environmental phenomena too, but for simplicity's sake, only animals, with special emphasis on human beings, will be discussed.)

In information terms, an information-rich world is one in which there are many variations from pure chaos or entropy, i.e., where there are many regularities, differentiations, bunchings and clusterings of matter and energy. To be more specific, let us take as an example a typical chair in a university classroom. The chair is made of molded blue plastic with chrome metal legs. The physical materials comprising the plastic chair have a different pattern of organization from that of the air around it. The chair and the air consist of different physical elements and so reflect light differently from each other. We happen to be animals that have developed a perceptual apparatus to detect those light differences, which differences handily correspond to physical differences in matter that are likely to be of importance to us as physical organisms moving about the world. Being able to discern the chair through sight, before bumping into it and getting a bruise, enables us to be more successful animals in the world than otherwise.

These perceptual apparatuses differ enormously from species to species in their character, complexity and power. The amoeba can perceive light, the bat can bounce signals off of the walls of a cave, the hound can smell a scent days after the source of the scent has moved on and the human can read a freeway sign from fifty meters away. Through the evolutionary processes of natural selection, we develop these senses to take advantage of the differentiations around us, which, in turn, promotes the survival of ourselves and our offspring. Animals in very deep ocean, where there is no light at all, lose, or never develop, light-sensing capabilities.

We humans are even so clever that we have found a number of ways to enhance our perceptual apparatus. We do it through direct magnification, as with a hearing aid or telescope, or through more indirect means, whereby we find ways of representing patterns of organization that would otherwise be undetectable by us. For example, the astronomer develops an apparatus to detect radio waves coming from distant stars. A human being cannot detect those waves directly even when they are magnified. So we find ways to convert the features of those waves into something we can detect and decipher, such as coded numbers on a screen.

Patterns of organization are not limited to perceptions, however. In our brains we create and store our own patterns of thought, feeling and memory in the neurons, which we then subsequently draw on for further thought and action. Further, we mold the world around us by imposing patterns of organization on the world, whether by intentionally producing houses, tools, books and the like, or unintentionally by beating a path through the woods by repeated use. In our own species, we generate enormously complex social systems and cultural expressions, which are, in turn, stored as information in our minds and in the patterns of organization of the objects and infrastructure around us.

Over the billions of years of life on this planet, it has been evolutionarily advantageous for living organisms to be able to discern distinctions and patterns in their environment and then interact knowingly with that environment, based on the patterns perceived and formed. In the process of natural selection, those animals survive that are able to feed and reproduce successfully to the next generation. Being able to sense prey or predators and to develop strategies that protect one and promote the life success of one's offspring, these capabilities rest on a variety of forms of pattern detection, creation and storage. Consequently, organisms, particularly the higher animals, develop large brains and the skills to discern, cognitively process and operationally exploit information in the daily stream of matter and energy in which they find themselves.

Why should animals experience *patterns* of organization? Why not just process each bit of information, whether a pulse along a neuron in the brain, or a pixel of light on a computer screen, as itself, as just one bit? The answer is that detecting a pattern can be vastly more efficient in processing and storage than detecting and storing all the individual bits of information that comprise the pattern.

Storage efficiency is not a trivial matter; brains use a huge part of the energy an animal takes in as nourishment. As Allman (2000, p. 175) notes, "in a newborn human the brain absorbs nearly two thirds of all the metabolic energy used by the entire body." Even in human adults, the brain uses about 20 percent of the body's metabolic energy (Elia, 1992, p. 63).

Allman also states:

> Brains are informed by the senses about the presence of resources and hazards; they evaluate and store this input and generate adaptive responses executed by the muscles. Thus, when the required resources are rare, when the distribution of these resources is highly variable, when the organism has high energy requirements that must be continuously sustained and when the organism must survive for a long period of time to reproduce, brains are usually large and complex. In the broadest sense then, brains are buffers against environmental variability (2000, p. 2–3).

Calvin (2002) has argued that we human beings developed our large brains and capacity for generativity, planning and cooperation because of the whipsaw effect of being subjected to hundreds of cooling and warming periods during the Ice Ages. Substantial climatic changes could occur in as little as one to ten years; consequently, gradual evolution of genetic adaptation was too slow for survival. Our species may have survived because, with the development of larger and larger brains, we were able to adapt *in real time* to drastic environmental shifts. Stanley (1996), another respected scientist, makes a similar argument.

The kind and nature of patterns that we can recognize vary from species to species, and, to some degree, from individual to individual within a species. These differences have arisen in response to evolutionary pressures, including the character of ecological niches that species inhabit. See, for example, the description by Korpimäki et al. (1996) of the different responses of voles (small, mouse-like rodents) to weasel predation vs. kestrel predation.

Human beings, with our large brains and extensive processing and storage capacity, can process patterns of great sophistication. Let us consider a human example. Consider the familiar concept of *bait and switch* in retail sales. In this pattern of behavior, a retailer offers an inexpensive product for sale through advertising, then when the prospective customer arrives at the store, the retailer says that the cheap item has been sold, but has some more expensive items for the customer to consider. The customer has been baited with an attractive product, then the product switched for actual sale.

Now, the person unfamiliar with bait and switch may be fooled once or twice, until she goes from processing the individual experiences to drawing the broader, categorizing conclusion that there is *a pattern across the instances*. She has now moved up from a simple collecting of retail experiences to seeing a broader, more sophisticated, emergent pattern in the interactions. From then on, she can think of a certain type of interaction with retailers by this common phrase and concept of *bait and switch*. There is a tremendous coding efficiency in this ability to see the broader patterns and she is also better protected against deceit because she can more quickly make a pattern match with her new understanding and avoid being cheated.

It is hard to imagine the use of language without the use of these categorizing capabilities. If every chair, being a different physical object from every other chair, had to be separately studied, remembered and named, in order for people to talk about it, we would quickly be bogged down in horrendous mental processing and storage overloads. Furthermore, we would not be able to call upon past experience when we encounter a new chair, because we could not group it with known objects with common features. Our ability to find useful commonalties between objects, which we then use to group related objects together and treat similarly under the same mental concept and same linguistic term, produces massive savings in cognitive processing and storage needs.

When an information definition such as this one is proposed, that is, one that starts with rudimentary physical and biological elements, a common reaction is to reject the definition on grounds that it is reductive, that is, that the purpose of the definition is to reduce processes at higher levels, such as the social and aesthetic, to their mere physical components. But as the argument above demonstrates, the *approach taken here is not only not reductive, it is actively constructive and emergent*. It is assumed that animals and human beings, particularly, have the ability to construct and remember global patterns perceptually, cognitively, kinesthetically, emotionally, etc. Rather than being reductive, this view of information is *productive*, because it assumes that processes at lower levels make possible the development of more and more sophisticated emergent patterns at higher levels, that is, patterns that are more than the sum of their parts, which we can then use for further development and understanding. The long-term history of our species has been one of understanding patterns of greater and greater levels of sophistication, which, in turn, produces ever-faster cycles of development and understanding.

Some readers of this essay in manuscript have argued that this definition of information simply equates *information* with *pattern* and that the definition is therefore trivial. I do not follow the logic: even if I were doing that, if I could make a good case for that definition and its usefulness to the field, would that not be a step forward?

In any case, my use of *pattern of organization of matter and energy* as the definition does more than define information as simply pattern. Instead, the creation in mind and actions, in perceptions and interactions, of patterns of organization at *many emergent levels* indicates the power of animal cognition and behavior. By seeing a wide range of different objects, each with somewhat similar, but by no means identical characteristics as all being chairs, for example, we humans are able to engage in powerful and efficient cognition and behavior. By seeing a similar but by no means identical sequence of actions as *bait and switch,* we are able to cope more effectively with our world and protect ourselves. Information is thus not just the literal atom-by-atom pattern of matter and energy; rather we can study information as existing at emergent levels as well. Surely, it enriches our understanding in information science to begin with this basic understanding of information, then see how this elemental definition contributes to ever-more-sophisticated understandings of patterns of organization, of information, in animal and human lives.

So, however emergence happens at a physical and biochemical level in mental processing, we accept that it does happen. Further, whatever else is going on in a human being's neural circuits, from the standpoint of information science, these mental abilities to detect emergent ideas and objects are immediately understandable in terms of coding efficiency and as effective and powerful means of representation. We in information science are accustomed to creating digital representations and compressing them for efficient storage and transmission through information technology; we should not be surprised that nature got there before us in making efficient use of our limited neurons.

In sum, animals evolve to be able to detect, process and store patterns of organization in their environment and experience. Such capabilities protect them against predators and enable them to find food and mates. Brain processing power is expensive for animals to support, however;

consequently brain size has generally grown under conditions of extreme and rapid change or variability, where the high costs of a large brain have been repaid in offspring survival. The large brains of humans permit the detection, processing and storage of patterns of organization of great sophistication, with many emergent qualities. An understanding of information as the pattern of organization of matter and energy thus makes sense in the context of the long-term evolutionary development of human beings.

Revisiting the definition

Now that we have seen the ways in which this definition fits within an evolutionary context, let us revisit the definition and consider its role in information science.

First, why should we in information science want to place information in an evolutionary context? After all, after World War II, Shannon's work (Shannon & Weaver, 1975) led many to try to conceptualize information in a physical way. Yet that application has never carried over well into the social and psychological senses in which we like to think about information. Will we not have the same problems with a biologically-based sense of information?

In response, Shannon's view of information greatly deepened and enriched our understanding of information. Space does not permit me to make the argument at length, but much of computer science and technical communication, as well as the information processing model of human cognition drew on Shannon's work. (See Gardner, 1985; Lachman et al., 1979; and Miller, 1951.) So the route science took through Shannon's information theory, while not ultimately wholly satisfying, was immensely enriching to our understanding of communication, information processing, and information transfer at the time. Indeed, the much deeper understanding provided by Shannon's work of how coding and combinatorial possibilities function in communication very probably made much easier the understanding of how coding works in DNA.

Similarly, I believe that the pass through evolutionary theory will greatly enrich our understanding of information as well. If we start with a physical sense that nature is not in a state of total entropy, but instead contains many differentiations of various kinds, then, we can see that once life began, it became advantageous for living things that move, particularly animals, to have a sensory capability to detect and construct patterns out of the differentiations existing in nature. In other words, if we accept that there is some world out there, outside of animals and that animals' survival is promoted, in ways well understood in evolutionary biology, when they can detect and process inputs about that exterior world, then we can expect such things as perceptual apparatuses, nervous systems and brains to develop. Such a conception holds out the possibility for information science of an understanding of information fully as broad as the world of things that people and other animals might experience as informative.

Thus, I would argue that this conception of information marks the next step up from the earlier understanding of information developed in the era of physical information theory. A purely physical information processing model of information for human beings fails to take into account the immense number of short-cuts, storage efficiencies and processing efficiencies that have evolved over the millennia in animals.

Evolutionary psychologists argue that the characteristics of modern humans evolved and were honed during the hundreds of thousands of years that we and our predecessor hominin species lived as hunter-gatherers (Barkow et al., 1992). These characteristics include brain organization and information processing. Thus, this evolutionary understanding is arguably a necessary, though not necessarily sufficient, part of a more general understanding of information for the purposes of information science.

In the next sections, we shall find that when we begin with the physical and biological definition provided herein, we are able to think about information in ways that are novel, productive and surprisingly relevant for information science. There may well be still more layers above the physical and biological levels that need to be developed for a full theory of information for the field, but we shall find that the understanding provided herein penetrates much further into the socially and psychologically important aspects of information than we might have imagined at the beginning.

So what is knowledge?

The principal intent of this essay is to define information. However, questions about the relationship of information to knowledge and data inevitably arise in the context of such a discussion. For that reason, knowledge and data are considered here and in the next section, relatively briefly, in order to demonstrate the relationship they are understood to have with information.

In the definition of information given above, information is seen to have no inherent meaning. The pattern of organization of matter and energy is just that; no more, no less. In living systems, however, things are always more complicated. Hundreds of millions of years of evolution have laid down structures associated with survival in animal brains that, in effect, give meaning to a stimulus even as the animal perceives the stimulus.

How might meaning have evolved? Mutations and other sources of genetic variation will have provided the raw material for natural selection. The animal that happens to have a more capable brain in information processing and, therefore, is able better to detect patterns of the sort discussed in this essay, will be more likely to survive and reproduce.

Earlier sections discussed how valuable it is for animals to be able to sense predators, or, using a human example, for a person to be able to detect a bait-and-switch pattern in the behavior of another person in a sales situation. The amount and type of meaning that the vole can assign to the shadow of a predator bird overhead is presumably limited compared to what humans can ascribe. We can recognize an eagle or a hawk; to the vole, the shadow may not even be understood as coming from a bird. Rather, the shadow may simply be processed as *mortal danger.*

When I say that an animal "assigns meaning," I mean it in this limited sense; it is in no way a conscious act of labeling, as it can be for humans. In most animals meaning assignment has developed to promote the survival of the animal, and many of its responses (including many of our own, such as "fight or flight"), have become instinctual and automatic. The findings of evolutionary psychology suggest that more of human behavior than we have generally been willing to acknowledge can be traced to fundamental biological survival needs and trade-offs (see Wright, 1994, for an excellent introduction).

However, in higher animals and above all in human beings, there is far greater flexibility in meaning assignment than in many other animals. The squirrel, for example, can recognize the food value of the scraps in a left-over lunch sack, while other creatures may be unable to see the value in any food that comes in an unconventional form.

Much of our human brain's huge general-purpose capacity, including our unique language capacity, can be devoted to a wide range of miscellaneous contents associated with our life knowledge, our cultural and social knowledge, our family experiences, our memories and so on. A great deal of human culture and meaning assignment varies by culture or group and can take on an extraordinary complexity and variety, though rooted in fairly stable cognitive equipment. *Bait and switch* may be a pattern I should be able to recognize to promote my survival in a capitalist economy and it may be completely unnecessary and meaningless in another kind of culture.

Furthermore, over the millennia, through natural selection, life-promoting behavior has been reinforced in various ways within surviving animals. Victory over a rival is experienced not just as the brute fact of victory. That victory, and the desire for more victories, may be reinforced in the animal by increases in hormones that promote aggressiveness, as a means of consolidating victories. In the case of human beings, the roles of various hormones are highly complex. (See, e.g., Bernhardt et al., 1998; Gray et al., 2002.) These hormonal reinforcements have evolved to the point where we can now say that we feel good when we win. In other words, a subtle mixture of various biochemical influences affects our mood and reinforces reactions in ways that were life-supporting over the history of our development as a species.

Whether it is physical defeat of a rival in a fight, or the delivery of the perfect *mot juste* at a conference that gets a laugh from the audience, humans ascribe meaning to the events of their lives, meaning that ultimately rests in biological survival, though that meaning is very complex and heavily socially mediated in human beings. (For debates on meaning formation in biochemical and neurological senses, see Atlan & Cohen, 1998; Barham, 1996; Freeman, 2000a, b; Langman & Cohn, 1999; and Wills, 1994.)

In general parlance, as well as in information studies, when we receive information from someone or something else ("Your package has arrived." "The Bruins won their game.") we consider ourselves *informed*. That is, we receive the natural pattern of organization of matter and energy that consists of the air moving with the sounds of someone's voice, or we read the pattern of organization of written words, or in some other way receive novel information. We then relate the sounds or letters to words and meaning in our mind and assign meaning to the communication.

In Bates (2006), to account for this process, two definitions of information are provided:

> *Information 1* The pattern of organization of matter and energy.

> *Information 2* Some pattern of organization of matter and energy given meaning by a living being.

In this way, we account for both the foundational sense of information, in which information is understood to exist throughout the universe, whether or not humans or other living beings perceive or use it, *and* the social and communicatory sense in which living beings interpret and give meaning to information. Once some interpretation has occurred, the information 2 may be integrated with the rest of the creature's neural stores or understanding. Thus, we define:

> *Knowledge* Information given meaning and integrated with other contents of understanding. (Compare Kochen, 1983, who simply defines knowledge as information given meaning.)

Less developed animals are capable of assigning less meaning to their environments, but from very low on the developmental scale, animals do assign meaning of some sort. The goat knows good moss when it sees it and knows to move toward it. Once our species developed enough to have a general-purpose and more or less conscious and deliberate meaning-assignment capability and, especially, developed language, then we could and do assign meaning to all kinds of things. So we developed religion and rituals, science and art and all the other cultural forms.

So how can we think about the relationship between information and knowledge? By the above definitions, for a batch of information 1 to become knowledge, it must first be assigned meaning by a living animal, at least, within the limit of that animal's capacity to assign meaning. This we call information 2. When the books are sitting in the closed library over the holiday and no one is reading them, they all contain information 1. Only when an understanding human (since no sub-human animal can understand books) is actually reading the information 1, does it become information 2. Its contents are assigned meaning in the act of reading. Thus, the reader is *informed* in the act of reading. Some or all of that information 2 may ultimately be integrated more or less permanently into the pre-existing knowledge stores of the animal, i.e., become, simply, knowledge.

Knowledge in inanimate objects, such as books, is really only information 1, a pattern of organization of matter and energy. When we die, our personal knowledge dies with us. When an entire civilization dies, then it may be impossible to make sense out of all the information 1 left behind; that is, to turn it into information 2 and then knowledge.

Finally, what are data?

Outside of information science, the term *data* conventionally refers to information gathered for processing or decision-making. Within the field, data are commonly seen as something raw, on a path to being fully *cooked* or distilled (Hammarberg, 1981), that is, in a sequence that goes from data to information to knowledge to wisdom. There are numerous versions of this sequence; see Houston and Harmon (2002) for a recent one.

Within the context provided by this essay, we can see data in two more precise ways. First, data 1 may be seen as that portion of the entire information environment available to a sensing organism that is taken in, or processed, by that organism. What is too distant or otherwise outside the sensing purview of the animal cannot be data, it is simply information 1. Thus, strictly speaking, in the terms used here, animals perceive data, not information. Data 1 become information 2 when they are given meaning, then become knowledge when they are integrated (to at least some degree) with pre-existing knowledge in the mind or brain.

The second sense of data, or data 2, refers to information selected or generated by human beings for social purposes. Human beings, especially in the last several hundred years and, especially, using scientific methods, have developed a remarkable capacity to manipulate portions of the world in such a way as to deliberately generate additional information, which they experience as data 2, from which they learn immense amounts. Data in this sense may be information generated by as minimal an act as a child poking a grasshopper to see what it will do, all the way through many more sophisticated attempts to generate new understanding. The initial products of all of research and scholarship fall under this sense of data. These data are entirely generated through the actions of the researcher, whether it is the astronomer aiming a telescope and capturing the resulting images, or a social scientist observing a group of people and recording those observations. Though the observed stars may have been there all along, or the behavior of the studied type may have been going on at other times as well, this information is not data (of the second type) until some human being has generated or captured it for purposes of gaining knowledge.

Data 2 may also be selected or generated for other social purposes. We create databases to store collections of facts about people, artifacts, etc. for the purposes of business, government, or other organizations. All the variants of this second sense of data have in common the fact that they are socially generated and socially meaningful.

Part 2: Theoretical and philosophical issues

Various theoretical and philosophical issues relating to this definition of information are discussed below, as responses to a series of questions.

Is information material?

Two questions: 1) Is information an abstract entity, independent of the physical materials that compose it? In other words, does information exist on a sort of Platonic abstract plane, independent of physical reality? 2) Or, is information inextricable from the material that composes it? In other words, if a piece of granite contains a pattern of organization, would we then say that the granite *is* information?

It is argued here that neither of these characterizations is accurate. With regard to the first point above, the position taken here is fully materialist, that is, no abstract plane is assumed to house or manifest the information associated with the physical realities we experience. If the information is anywhere, it resides in the physical realities of nature, whether in the structure of a piece of granite, or in the neural pathways of the brain.

To say this, however, is not to say that information is *identical* with the physical materials or waves that make up the pattern of organization. *The information is the pattern of organization of the material, not the material itself.* As Wiener (1961, p. 132) has said: "Information is information, not matter or energy."

The ability to discern that aspect which we, as human beings, choose to label *pattern of organization* comes about because we are animals with sufficiently developed nervous systems and sensory apparatuses that we can think about the patterns of the sensory inputs we are receiving independently of their material. Thus, we can, for example, make a drawing of something we are looking at. We can, further, evaluate whether our drawing is a good representation of the item we have depicted, i.e., does the drawing constitute a reasonable mapping of the object from three to two dimensions? The information may now be manifested in three places—the physical object, in our mind and on the drawing essay. But it is still always in the physical world and we are dealing with the pattern in some independence from the material. (See Hektor, 1999, for a discussion of the immateriality of information and Hayles, 1999, for an insightful analysis of the consequences of separating information from physical embodiment in the intellectual fashions of our culture.)

These patterns are known to us, because we, like other organisms, have developed evolutionarily to be able to discern and construct patterns of organization and take advantage of them. We can recognize the pattern of organization of a rock, or the pattern of behavior of a lemur we are studying in the wild, because we have developed the sensory and cognitive apparatus to have observational experiences and to articulate our experiences. The rock does not know of information at all. The lemur uses information, but does not know it is doing so. With few exceptions, only humans are aware that we are using information from an observed structure or event separately from the observed structure or event.

Stonier (1997), also working from an evolutionary perspective, defines information differently:

> [I]nformation, like energy, is conceived of as a basic property
> of the universe; and like energy, which is traditionally defined
> operationally as possessing the capacity to perform work, so

information is defined operationally as possessing the capacity to organize a system (p. 1).

I would say that the information is the order in the system, not the capacity to create it. Information is not a force that does things to natural objects. Rather, it is the pattern of organization to be found in any matter or energy, moving or still.

Is the definition too inclusive?

Some have said that this definition of information includes "everything" and that the definition is, therefore, meaningless. First, it does not include everything, because it does not include total entropy, undifferentiated chaos, which, by definition, has no pattern or order. Second, information has not been defined as "everything." Rather, information is *the pattern of organization of everything*. As argued in the previous section, information is distinct from matter and energy (Wiener, 1961, p. 132). The value and conceptual productivity of this more specific definition was discussed at length in the section, "The evolutionary context," above. Third, if animals and humans process all the things that this definition of information includes and derive meaningful interpretations of them, then, surely, anything that they can so interpret should be included in a comprehensive and fundamental definition of information for information science.

Do not human beings potentially remember, respond, or act on any conceivable bit of information encountered? Hitting your knee on a rock, or hearing your friend's question, as well as countless other things and experiences can all be informative. So why should any of these information structures be eliminated in a definition of information?

What really matters for the development of information science as a discipline is whether such a definition can be successfully employed and built upon in the field's theory and practice. In effect, this entire essay and its companion essay, "Fundamental forms of information" (Bates, 2006), constitute an effort to demonstrate the potentialities within this definition.

It has also been argued by some that only patterns of organization that have actually been experienced by a person should be considered information, that is, information is a meaningful concept only in relation to subjective experience. By that line of reasoning, if some signal or data has no impact on a person somewhere sometime, then there is no information in that signal or data. But are not many things *potentially* informative? Should we eliminate those from our definition of information because they have not actually informed a person yet? Further, how can we realistically study information-seeking behavior, if we cannot consider as at least potentially informative all those things a person has *not yet* been informed by?

How does this information definition relate to emergence?

Following Claude Shannon, who produced his ground-breaking work on means to measure the amount of information after World War II, many efforts went into adapting that understanding to more cognitively- and socially-based perspectives on information. Though much was learned from those efforts, on the whole, they did not succeed in forming a basis upon which information science could build a broad understanding of information and knowledge. The approach taken in this essay may take us a step closer to that development, if not all the way.

Decades of research in evolutionary biology and psychology on human development over hundreds of thousands of years has made clear the extent to which humans are *not* simple straightforward information processing machines, processing so and so many bits of information, as a computer does, to get through life. Instead, there has been more and more recognition that human beings are mixtures of capabilities, some very highly developed and efficient, others slower and clumsy. For example, we are a very social animal and can detect extremely small distinctions in the appearance of people's faces and have a part of our brain devoted to that capability, while we cannot distinguish people nearly so well by the look of their knees or hands. The latter may be nearly as large as faces, but we do not have the mental equipment to make the key distinctions we need using those as indicators of identity.

We have evolutionary short-cuts built into our cognitive processing, so we know to flee fire, for example, without thinking practically at all. But in the case of a slower-developing crisis, one not typically a part of our hunter-gatherer inheritance, we may not take it seriously or miss the warning signs altogether until it is too late.

Information 2 has been defined as the pattern of organization of matter and energy given meaning. This initial meaning assignment, before the information is integrated with the stores of knowledge, can occur relatively instantly in cases where the association among elements is well developed. Human language capabilities are so well developed that most of the time when we look at a chair, we think and speak of it as a chair *first* and not as a pattern of edges and light and dark, though we can look at it in the latter sense too, if we want. As an experienced adult in a capitalist society, I can recognize certain sequences of behavior and circumstances as *bait and switch* shortly after entering a store. I experience my interactions with the salesperson with this extremely compact, efficient, *emergent* characterization of *bait and switch,* not as a long sequence of vocal and behavioral interactions.

In other cases, emergent understandings may be inherently more difficult or for some other reason take longer to form. For example, in dealing with someone who has a clever, subtly manipulative personality, it may take numerous interactions over years before we grasp how they relate to us and others and conclude that we should not trust them. Many new ideas in science and the arts took a long time to emerge because, despite our great intelligence as a species, something about those ideas so violated conventional thinking, that we were unable to develop that emergent understanding. With many such ideas, however, once developed, they seem terribly obvious: "Why didn't we think of that sooner?!"

Thus, it is argued that this understanding of information, drawing on an evolutionary psychology approach, takes information science's understanding of information from the physical up to the biological and anthropological. Further, by thinking in terms of emergence, we can see that many actions and expressions can be efficiently stored and talked about in terms of their emergent meanings, while still using the key definitions for information, data and knowledge that were presented in this essay.

Can social science metatheories in information studies be reconciled with the approach taken here?

In recent years, there has been a healthy debate in information studies over what the philosophical underpinnings of the field should be (Bates, 1999; Budd, 2001; Dick, 1995; Frohmann, 1994; Hjørland, 1998, 2002a, b; Hjørland & Albrechtsen, 1995; Talja et al., 1999; Tuominen et al., 2002; Wiegand, 1999). Over the history of this field, people from many different disciplinary traditions have brought their worldviews and interests to bear on information-related questions. Humanities, social science and physical science and engineering approaches all have deep roots in information studies. Though sophisticated social science research dates back to the 1930s in the field, the growing interest in studying people and their interactions with information over the last thirty years has eventually led to a very large social science presence. The immediate reaction of many people to the ideas in this essay is that those ideas are purely physical and sensory and do not have much to do with the core interests in information seeking and other social science questions in the field. It is to be hoped that the arguments and examples provided here and in the companion essay, (Bates, 2006), have gone some way to dispel that assumption.

However, there are deeper questions lurking here as well, regarding the underlying "metatheory," or fundamental theoretical and philosophical outlook of this essay. Recent papers by Tuominen et al. (2002) and Talja et al. (2005) have identified a total of four principal world views that they find to be operative in this field over the last thirty years or so. They name these the information transfer model, constructivism, collectivism and constructionism. They view these four metatheories as emphasizing, respectively, the information, the individual's cognitive viewpoint, the combined socio-cognitive viewpoint, and the social and dialogic aspects of knowledge formation.

Talja et al. (2005) argue that each metatheory has distinct applications that are most useful in tackling particular kinds of questions. I would agree and would add that even though there may be conflicts and disagreements, it is worthwhile to work at integrating the research results of the applications of these metatheories and even, perhaps, ultimately, the metatheories themselves.

Cosmides et al., working from the standpoint of evolutionary psychology, argue for what they call conceptual integration in the natural and social sciences. In the context of this essay, theirs constitutes a fifth metatheoretical perspective. Specifically, they state:

> *Conceptual integration.* . . [italics in original] refers to the principle that the various disciplines within the behavioral and social sciences should make themselves mutually consistent and consistent with what is known in the natural sciences as well. . . . The natural sciences are already mutually consistent: the laws of chemistry are compatible with the laws of physics, even though they are not reducible to them. . . .
>
> Such is not the case in the behavioral and social sciences. Evolutionary biology, psychology, psychiatry, anthropology, sociology, history and economics largely live in inglorious isolation from one another (Cosmides et al., 1992, p. 4).

The authors provide a number of examples of ways in which the behavioral and social sciences can help each other and, in some cases, prevent dead-end research paths in one discipline based on presuppositions that have been definitively refuted in other disciplines. They go on to argue:

Conceptual integration generates this powerful growth in knowledge because it allows investigators to use knowledge developed in other disciplines to solve problems in their own (Cosmides et al., 1992, p. 12).

They continue:

> At present, crossing such boundaries is often met with xenophobia, packaged in the form of such familiar accusations as 'intellectual imperialism' or 'reductionism'. But. . . we are neither calling for reductionism nor for the conquest and assimilation of one field by another. . . . In fact, not only do the principles of one field not reduce to those of another, but by tracing the relationships between fields, additional principles often appear.
>
> Instead, conceptual integration simply involves learning to accept with grace the irreplaceable intellectual gifts offered by other fields (Cosmides et al., 1992, p. 12).

In short, they are arguing for working toward the goal of uniting all human knowledge, and doing so by examining what each field has to offer. To date, the exciting new work that is going on in a cluster of new fields: evolutionary culture theory (Durham, 1991), evolutionary psychology (Barkow et al., 1992), evolutionary neuroscience (La Cerra & Bingham, 2002) and human behavioral ecology (Winterhalder & Smith, 2000), have been largely ignored in many of the social sciences, including within information science. (In our field, Brier, 1996; Cooper, 2001; Madden, 2004; Sandstrom, 1994, 1999, 2001 and Sandstrom & Sandstrom, 1995, are exceptions, having drawn on this literature.) This essay has constituted a very preliminary and novice effort to draw on that perspective, but within an orientation that is meaningful for our field.

Some have insisted that the integrative perspective is fundamentally at odds with the several metatheories described by Tuominen et al. (2002) and Talja et al. (2005), that these metatheories cannot be reconciled with a scientific view. Cosmides et al. (1992) would suggest that such a claim of incompatibility would speak against, rather than for, those other metatheories. They would argue that the long-term objective is surely to unite all human knowledge, not to argue that certain perspectives are so unique and special that they cannot be held in mind at the same time as perspectives and knowledge from other disciplines.

Again, having an understanding of psychology that is consonant with biology does not reduce psychology to biology, nor anthropology to psychology, nor sociology to anthropology, etc. Key findings about cultures and societies cannot be predicted from psychology, nor can psychology be predicted from biology, any more than key findings in biology can be predicted from chemistry or physics. Each discipline rests within the discipline(s) at a more fundamental level, but does not consist solely of, nor is reducible to, the discipline(s) below it.

In information studies, we came late to seeing the value in the several newer social science metatheories described by Talja et al. (2005); these perspectives were well developed in other fields decades before we began to draw on their literatures. We have much yet to absorb.

But evolutionary psychology and related new initiatives have already been developing for decades as well. See references in Barkow et al. (1992); Clark (1997); Damasio (1999); Durham (1991); Jackendoff (2002); Johnson (1987); Knight et al. (2000); La Cerra and Bingham (2002); Sperber (1996); Winterhalder and Smith (2000). These ideas have also seen application in many fields both inside and outside the sciences. See, for example, a sampling of recent papers drawing on this perspective in literature, history, social science, and philosophy (Carroll, 1999; Kenrick et al., 2002; Lachapelle, 2000; Mallon & Stich, 2000; Plotkin, 2003; Siegert & Ward, 2002; Stuart-Fox, 1999; Sugiyama, 2001; Waal, 2002). Indeed, the book edited by Barkow et al. (1992) and its nineteen component papers have already been cited collectively over 1600 times. Shall we fall decades behind on this evolutionary approach as well, as we did with the other social science metatheories, or do we begin to draw from it now?

Summary and conclusions

Summary

Many definitions of information have been suggested throughout the history of information science. In this essay, the objective has been to provide a definition that is usable for the physical, biological and social meanings of the term, covering the various senses important to our field.

Information has been defined as the *pattern of organization of matter and energy*. Information is everywhere except where there is total entropy. Living beings process, organize and ascribe meaning to information. Some pattern of organization that has been given meaning by a living being has been defined as information 2, while the above definition is information 1, when it is desirable to make the distinction. Knowledge has been defined as information given meaning and integrated with other contents of understanding. Meaning itself is rooted ultimately in biological survival. In the human being, extensive processing space in the brain has made possible the generation of extremely rich cultural and interpersonal meaning, which imbues human interactions. (In the short term, not all meaning that humans ascribe to information is the result of evolutionary processes. Our extensive brain processing space also enables us to hold beliefs for the short term that, over the long term, may actually be harmful to survival.)

Data 1 has been defined as that portion of the entire information environment (including internal inputs) that is taken in, or processed, by an organism. Data 2 is that information that is selected or generated and used by human beings for research or other social purposes.

This definition of information is not reductive—that is, it does not imply that information is all and only the most microscopic physical manifestation of matter and energy. Information principally exists for organisms at many *emergent* levels. A human being, for example, can see this account as tiny marks on a piece of paper, as letters of the alphabet, as words of the English language, as a sequence of ideas, as a genre of publication, as a philosophical position and so on.

Thus, patterns of organization are not all equal in the life experience of animals. Some types of patterns are more important, some less so. Some parts of patterns are repetitive and can be compressed in mental storage. As mental storage space is generally limited and its maintenance costly to an animal, adaptive advantage accrues to the species that develops efficient storage. As a result, many species process elements of their environment in ways efficient and effective for their particular purposes; that is, as patterns of organization that are experienced as emergent wholes.

We see a chair as a chair, not only as a pattern of light and dark. We see a string of actions by a salesperson as *bait and switch,* not just as a sequence of actions. We understand a series of statements as parts of a whole philosophical argument, not just as a series of sentences. The understanding of information embraced here recognizes and builds on the idea that these emergent wholes are efficient for storage and effective for the life purposes of human beings as successful animals (to date) on our planet.

Thus, people experience their lives in terms of these emergent objects and relations, for the most part. Likewise, information is stored in retrieval systems in such a way that it can be represented to human beings in their preferred emergent forms, rather than in the pixels or bits in which the information is actually encoded within the information system.

Conclusions

I have long felt that to succeed in the process of developing a broadly applicable, encompassing understanding of information for our field, we must begin at the physical and biological levels and move up to the cultural, social, cognitive and aesthetic. Beginning with the physical substrates of our bodies and of our cultures in defining information does not mean stopping there. Instead, we are thereby enabled to develop a truly grounded understanding of the core concept of our discipline—an understanding that can be built on and enriched indefinitely much more by drawing on the learning from the several other metatheoretical approaches of interest in information studies.

Thus, to take specific examples from the thinking of information studies, the construction of emergent patterns of organization can be seen as the general process at the root of Dervin's (1999) concept of "sensemaking," in which people are seen as seeking information in order to make sense of situations in their lives. 'Making sense' consists operationally in constructing an understanding, an emergent, simplifying organization of ideas and experience that satisfies people's sense of completion, of wholeness. Likewise, the construction of such emergent patterns is also at the heart of Kuhlthau's (2004) information search process, in which the searcher gradually forms a more and more coherent conception of the focus, theme, or argument of the planned paper. The formulation of an 'Anomalous State of Knowledge' (Belkin et al., 1982) is in response to a situation where the information seeker feels a lack of success in creating a coherent, emergent, understanding around some phenomenon of interest and attempts to describe the shape of the gap in knowledge.

Similarly, at the social, rather than individual level, people create coherent, social groupings utilizing emergent constructions of their situations. Drawing on Hjørland and Albrechtsen's (1995) sense of *domain,* we can imagine a group of sociologists studying, say, the intersection of class and race. They create a social and intellectual domain, a specialty, that is structured socially in various forms useful to researchers, such as in professional organizations and listservs. There is a body of agreed-upon learning, as well as a research philosophy and methodology, that is, a paradigm (Kuhn, 1970), motivating and uniting the group of researchers, as well as publications, laboratories and other social institutions. This entire complex of social arrangements and information forms is a domain. We are now recognizing that to understand information seeking and use, we must examine all the forms of information that compose the emergent entity we call a domain—the social arrangements, information genres and behavioral patterns.

Finally, the discourses of the constructionist metatheory that Talja et al. (2005) refer to, that is, the social and intellectual arrangements of language and language patterns used by a group of people, can be understood, in the terms used here, as one means of constructing a useful emergent whole, a world view about some part of life expressed through language.

I see no conflict between a definition of information as the pattern of organization of matter and energy and our social understanding of human beings constructing all manner of emergent patterns of importance to themselves. Whether called sense, focus, or ASK, whether called domains, paradigms, or discourses, these information structures constitute coherent, emergent wholes, built up out of the information described herein.

Much remains to be done to develop the fullness of an understanding of information for this field. This essay will, I hope, contribute to that effort.

ACKNOWLEDGEMENTS

My thanks to Suresh Bhavnani, Terry Brooks, Michael Buckland, Michele Cloonan, William Cooper, Laura Gould, Jenna Hartel, Robert M. Hayes, P. Bryan Heidorn, Anders Hektor, Birger Hjørland, Peter Ingwersen, Jarkko Kari, Jaana Kekäläinen, Kimmo Kettunen, Erkki Korpimäki, Poul S. Larsen, Mary N. Maack, Marianne L. Nielsen, Pamela Sandstrom, Reijo Savolainen, Dagobert Soergel, Sanna Talja, Pertti Vakkari and the anonymous reviewers for their thoughtful and insightful comments on this and the companion essay (Bates, 2006).

REFERENCES

Allman, J.M. (2000). *Evolving brains.* New York, NY: Scientific American Library.

Aspray, W. (1985). The scientific conceptualization of information: A survey. *Annals of the History of Computing, 7*(2), 117–140.

Atlan, H., & Cohen, I.R. (1998). Immune information, self-organization and meaning. *International Immunology, 10*(6), 711–717.

Barham, J. (1996). A dynamical model of the meaning of information. *BioSystems, 38*(2–3), 235–241.

Barkow, J.H., Cosmides, L., & Tooby, J. (Eds.). (1992). *The adapted mind: Evolutionary psychology and the generation of culture.* New York: Oxford University Press.

Bates, M.J. (1987). Information: The last variable. *Proceedings of the 50th ASIS Annual Meeting, 24,* 6–10.

Bates, M.J. (1999). The invisible substrate of information science. *Journal of the American Society for Information Science, 50*(12), 1043–1050.

Bates, M.J. (2006). Fundamental forms of information. *Journal of the American Society for Information Science and Technology, 57*(8), 1033-1045.

Belkin, N.J. (1978). Information concepts for information science. *Journal of Documentation, 34*(1), 55-85.

Belkin, N.J., & Robertson, S.E. (1976). Information science and the phenomenon of information. *Journal of the American Society for Information Science, 27*(4), 197-204.

Belkin, N.J., Oddy, R.N., & Brooks, H.M. (1982). ASK for information retrieval: Part 1: Background and theory. *Journal of Documentation, 38*(2), 61-71.

Bernhardt, P.C., Dabbs, J.M., Fielden, J.A., & Lutter, C.D. (1998). Testosterone changes during vicarious experiences of winning and losing among fans at sporting events. *Physiology & Behavior, 65*(1), 59-62.

Boulding, K.E. (1961). *The image: Knowledge of life and society.* Ann Arbor, MI: University of Michigan Press.

Brier, S. (1996). Cybersemiotics: A new interdisciplinary development applied to the problems of knowledge organisation and document retrieval in information science. *Journal of Documentation, 52*(3), 296-344.

Brier, S. (1998). Cybersemiotics: A transdisciplinary framework for information studies. *Biosystems, 46*(1-2), 185-191.

Brookes, B.C. (1975). The fundamental problem of information science. In V. Horsnell (Ed.), *Informatics 2*: Proceedings of a Conference Held by the ASLIB Coordinate Indexing Group (pp. 42-49). London: ASLIB.

Brookes, B.C. (1980). The foundations of information science. Part 1. Philosophical aspects. *Journal of Information Science, 2*(3-4), 125-133.

Buckland, M.K. (1991). Information as thing. *Journal of the American Society for Information Science, 42*(5), 351-360.

Budd, J.M. (2001). *Knowledge and knowing in library and information science: A philosophical framework.* Lanham, MD: Scarecrow Press.

Calvin, W.H. (2002). *A brain for all seasons: Human evolution and abrupt climate change.* Chicago, IL: University of Chicago Press.

Capurro, R., & Hjørland, B. (2003). The concept of information. *Annual Review of Information Science and Technology, 37*, 343-411.

Carroll, J. (1999). The deep structure of literary representations. *Evolution and Human Behavior, 20*(3), 159-173.

Clark, A. (1997). *Being there: Putting brain, body and world together again.* Cambridge, MA: MIT Press.

Cooper, W.S. (2001). *The evolution of reason: Logic as a branch of biology.* Cambridge, England: Cambridge University Press.

Cornelius, I. (2002). Theorizing information for information science. *Annual Review of Information Science and Technology, 36*, 393-425.

Cosmides, L., Tooby, J., & Barkow, J.H. (1992). Introduction: Evolutionary psychology and conceptual integration. In J.H. Barkow, L. Cosmides, & J. Tooby (Eds.), *The adapted mind: Evolutionary psychology and the generation of culture*, (pp. 3-15). New York: Oxford University Press.

Damasio, A. (1999). *The feeling of what happens: Body and emotion in the making of consciousness.* San Diego, CA: Harvest.

Day, R.E. (2001). *The modern invention of information.* Carbondale, IL: Southern Illinois University Press.

Derr, R.L. (1985). The concept of information in ordinary discourse. *Information Processing & Management, 21*(6), 489-499.

Dervin, B. (1977). Useful theory for librarianship: Communication, not information. *Drexel Library Quarterly, 13*(3), 16-32.

Dervin, B. (1983). Information as a user construct: The relevance of perceived information needs to synthesis and interpretation. In S.A. Ward & L.J. Reed (Eds.), *Knowledge structure and use: Implications for synthesis and interpretation* (pp. 155-183). Philadelphia: Temple University Press.

Dervin, B. (1999). On studying information seeking methodologically: The implications of connecting metatheory to method. *Information Processing & Management, 35*(6), 727-750.

Dick, A.L. (1995). Library and information science as a social science: Neutral and normative conceptions. *The Library Quarterly, 65*(2), 216-235.

Dretske, F.I. (1981). *Knowledge and the flow of information.* Cambridge, MA: MIT Press.

Durham, W.H. (1991). *Coevolution: Genes, culture and human diversity.* Stanford, CA: Stanford University Press.

Elia, M. (1992). Organ and tissue contribution to metabolic rate. In J.M. Kinney & H.N. Tucker (Eds.), *Energy metabolism: Tissue determinants and cellular corollaries* (pp. 61-79). New York: Raven Press.

Ellis, D. (1992). The physical and cognitive paradigms in information retrieval research. *Journal of Documentation, 48*(1), 45-64.

Fischer, R. (1993). From transmission of signals to self-creation of meaning: Transformations in the concept of information. *Cybernetica, 36*(3), 229-243.

Floridi, L. (2002). What is the philosophy of information? *Metaphilosophy, 33*(1-2), 123-145.

Fox, C.J. (1983). *Information and misinformation: An investigation of the notions of information, misinformation, informing and misinforming.* Westport, CT: Greenwood Press.

Freeman, W.J. (2000a). *How brains make up their minds.* New York: Columbia University Press.

Freeman, W.J. (2000b). A neurobiological interpretation of semiotics: Meaning, representation and information. *Information Sciences, 124*(1-4), 93-102.

Frohmann, B. (1994). Discourse analysis as a research method in library and information science. *Library & Information Science Research, 16*(2), 118-138.

"Frost." (2005). *Encyclopædia Britannica.* Retrieved 1 June, 2005, from Encyclopædia Britannica Online. http://search.eb.comebarticle?eu=36165 [Subscription required].

Gardner, H. (1985). *The mind's new science: A history of the cognitive revolution.* New York: Basic Books.

Goonatilake, S. (1991). *The evolution of information: Lineages in gene, culture and artefact.* London: Pinter.

Gray, P.B., Kahlenberg, S.M., Barrett, E.S., Lipson, S.F., & Ellison, P.T. (2002). Marriage and fatherhood associated with lower testosterone in males. *Evolution and Human Behavior, 23*(3), 193-201.

Hammarberg, R. (1981). The cooked and the raw. *Journal of Information Science, 3*(6), 261-267.

Hayles, N.K. (1999). *How we became posthuman: Virtual bodies in cybernetics, literature and informatics.* Chicago, IL: University of Chicago Press.

Hektor, A. (1999). Immateriality: On the problem of information. *KFB—Rapport, (24)* 134-162.

Hjørland, B. (1998). Theory and metatheory of information science: A new interpretation. *Journal of Documentation, 54*(5), 606-621.

Hjørland, B. (2002a). Epistemology and the socio-cognitive perspective in information science. *Journal of the American Society for Information Science and Technology, 53*(4), 257-270.

Hjørland, B. (2002b). Principia informatica: Foundational theory of information and principles of information services. In H. Bruce, R. Fidel, P. Ingwersen, & P. Vakkari (Eds.), *Emerging frameworks and methods: Proceedings of the Fourth International*

Conference on Conceptions of Library and Information Science (CoLIS4), (pp. 109–121). Greenwood Village, CO: Libraries Unlimited.

Hjørland, B., & Albrechtsen, H. (1995). Domain analysis: Toward a new horizon in information science. *Journal of the American Society for Information Science, 46*(6), 400–425.

Houston, R.D., & Harmon, E.G. (2002). Re-envisioning the information concept: Systematic definitions. In H. Bruce, R. Fidel, P. Ingwersen, & P. Vakkari (Eds.), *Emerging frameworks and methods: Proceedings of the Fourth International Conference on Conceptions of Library and Information Science* (CoLIS4) (pp. 305–308). Greenwood Village, CO: Libraries Unlimited.

Jackendoff, R. (2002). *Foundations of language: Brain, meaning, grammar, evolution.* New York: Oxford University Press.

Johnson, M. (1987). *The body in the mind: The bodily basis of meaning, imagination and reason.* Chicago, IL: University of Chicago Press.

Kenrick, D.T., Maner, J.K., Butner, J., Li, N.P., Becker, D.V., & Schaller, M. (2002). Dynamical evolutionary psychology: Mapping the domains of the new interactionist paradigm. *Personality and Social Psychology Review, 6*(4), 347–356.

Knight, C., Studdert-Kennedy, M., & Hurford, J.R. (Eds.). (2000). *The evolutionary emergence of language: Social function and the origins of linguistic form.* Cambridge, England: Cambridge University Press.

Kochen, M. (1983). Library science and information science: Broad or narrow? In F. Machlup & U. Mansfield (Eds.), *The study of information: Interdisciplinary messages* (pp. 371–377). New York: Wiley.

Korpimäki, E., Koivunen, V., & Hakkarainen, H. (1996). Microhabitat use and behavior of voles under weasel and raptor predation risk: Predator facilitation? *Behavioral Ecology, 7*(1), 30–34.

Kuhlthau, C.C. (2004). *Seeking meaning: A process approach to library and information services* (2nd ed.). Westport, CT: Libraries Unlimited.

Kuhn, T.S. (1970). *The structure of scientific revolutions* (2nd ed.). Chicago, IL: University of Chicago Press.

La Cerra, P., & Bingham, R. (2002). *The origin of minds: Evolution, uniqueness and the new science of the self.* New York: Harmony Books.

Lachapelle, J. (2000). Cultural evolution, reductionism in the social sciences and explanatory pluralism. *Philosophy of the Social Sciences, 30*(3), 331–361.

Lachman, R., Lachman, J.L., & Butterfield, E.C. (1979). *Cognitive psychology and information processing: An introduction.* Hillsdale, NJ: Lawrence Erlbaum Associates.

Langman, R.E., & Cohn, M. (1999). Away with words: Commentary on the Atlan-Cohen essay 'Immune information, self-organization and meaning'. *International Immunology, 11*(6), 865–870.

Losee, R.M. (1997). A discipline independent definition of information. *Journal of the American Society for Information Science, 48*(3), 254–269.

Losee, R.M., Jr. (1990). *The science of information: Measurement and applications.* San Diego, CA: Academic Press.

Machlup, F., & Mansfield, U. (Eds.). (1983). *The study of information: Interdisciplinary messages.* New York: Wiley.

MacKay, D.M. (1969). *Information, mechanism and meaning.* Cambridge, MA: MIT Press.

Madden, A.D. (2004). Evolution and information. *Journal of Documentation, 60*(1): 9–23.

Mallon, R., & Stich, S.P. (2000). The odd couple: The compatibility of social construction and evolutionary psychology. *Philosophy of Science, 67*(1), 133–154.

Meadow, C.T., & Yuan, W. (1997). Measuring the impact of information: Defining the concepts. *Information Processing & Management, 33*(6), 697–714.

Miller, G.A. (1951). *Language and communication.* New York: McGraw-Hill.

Parker, E.B. (1974). Information and society. In C.A. Cuadra & M.J. Bates (Eds.), *Library and information service needs of the nation: Proceedings of a conference on the needs of occupational, ethnic and other groups in the United States* (pp. 9–50). Washington, DC: U.S.G.P.O. (ERIC #ED 101 716).

Plotkin, H. (2003). We-intentionality: An essential element in understanding human culture? *Perspectives in Biology and Medicine, 46*(2), 283–296.

Pratt, A.D. (1977). The information of the image: A model of the communications process. *Libri, 27*(3), 204–220.

Raber, D. (2003). *The problem of information: An introduction to information science.* Lanham, MD: Scarecrow Press.

Random House Unabridged Dictionary (2nd ed.) (1993). New York: Random House.

Rayward, W.B. (1992). Restructuring and mobilising information in documents: A historical perspective. In P. Vakkari & B. Cronin (Eds.), *Conceptions of library and information science: Historical, empirical and theoretical perspectives.* Proceedings of the International Conference Held for the Celebration of 20th Anniversary of the Department of Information Studies, University of Tampere, Finland, 26–28 August, 1991, (pp. 50–68). London: Taylor Graham.

Sandstrom, A.R., & Sandstrom, P.E. (1995). The use and misuse of anthropological methods in library & information science research. *The Library Quarterly, 65*(2), 161–199.

Sandstrom, P.E. (1994). An optimal foraging approach to information seeking and use. *The Library Quarterly, 64*(4), 414–449.

Sandstrom, P.E. (1999). Scholars as subsistence foragers. *Bulletin of the American Society for Information Science, 25*(3), 17–20.

Sandstrom, P.E. (2001). Scholarly communication as a socioecological system. *Scientometrics, 51*(3), 573–605.

Shannon, C.E., & Weaver, W. (1975). *The mathematical theory of communication.* Urbana, IL: University of Illinois Press.

Siegert, R.J., & Ward, T. (2002). Evolutionary psychology: Origins and criticisms. *Australian Psychologist, 37*(1), 20–29.

Sperber, D. (1996). *Explaining culture: A naturalistic approach.* Oxford, England: Blackwell.

Stanley, S.M. (1996). *Children of the ice age: How a global catastrophe allowed humans to evolve.* New York: Harmony Books.

Stonier, T. (1997). *Information and meaning: An evolutionary perspective.* Berlin: Springer.

Stuart-Fox, M. (1999). Evolutionary theory of history. *History & Theory, 38*(4), 33–51.

Sugiyama, M.S. (2001). Narrative theory and function: Why evolution matters. *Philosophy and Literature, 25*(2), 233–250.

Talja, S., Keso, H., & Pietiläinen, T. (1999). The production of 'context' in information seeking research: A metatheoretical view. *Information Processing & Management, 35*(6), 751–763.

Talja, S., Tuominen, K., & Savolainen, R. (2005). 'Isms' in information science: Constructivism, collectivism and constructionism. *Journal of Documentation, 61*(1), 79–101.

Thompson, F.B. (1968). The organization is the information. *American Documentation, 19*(3), 305–308.

Tuominen, K., Talja, S., & Savolainen, R. (2002). Discourse, cognition and reality: Toward a social constructionist metatheory for library and information science. In H. Bruce, R. Fidel, P. Ingwersen, & P. Vakkari (Eds.), *Emerging frameworks and methods: Proceedings of the Fourth International Conference on Conceptions of Library and Information Science* (CoLIS4). (pp. 271–283). Greenwood Village, CO: Libraries Unlimited.

Vakkari, P., & Cronin, B. (Eds.). (1992). *Conceptions of library and information science: Historical, empirical and theoretical perspectives*. Proceedings of the International Conference Held for the Celebration of 20th Anniversary of the Department of Information Studies, University of Tampere, Finland, 26–28 August, 1991. London: Taylor Graham.

Waal, F.B.M. de (2002). Evolutionary psychology: The wheat and the chaff. *Current directions in Psychological Science, 11*(6), 187–191.

Wagner, D.G., & Berger, J. (1985). Do sociological theories grow? *American Journal of Sociology, 90*(4), 697–728.

Wellisch, H. (1972). From information science to informatics: A terminological investigation. *Journal of Librarianship, 4*(3), 157–187.

Wersig, G., & Neveling, U. (1975). The phenomena of interest to information science. *The Information Scientist, 9*(4), 127–140.

Wiegand, W.A. (1999). Tunnel vision and blind spots: What the past tells us about the present; reflections on the twentieth century history of American librarianship. *The Library Quarterly, 69*(1), 1–32.

Wiener, N. (1961). *Cybernetics: Or control and communication in the animal and the machine* (2nd ed.). Cambridge, MA: MIT Press.

Wills, P.R. (1994). Does information acquire meaning naturally? *Berichte der Bunsen-Gesellschaft für Physikalische Chemie, 98*(9), 1129–1134.

Winterhalder, B., & Smith, E.A. (2000). Analyzing adaptive strategies: Human behavioral ecology at twenty-five. *Evolutionary Anthropology, 9*(2), 51–72.

Wright, R. (1994). *The moral animal: Evolutionary psychology and everyday life*. New York: Vintage Books.

2

Fundamental forms of information

ABSTRACT

Fundamental forms of information, as well as the term *information* itself, are defined and developed for the purposes of information science/studies. Concepts of natural and represented information (taking an unconventional sense of representation), encoded and embodied information, as well as experienced, enacted, expressed, embedded, recorded, and trace information are elaborated. The utility of these terms for the discipline is illustrated with examples from the study of information-seeking behavior and of information genres. Distinctions between the information and curatorial sciences with respect to their social (and informational) objects of study are briefly outlined.

Introduction

The objective of this article is to present and justify a definition of information and several fundamental information forms. These forms are capable of, and suitable for, use in the research and theory development of information science. I address common questions and reactions to the definitions in the latter part of the article. At the end of the article, I present several examples to illustrate how these forms can be utilized in the field's thinking, and form the basis for further thought on the fundamental questions of the field.

First published as Bates, M. J. (2006). Fundamental forms of information. *Journal of the American Society for Information Science and Technology, 57*(8), 1033–1045.

First, a definition of information itself is presented and briefly justified. Then the various fundamental forms of information are defined and elaborated. These forms are presented in relation to the work of Susantha Goonatilake (1991), a writer much overlooked both in information science and in the larger scientific community. Goonatilake presents a model of "information flow lineages," which complements the theory being developed here around information forms.

In a companion article (Bates, 2005), I review the literature and present and justify the definition of information in more detail, and in an evolutionary context. There, various theoretical and philosophical issues that inevitably surround such a fundamental concept as information are developed and argued in greater detail, including a discussion of the meaning of knowledge and data as well. Here, the information definition is presented more or less as a given; only the most common concerns expressed by readers and respondents at talks are briefly addressed, in order to focus the discussion primarily on the several fundamental forms.

A definition of information

The word *fundamental* is as important as the other words in this article's title. The effort here is to begin a consideration of information at the most fundamental levels possible. This is done on the grounds that for a word so basic that it defines our field, we need, for a satisfying theory of information, to begin at the beginning, at the root meaning of the term, and build up from there to the more social and other common meanings.

We know that we are continually subjected to a huge range of sensory inputs and internal experiences of sensations and thoughts. In fact, almost anything existing in the universe, that can come into human and other animals' purview, can be experienced as information—a bird call, our friend's "hello," the rock we trip over, the intuition we have about the honesty of someone we are talking to, a book we read. The definition of information used here, therefore, goes to the very basis of any living being's awareness: "Information is the pattern of organization of matter and energy." Though this definition is quoted from Edwin Parker (1974, p. 10), this approach to the concept was endemic at the time. Parker does not elaborate his definition, and no more recent theoretical development of this approach to information has been found. I believe this definition has much undeveloped potential.

Information is the pattern of organization of the *matter* of rocks, of the earth, of plants, of animal bodies, or of brain matter. Information is also the pattern of organization of the *energy* of my speech as it moves the air, or of the earth as it moves in an earthquake. Indeed, the only thing in the universe that does not contain information is total entropy; that alone is pattern-free. Because human beings can potentially act on or be influenced by virtually any imaginable information in the universe, if we want a truly fundamental and broadly applicable definition of information for our discipline, we must begin with one just this broad in meaning and application.

Applications of the term

First, information exists independently of living beings in the structure, pattern, arrangement of matter, and in the pattern of energy throughout the universe, and would do so whether or not any living being were present to experience the information. This is not to claim that we humans have a complete, clear, or "true" understanding of that independent universe, or that it is ultimately understandable in only one way, just that it exists in some form.

Second, the term *information*, as used here, also includes all the patterns of organization of matter and energy in living matter, including in the brains and bodies of human beings and other animals. This information arises from their genetic heritage and is further constructed by living beings interacting with the world, and stored in their sensory, nervous, and biochemical systems. Thus, our subjectively constructed understanding of the world, stored in our minds and feelings, can be viewed from the exterior as well, as one more body of information with a particular pattern of organization. These patterns of organization exist just as surely as the inanimate ones do, except that they are manifested in neuronal connections in the brain, action potentials, and the like. Each construction is conditioned by the animal's current experience and environment, genetic make-up, life history, and information-processing characteristics and limitations. Consequently, any construction animals may make of a given situation may vary considerably from animal to animal.

Thus, the argument presented here is that we can talk about information as an objectively existing phenomenon in the universe, which is also constructed, stored, and acted upon by living beings in countless different subjective ways, each way distinctive to the individual animal having the experience. At the same time, the selection and shaping of the information to be stored and acted upon by any individual animal or animal species is environmentally and evolutionarily shaped too, so experiences stored across a group of animals will also have many similarities. (Plants can respond to light and other environmental phenomena too, but for simplicity's sake, only animals—with special emphasis on human beings—will be discussed.)

One approach to information (Brier, 1996) draws on Bateson's definition of information as "a difference that makes a difference" (Bateson, 1972, p. 453). A difference to whom or what? Here, I argue that we must begin prior to that understanding, begin even before a sensing animal detects or assigns meaning to an experienced difference. As we shall see, humans and other animals can usefully identify a number of distinct types of information even prior to meaning assignment. Later, we will relate this definition of information to a more familiar understanding of the term, addressing what happens when we *become informed*.

In the end, the fundamental stance taken here is one of scientific observation. The phenomenon being observed is information, the pattern of organization of matter and energy as it exists in the universe and in living beings. The fact that we are observing, however, and claiming the objective existence of patterns of organization such as neurally stored memories, does not imply that *our understanding or construction of that objective existence is true, complete, correct, or the only possible understanding.* Nor does this claim imply that we deny the subjective variations and uniqueness in each individual's perception, extraction, and use of information in their minds and surroundings.

The senses of "pattern"

In defining information in this way, "pattern" is understood to refer both to (1) any kind of arrangement that is not pure chaos or disorganization, such as *"patterns of frost on the window"* and (2) *"a combination of qualities, acts, tendencies, etc., forming a consistent or characteristic arrangement: the behavior patterns of teenagers"* (*Random House Unabridged Dictionary*; Random House, 1993, p. 1423). (These are just two of many senses of "pattern" as typically defined in dictionaries.)

The first definition above implies a first-order pattern. The design we see on the frosted window is not part of any larger whole; it is simply a grouping or spotting of frost on the window that is not totally chaotic—so it does contain information—but nor is it a system. The second definition implies a second-order kind of pattern, one in which a variety of features are knitted together in an overall system or integral design. A pattern of behavior implies repeated similar cycles of activity, that is, some coherence above and beyond the bunching and scattering of the frost on the window.

These latter patterns may be characterized as "emergent," meaning that the sum of the elements constitutes something new, a whole with its own distinct qualities. Emergent phenomena are often dramatically different in character from the component elements that go into them. When we look at a chair, we see a (first-order) pattern of light and dark, solid colors and edges. We also are capable, however, of recognizing the chair *as a chair*, an emergent pattern that we can recognize quickly as a whole, because it possesses certain features we have learned to recognize through experience in a culture that uses chairs, and with the help of our inherent cognitive abilities.

This is not to imply that a sharp distinction between first- and second-order patterns is being proposed. Patterns form, dissolve, fragment, etc., in many ways continually, and *are seen* to dissolve, fragment, etc. continually. Rather, both definitions are provided to claim both senses for "pattern" in the above definition. A "pattern of organization," as used here, does not have to imply coherence, though it often does—just something other than pure, pattern-less entropy. (These last four paragraphs taken largely verbatim from Bates, 2005.)

Does this mean, then, that information is merely another word for pattern, or form? No, the crucial phrase in the definition is "pattern of organization." At the physical level, form is often thought of as the outer shape of an object, a three-dimensional concept. Pattern, on the other hand, is often thought of as two-dimensional—though frequently also emphasizing the outer shapes, but in two dimensions. For example, a checkerboard is seen as consisting of 64 squares of alternating colors; thus, it has a *checkerboard pattern*. That is certainly a *pattern*. But, as used here, the pattern of organization of the checkerboard includes everything in and around the board, not just its surface—the pattern of organization of the atoms and molecules of the material of the board itself, i.e., its internal structure, the waves/particles of different colors of light of the checkerboard surface, the pattern of differentiation of the edge of the board with the table upon which it sits and the air around it. The patterns of organization associated with this one checkerboard are multifarious in terms of the physical existence of the board alone.

Numerous additional patterns of organization are involved when we consider the ways observing animals may perceive and give meaning to the sight or touch of the board. Perhaps an insect crawling across the board feels slight indentations in the surface of the board that are invisible to humans, yet may not even observe the color shift from one square to the next. Someone from a society that does not play board games may see the board and its squares, but not think that they have any imaginable use. An advanced chess player, on the other hand, may ignore the color alterations of the squares and think only of regions of the board that are typically involved in different stages of play of chess.

The patterns of organization of everything in the universe (other than pure entropy or "patternless-ness") involve every physical, biological, perceptual, and cognitive pattern of organization that exists or is extracted by sensing beings. Information is thus not just the outer form, shape, or pattern of something as interpreted by human beings; rather, it includes the physical and biological patterns of organization not sensed by us as well, from the atomic to the galactic, from the virus to the ecosystem. Information, as defined here, includes all physical patterns of organization, all biological patterns of organization of life forms, and all constructed (and emergent) patterns of organization as extracted, stored, and used by living beings.

Breadth of the concept

Information science, or information studies, as it is variously known, is concerned with both animate and inanimate information in a very wide array of forms. Anything human beings interact with or observe can be a source of information. As information scientists, we accept that people create subjective constructions of their experience, and those constructions of information also have an objective existence in the nervous system.

Further, we collect and manage huge quantities of a wide array of kinds of information (patterns of organization of matter and energy), in an ever-growing set of media forms. For this field then, we need a basic information definition that incorporates all these various forms that we research and work with. The definition presented here is broad enough to cover all these kinds of information.

However, it might be argued that libraries and other information institutions do not collect rocks, bird calls, or intuitions. Therefore, is this definition *too* broad? We shall see below that within this encompassing definition, there are fundamental forms of information that distinguish well the focus of the professional activities of our discipline.

Considering the breadth of this definition in another respect, does the definition then imply that information is everything, and therefore not a particularly meaningful concept? The answer here is that no, information is *not* everything—rather, it is the *pattern of organization of everything*—except for total entropy or chaos, which is assumed to be pattern-free. This distinction between the pattern of organization and the material or energy that constitutes the pattern is crucial. As Wiener has said, "Information is information, not matter or energy" (1961, p. 132). Ours is the discipline that takes this phenomenon—the patterns of organization of matter and energy—as our central focus. (See Bates, 1999, for a detailed discussion of the distinctive character of information science, and Bates, 2005, for more discussion of the philosophical issues embedded in this approach to information.) We shall see that we can build effectively on this foundation.

Using the above definition of information as a basis, we turn now to consider several fundamental forms of information that may be understood to build on the basic definition, and which, it is argued, are useful for the disciplinary needs of information science/studies.

Natural and represented information

Here I introduce and develop the concepts of natural information, represented information, encoded information, and embodied information. See Table 1 for a glossary of terms with concise definitions of all major terms used or introduced in this article.

All information is *natural information*, in that it exists in the material world of matter and energy. Some natural information is distinctive, in that it is involved in representation at some moment of observation. *Represented information* is natural information that is encoded or embodied. Represented information can only be found in association with living organisms.

Encoded information is natural information that has symbolic, linguistic, and/or signal-based patterns of organization. *Embodied information* is the corporeal expression or manifestation of information previously in encoded form.

In the genetic, neural, and biochemical information of living organisms, and in information produced by living organisms, the information exists, actually or potentially, in a duality of embodiment and encoding. Specifically, encoded information may become embodied, and embodied information may become encoded.

Represented information, i.e., encoded or embodied information, can be created only by living organisms. Turning this around, with life begins representation. No effort will be made here to solve the "chicken or egg" problem of how represented information began. Suffice it to say that life does exist on this planet, and with all life comes the *encoded* genetic

TABLE 1. *Glossary of terms*

Embedded information: The pattern of organization of the enduring effects of the presence of animals on the earth; may be incidental, as a path through the woods, or deliberate, as a fashioned utensil or tool.

Embodied information: The corporeal expression or manifestation of information previously in encoded form.

Enacted information: The pattern of organization of actions of an animal in, and interacting with, its environment, utilizing capabilities and experience from its neural stores.

Encoded information: Natural information that has symbolic, linguistic, and/or signal-based patterns of organization.

Exosomatic information: Information stored in durable form external to the body (idea drawn from Goonatilake, 1991).

Experienced information: The pattern of organization of subjective experience, the feeling of being in life, of an animal.

Expressed information: The pattern of organization of communicatory scents, calls, gestures, and ultimately, human spoken language used to communicate among members of a species and between species.

Genetic information: Information contained in the genotype.

Genotype: The genetic constitution of a living thing (drawn from standard definitions in the biological literature).

Information 1: The pattern of organization of matter and energy.

Information 2: Some pattern of organization of matter and energy given meaning by a living being.

Knowledge: Information given meaning and integrated with other contents of understanding.

Natural information: All information is natural information, in that it exists in the material world of matter and energy. Represented information (below) is an important subset of natural information.

Neural-cultural information: Information that has been created by, processed in, or disseminated from animal nervous systems, especially human nervous systems. (**Neural information** may also be used for lower animals and/or specifically for the nervous system structures that make memory and action possible in general in animals and humans.)

Phenotype: The genetically and environmentally determined embodiment of a genotype. (Drawn from standard definitions in the biological literature.)

Recorded information: Communicatory or memorial information preserved in a durable medium.

Represented information: Natural information that is encoded or embodied.

Trace information: The pattern of organization of the residue that is incidental to living processes or which remains after living processes are finished with it.

Note. Definitions are the author's unless otherwise stated.

material, or genotype, that can be *embodied* in a living being, known to biologists as the phenotype.

So genetic information is encoded as DNA in the genotype and embodied in the phenotype, the living animal. Neural-cultural information is encoded in the brain and nervous system, and embodied in the experience, actions, and expressions of animals (more below). Exosomatic information, that is, information stored externally to the body, has been developed in complex ways by human beings, and may appear in many combinations of re-encoded and re-embodied forms. Humans may re-encode information as memory or as writing or by other means. These changes occur through natural processes, which may be automatic under certain conditions, or which may be carried out deliberately by a living being.

The use of the term *representation* here refers only to this duality of encoding and embodiment. The term is not used to refer to the various popular senses of an exact replica, model, or other common meaning of the term. In fact, rather than being alike, embodied information and encoded information usually look strikingly different. For example, the encoded strands of pig DNA—i.e., the genotype—do not look at all like the embodied piglets (phenotype) they may produce.

Goonatilake's Three Information Flow Lineages

Susantha Goonatilake (1991) has developed the concept of "information flow lineages," which I use in the following discussion. These lineages are first introduced, then related to the fundamental forms of information developed in this article.

The information flow lineages Goonatilake has identified and defined what he calls information flow lineages through the history of living matter on the planet. He argues that there have been, and continue to be, three lines of information transmission in association with life, which he calls the genetic, the neural-cultural, and the exosomatic "flow lines" of information transmission (1991, summary on pp. 118–120). Genetic information is transmitted through the usual processes of biological inheritance, influenced by natural selection. He states, "Beginning from prebiotic origins a continuous lineage of information and of organized complexity exists as a genetic flow system. As evolution proceeds through time, these lines of genetic information spread out and radiate into new environmental niches" (p. 118). Further, "Metaphorically one could say that the flow line has a 'conversation' with the environment, successful conversations becoming congealed in the genome" (p. 125).

Thus, the characteristics of organisms that promote their fitness, that is, that enable organisms to survive and reproduce in a given environment, are propagated through time in their genetic makeup.

Goonatilake describes the neural element as follows: "To adapt to the changing everyday environment feedback loops exist between the neural system and the environment, influencing the behavior of both" (p. 119). "In the phenotype which is created out of genotypal information, there are other information-producing 'devices' in the form of hormonal and neural circuits. . . . These extra-genetic devices provide also the means by which cultural information is transmitted from generation to generation" (p. 15).

He is careful to distinguish between neural systems, which "can exist without transmissions of acquired information from parent to offspring" (p. 119), and cultural systems in which "/s/uch transmissions across generations, however, do occur . . ." (p. 119). He treats the combined "neural-cultural" as the second flow line. Thus, information can be transferred between the animal and the environment, and from one animal to another through observation or communication in real time. He traces the history of encephalization, or growth of brain size relative to body size in mammals, and points out that some ecological niches demand more and some less brain development (p. 19).

Goonatilake further argues that a third flow line, the "exosomatic," has also developed. This exosomatic line consists of information stored outside the animal as the "externalization of memories" (p. 83). He uses as examples the pheromone trails laid down by ants to guide other ants to food, and even the beaten trails to a water hole that animals follow in a forest (p. 84). (There is some resemblance here to Dawkins' discussion of "the extended phenotype," 1982, not cited by Goonatilake.) The amount and complexity of exosomatic information has grown tremendously over the last several thousand years and has become extremely important for humans. As he points out, books were initially "a repository of men's memories" and later they became memory stores on which brains work (p. 122). Thus, in the third flow line, information can be transferred from one person to another, without the two people ever being in each other's presence, and therefore can skip generations.

He relates the three flow lines, and suggests, metaphorically, why they developed:

The rudimentary beginnings of these exosomatic information lines can be traced back to even the earliest animals. But they developed and expanded only with the primates and most elaborately only in association with humans. It is only when the adaptive limits of the neural-cultural information line begin to be reached that it, in at least some functions 'spills over' into the non-biological; in a similar way, the neural-cultural line developed after 'spilling over' from the genetic. (p. 83)

In other words, neural-cultural transmission is dependent on and arises out of genetic transmission, and exosomatic transmission develops out of the neural-cultural. As human beings used their sophisticated brains to develop dense, informationally rich cultures and learning that they wanted to retain and re-use, they first passed down stories and learning in person (neural-cultural flow line), then discovered and developed means of creating sophisticated external memory stores (exosomatic flow line), with which they could store and pass on vastly greater amounts of information.

Earlier, Brookes (1975) had also used the term *exosomatic* in the same sense to refer to external information stores. White developed the concept extensively also, calling it "external memory" (1992), and Brunk (2001) called a similar concept "exoinformation." At the genetic and neural-cultural levels, respectively, Dawkins distinguished genes and "memes" (1976), and Swanson (1983) defined "biogenes" and "sociogenes." However, I have found no one but Goonatilake who has incorporated these three paths of information transmission into a single model.

Goonatilake "flow lines" in relation to the "fundamental forms of information" The types of information to be discussed in the next three major sections are of my own devising, and were, in fact, developed largely prior to encountering Goonatilake's little-cited book. (Goonatilake defines information as "an organizing mechanism which provides an ability to deal with the environment. It is a symbolic description having modes of interpreting and interacting with the environment" 1991, p. 1.)

Table 2 shows the relationships between Goonatilake's flow lines and my fundamental forms of information.

Genetic information moves through the genetic flow line. My defined fundamental forms of experienced, enacted, and expressed information move through what Goonatilake called the "neural-cultural flow line," and embedded and recorded information move through Goonatilake's "exosomatic flow line." The emphasis in the following discussion will be on neural-cultural and exosomatic information, as well as residue, namely, the information left after animals are done with it—which was not addressed by Goonatilake.

More generally, I believe all pattern of organization in the universe is natural information. *Some* natural information is associated with life—either

TABLE 2. *Goonatilake's information flow lineages in relation to Bates' information forms.*

Bates: Natural information (in all things living and non-living)

    Bates: Represented information (associated with living things only)

GOONATILAKE'S FLOW LINES	BATES' TYPES OF INFORMATION PER FLOW LINE
Genetic flow line	Genetic information
Neural-cultural flow line	Experienced information Enacted information Expressed information
Exosomatic flow line	Embedded information Recorded information
Residue	Trace information

Note. Bates' represented information is a subset of natural information. It may be encoded or embodied at any moment of observation. For example, encoded capabilities stored in the brain may be embodied in the actions of an animal. Residue (not addressed by Goonatilake), consisting of trace information, is a flow line in which the information degrades from represented information to, simply, natural information (neither encoded nor embodied). Residue is the flow line of extinction and of the Biblical "dust to dust."

is living itself or was created by the living. That life-associated information is here called *represented information*, and includes genetic, neural-cultural, and exosomatic information, i.e., all the kinds of information that move through Goonatilake's three flow lines. Thus, I define information as both associated with and not associated with life, while Goonatilake dealt only with life-associated information. I further identify information in the process of degrading from life-associated to non-life-associated, that is, *trace information*.

Here, by definition, represented information is either encoded or embodied. Examples of *encoded* genetic, neural-cultural, and exosomatic information are, respectively, the genotype, the nervous system's links and structures laid down through experience (and deriving ultimately from genetic capabilities), and writing. Genetic *embodied* information is exemplified by the phenotype. Embodied neural-cultural information can be seen in the phenotype's experience, actions, and communications; these three all embody previously encoded information. (The interesting complexities of encoding and embodiment with respect to exosomatic information are not developed in this article.) The following sections provide more detail on several of the named types of information.

Types of neural-cultural information

As noted, neural-cultural information is *encoded* in the brain and nervous system. In animals and, in particular, in human beings, three fundamental modes of *embodied* information are identified: in experience, in actions in the world, and in communicatory expression. Each is discussed in turn in this section.

Experienced information

If we were to look into the brain of a person looking at a classroom chair, we would not see a miniature chair; rather, we would see only neurons firing. However, the person looking at the chair *does* see a chair. The neurons firing in the person's brain, therefore, create *an embodied subjective experience of seeing the chair*, an experience utterly unlike what is going on in the brain to create that experience. A stubbed toe, which produces another round of neuronal activity, is not felt as neurons firing by the person experiencing the incident; rather, that person feels the pain of a stubbed toe—*and in the toe*—not in the brain.

Thus, to feel our own experiences, the brain must create some pattern of neuronal firing that produces consciousness and the associated sense of experiencing life. The question of the nature of consciousness and the mind–brain relationship is one of the most hotly debated questions in all of science and philosophy currently (Chalmers, 1996; Damasio, 1999; Dennett, 1991; McCauley, 1996; Varela, Thompson, & Rosch, 1993, and many others). There is no agreed-upon understanding at this time in those fields, and no attempt is made to solve this challenging question here.

Here, for our purposes, we will simply consider subjective experience, including the experience of remembering, to be the first on a list of kinds of embodied information that result from neural encoded information. Again, all the stored knowledge, life experience, etc. that a person (or other animal) has is *encoded* in neural pathways of the brain. Life as experienced and remembered by the individual is *embodied* in whatever degree of consciousness or awareness that individual has.

One final point about experienced information: We experience our thoughts and activities as a conscious self, while, in ordinary practice, all of our other knowledge and memories are out of consciousness. Experienced information is not solely what we experience with conscious awareness, however. We can also experience a variety of kinds of out-of-consciousness information that are nonetheless active in creating our current experience. For example, when a pianist is asked by her music teacher to "play with more feeling," she can bring forth a variety of out-of-awareness knowledge and experience that will enrich her playing and give it more feeling. She could not articulate how she does this; she does not know what she draws upon. Yet she can do it—and the student with less training and experience cannot do it. Something in her encoded neural information, therefore, was brought to bear and was embodied in her playing.

Enacted information

We may experience our lives mentally in private, but when we start acting in the world, our genetically endowed talents and life knowledge become visible to the external world. When an animal enacts information, it acts in the world, utilizing whatever capabilities and experience it can from its neural stores. Fish cannot hide nuts and squirrels cannot breathe under water, but each type of animal is capable of embodying many other types of skills or behaviors, which it does lifelong. Animals enact their neural information by carrying out all the activities of their lives. Throughout the history of animals on the planet, much learning, especially that transferred from mother to offspring, has come about by observing and copying enacted information.

Human beings, who possess extraordinarily extensive knowledge, can enact vast numbers of different types of behaviors. Not only do we carry out the usual animal behaviors of eating, birthing, fighting, etc., but we have also developed a huge range of skills, from plumbing to brain surgery, as well as social institutions—religions, the arts, business, government, science. Aside from the physical buildings that often house these activities, the institutions themselves fully exist only when human beings use their knowledge and experience to enact the institutions in real time. Thus, enacted information can occur in isolation or in social contexts, where it becomes a part of the larger texture of social behavior.

Demonstrating how much power and other human relations and choices are embedded within social institutions has been one of the great achievements of the social sciences in the last 20 to 30 years. These social institutions exist by being renewed, reinforced, and gradually changed through time by the people involved in enacting them on a daily basis.

For example, the welfare office in a typical United States city exists physically only as a building. Nevertheless, what counts for the people working and seeking help there is the daily enacting of roles and relationships in real time in the interactions among people in that office. If the society sees welfare as unavoidable, and welfare recipients as people who are getting resources they do not deserve—the assumption in much of the United States today—then everyone involved will enact that relationship in countless ways in the daily activity of the welfare office. The office will be dirty and not air-conditioned. Supplicants will be required to wait many hours and will be treated rudely. In resentment, they will react with hostility, leading the staff to become still more negative in their relations, and so on. These unfortunate results do not happen by accident. They arise out of pre-existing collective social assumptions and attitudes, and the people involved carry out the social consensus about the institution of welfare in their daily enactment of work or supplication at the welfare office.

Expressed information

This form of embodied neural-cultural information consists of the pattern of organization of communicatory scents, calls, gestures, and ultimately, human spoken language used to communicate among members of a species and between species. Thus, expressed information has a quintessentially social function. Other than in a few cases, such as a spontaneous cry of pain or fear, all expressed information is intentionally communicative to others in the environment. Animals mark territory with scent, produce mating calls and danger calls; primates gesture expressively. Humans use spoken language and body language to communicate an extraordinarily rich variety of meaning. In humans, communication through expressed information is enormously important, and is supported by brain structures that make language possible. For all these reasons, I have set aside expressed information, technically a subset of enacted information, for independent consideration.

Types of exosomatic information

Exosomatic information, that is, information stored externally to the body of animals, is a type that is core to the interests of information science. Embedded and recorded information are described below.

Embedded information

If we survey all that results from the presence of living things on earth, we find many objects and other visible effects of the presence of animals. The spider makes its web; the bird builds a nest; the human being makes tools, utensils, and other artifacts. Embedded information is that enduring information created or altered by the actions of animals and people in the world. It may be incidental, as a path through the woods, or deliberate, as a fashioned artifact. The changes and added structure found in the nest, the cell phone, or the house all constitute embedded information—information that would not exist without the agency of animals. Because animals act, they leave evidence of their presence. (See also Dawkins' discussion of the "extended phenotype," 1982.)

The study of the embedded information of artifacts has been a prime means of learning about other cultures in the human sciences, especially about extinct cultures. Just as cultures develop socially shared attitudes and institutions that are enacted, so do people develop socially shared design styles and artifacts that are often remarkably stable in character through time and over wide geographical areas. We can learn much about people by studying these characteristics of the enduring physical remains of their cultures.

Though these objects may be seen to carry embedded information, there is only so much understanding the objects can provide. We can deduce, perhaps, how a flint knife was made, but may not be able to determine how a lost pottery-glazing technique was carried out. In short, the embedded information is generally not left by its creators to be informative, but rather is informative as an incidental consequence of the activities and skills of the people leaving the artifacts. We deduce what we can, and often must forego some other knowledge we might wish to extract.

Embedded information is not limited to earlier cultures, however. Quite the contrary, the impact, in embedded information, of the current human cultures on the planet is beyond measure. Every building, every object, every plowed furrow that human beings have left on the planet is a kind of embedded information.

Recorded information

Recorded information is communicatory or memorial information preserved in a durable medium. While an animal scent mark in the woods may be thought to be intended to communicate "This is my turf; stay off," I will limit the discussion of recorded information here to human products.

The use of symbols is primary to human beings (a symbol is "[s]omething that represents something else by association, resemblance, or convention...," *American Heritage Dictionary*; Houghton Mifflin, 2000), and constitutes a powerful and extensively used capacity on our part. Written language is an obvious form of communicatory information, but over the centuries, people have used our symbol-making capacity in countless ways. See, for example, Wilkinson (1994), who characterizes nine broad classes of symbolism in Egyptian art, only one of which is related to language (examples of others are form, size, color, action). Recorded information may have begun with drawings or carvings; however, the most revolutionizing form of recorded information was almost certainly the technology of writing, which was followed in later years by musical and mathematical notation, and other sorts of recorded information (compare Hjørland, 2002). Other forms of recorded information, such as photography, film, audio recordings, and many more, need to be incorporated in a general theory of recorded information for information studies.

Relationships among the forms of information

The crucial difference between embedded and recorded information is communicatory intent. The activities of the animal—enacted information—produce embedded information, the durable effects of action. (There are many nonenduring effects as well, such as the displacement of air past my legs as I walk.) The activity of using language, or other communicatory means—expressed information—has its enduring equivalent in recorded information. (Along these same lines, Heilprin and Goodman [1965] distinguished "short-duration" and "long-duration" messages. This is not to make the simplistic assumption that writing is simply spoken language written down. Written language is generally formulated differently from spoken language.)

Enacted information creates embedded information as its durable result, and expressed information leaves recorded information as its durable result. Further, just as expressed information is a communicatory subset of enacted information (as there are many noncommunicatory forms of action), so also is recorded information a communicatory subset of embedded information because there are many noncommunicatory (that is, not intentionally communicatory) forms of embedded information. Though written language is central as a form of recorded information, it is not the only form. A monument to a battle, for instance, may be intended primarily as a statement about the event, and only secondarily as a work of art. Embedded and recorded information may appear in or on the same artifact.

Recorded information is distinguished here from expressed information because the invention of writing and the development of the technologies to produce durable recorded information appear to have had an immeasurable impact on human cultures and on the speed of development of those cultures. No longer do humans have to try to memorize all that their culture knows; now a lot of that information can be kept in durable form outside the body. The durability and storage efficiency of such information have enabled a great leap in human information processing. The impact on human cognition of written records has been discussed at length by Ong (1982), Havelock (1980), and others.

Residue

In Table 1, *Residue* was listed as the fourth and final information flow channel (not named or discussed by Goonatilake). However, information does not cumulate in this channel as it does in the other three. Rather, residue represents the trace or deteriorating form of prior genetic, neural-cultural, or exosomatic information. The flow here is of a different sort—the Biblical "dust to dust"—in which structures previously associated with life recede back into their natural, inert forms. Trace information is that information that is degrading from being represented information (encoded or embodied) into being natural information only (neither encoded nor embodied). Trace information includes the no-longer-used wasps' nest, waste heaps, carrion, disintegrating ancient scrolls, and so on. Trace information is included here to acknowledge that all living processes produce waste, and degrade eventually, according to the law of entropy.

After a fire has leveled a house, for instance, most of the information in it—including the pattern of organization of the building materials, the arrangement of rooms, the structure of the furniture, the texts of the books, magazines, and so on—has been consumed and cannot be restored, without bringing in new materials. There may be traces of information left, but not anything like the amount of information—embedded and recorded information—that was there before the sudden degradation of the house by fire. Eventually, embedded information (as well as all other kinds of represented information) degrades to a residue of its former

Bradley times three—Los Angeles Mayor Tom Bradley speaks Friday during unveiling of a bronze bust of himself at Bradley International Terminal at Los Angeles Airport. The work by Russian-born artist S. Sarkis made its debut on third anniversary of terminal's opening and completion of other major improvements.

FIG. 1. *"Bradley times three," photo by George R.Fry,* Los Angeles Times, *June 20, 1987, part II, p. 1. ©1987. Los Angeles Times. Reprinted with Permission.*

self, then ultimately loses all trace of representation and becomes simply natural information.

Collecting the information forms

A summary example of information forms

Figure 1, "Bradley times three," illustrates the information forms discussed here. Former Mayor Bradley of Los Angeles and the other human beings in the photograph are the phenotypic expressions, or embodiments, of their genotypes, their encoded genetic information. We may assume that the Mayor is conscious of his own experiences in the moment (experienced information). Further, he is carrying out a physical action of standing at the podium, and a sociocultural action of dedicating the bust of himself (enacted information). He is speaking (expressed information). The statue, the terminal, the podium—all the human-made objects around him—carry embedded information. The words on the terminal entryway and on the plaque on the face of the statue's pedestal are recorded information. Had an empty candy wrapper been visible on the floor of the terminal entryway, it would be an example of trace information.

Other categorizations of information

Clusters of writings, principally in the management and knowledge management literatures, have dealt with different categorizations of knowledge. Tacit versus explicit knowledge is greatly discussed (Baumard, 1999; Davenport & Prusak, 1998), and some efforts have been made to create even finer distinctions. Blumentritt and Johnston (1999) have collected over 25 differently named types of knowledge, drawn from the literature. Among the terms included is "encoded knowledge" (Blackler, 1995, p. 1025), used similarly to "encoded information" here. "Embedded knowledge" and "embodied knowledge" are also used (Blackler, 1995; Collins, 1996), but with different meanings from those used here for information. The referenced definitions are, of course, all for the concept of knowledge, not of information, and are not structurally related to each other in the way they are here.

Common issues raised regarding the definition

The ideas presented here go against some common assumptions in the field, and evoke various reactions in readers and audience members. In this section, I will address the most common of these reactions. (Consult my 2005 article for more detailed discussion of some aspects.)

First, let us discuss three seemingly disparate issues:

- Is information a "sign," as used in the theory of semiotics?

- What is the relationship of information to knowledge?

- In this article's terms, what does it mean to "be informed"?

Briefly, *semiotics* is "the study of signs and sign-using behavior" (Encyclopædia Britannica Online, n.d.). The American father of the field was C.S. Peirce, who conceptualized the world of signs in the following way: There is (a) a "representamen," that is, a sign vehicle, or form, which the sign takes, (b) an interpretation, or sense, made of the sign, and (c) an object (material or conceptual) to which the sign refers (Chandler, 2004).

For example, to most English speakers, the letters "tomato" on a page are a sign. This comes about because we *interpret* ("b" above) the *sign vehicle* of the letters ("a" above) to represent tomatoes (*object*—"c" above). Thus, to us, "tomato" is a sign.

Chandler also notes that "[t]he sign is more than just a sign vehicle. The term 'sign' is often used loosely, so that this distinction is not always preserved." However, the *"representamen* is the *form* in which the sign appears (such as the spoken or written form of a word) whereas the *sign* is the whole meaningful ensemble" (Chandler, 2004). Nothing is a sign unless it is interpreted as a sign (Chandler, 2004). To Peirce, a sign thus stands for something to somebody in some respect (Chandler, 2004; Hoopes, 1991).

The interaction of sign vehicle, sense, and object is called *semiosis* (Chandler, 2004). Semioticians emphasize the dynamic nature of semiosis, as a social and linguistic process in which meanings and associations shift through time and with changing circumstances (Lidov, 1998; Taborsky, n.d.). Semiosis is thus a broader and more encompassing process than individual acts of interpretation of signs.

We can mark out a relationship between signs and information. To interpret a pattern of organization as a sign, an animal must have some association, either derived instinctively or through learning and experience, between that sign vehicle and the object to which the vehicle is interpreted as referring. The marks on a page that make up the sign vehicle "tomato" do not automatically or inherently refer to the object to which we English speakers give that name; rather the object, the sign vehicle, and the interpretation of the vehicle have been linked through time in the development of one particular language. Likewise, a certain kind of strutting in the male birds of a species is not automatically and inherently a courting display, i.e., a sign of a desire to mate, by the male to females. Rather, that association has been developed over time through natural selection (see also Hoffmeyer, 1997).

In the language used here, information can be seen as the raw material, the fodder, that goes into the process of semiosis, as well as into individual acts of interpretation. For example, in various cultures over the years, waving a hand toward oneself has come to mean "come here." Thus, a long-term semiotic process has resulted in that association being present in many human beings' minds. My culture shares that association. When, at a party, I interpret that hand motion from a friend across the room as a sign to come join him, I am selecting out a certain subset of all the information around me. From all the patterns of organization of sound, sight, smell, and touch that I experience in the room, I separate out the hand motion and read it as "come here," and then heed the call.

In the conventional usages of information studies, we take the above as an example of *being informed*. I now know something I did not know before—he wants me to come join him. We can make countless other such examples: Jane tells Joe that his suitcase has arrived, that the ship has run aground, that the Red Sox have won the World Series, that he got 100 on his test, etc. All these involve before and after moments, not knowing something, learning it, then knowing that something.

Looking on from the outside, we can say that something in my or Joe's nervous systems changed, the neural patterns of organization changed after we were informed. So the nervous system information changed in the sense in which information has been discussed to this point in this article.

In ordinary parlance, however, and in information studies as well, we use "informed" to refer not only to a new pattern of brain matter, but also to a change in meaning or understanding in our experience. Although not stated explicitly, information as defined to this point is without meaning, is meaning-free. Living beings can *assign meaning* to information, but patterns of organization of matter and energy are not inherently meaningful. Thus, how do we account for the "meaning-free" sense of information as well as the "meaning-full" sense of information given meaning, as in the above examples?

As noted earlier, any information in the universe can potentially be informative, so we began with a definition of information broad enough to encompass all that potential information. At the same time, we study "information-seeking," where people have a specific need that has meaning, and seek to meet that need with information that likewise has meaning for them, i.e., it matches or fills the stated need.

Let us then define the following terms for use as needed in the field:

- Information 1: The pattern of organization of matter and energy.

- Information 2: Some pattern of organization of matter and energy given meaning by a living being (or its constituent parts).

- Knowledge: Information given meaning and integrated with other contents of understanding.

We can thus unite the several issues discussed in this section by saying that Information 1 is the basis upon which semiosis acts over the long term, as well as the basis upon which individual acts of interpretation of sign vehicles operate. Taking the above example again, I have learned through participation in my society that a certain hand wave means "come here." Assuming my friend succeeds in moving his hand in a pattern sufficiently close to the generally understood features of that hand movement, and I see him when he is waving his hand, then I am able to read the movement as a sign with that meaning. I now feel informed that my friend wants me to join him. My interpretation of the hand movement—Information 1—has yielded Information 2, some pattern of organization of matter and energy given meaning in my mind.

As the receiver, I have made the linkages, i.e., I have constructed an understanding and am informed. As many have pointed out in recent years, information does not move, like a physical substance, unaltered, in a pipeline from the mind of a sender to the mind of a receiver. Rather, meaningful Information 2, assembled by a sender from knowledge stores, is disseminated via speaking, writing, or other means, and directed to a receiver. The receiver initially receives the message as Information 1, perhaps as sounds, or marks on a page. These sign vehicles remain meaningless until given meaning by the receiver through interpretation, thus constructing an understanding of the message as Information 2. As we know, the Information 2 constructed by the receiver may be very different from the Information 2 that the sender intended to communicate. Eventually, the constructed Information 2 is integrated more or less permanently into the receiver's knowledge stores, and disappears as a separate entity in the individual's experience.

This is not to imply that "being informed" is a situation that occurs solely at the level of a single act of interpretation in relation to a single sign. Rather, in most cases when a person feels informed, numerous interrelated acts of semiosis have occurred, both simultaneously and sequentially, within a context of complex social and cultural meaning-codes that can be very elaborate in the case of human beings (Chandler, 2004).

When needed, Information 1 and 2 can be distinguished in discussion in the field; otherwise, the term *information* is understood to refer to one or both senses.

Meaning is *ascribed* to some of the Information 1 in the world by living beings, that is, living beings interpret some of the information in the universe as signs. An enormous part of all the Information 1 in the universe, however, has never been interpreted as a sign by any living being.

By the way, "informing" always takes place between living beings, because, as far as we know, only living beings (or their constituent parts) can engage in sign production and interpretation. The tomcat marks its territory to communicate to and ward off competitors; however, the waves do not "inform" the beach by moving sand particles.

Note that truth is not a requirement of knowledge as described here; knowledge is a kind of meaningful belief. We may or may not be able to offer various kinds of evidence to support such beliefs, and we may or may not be able to claim fairly, by the understandings of our culture, that these beliefs are true to the reality of the world.

There is just one final point; I have included the discussion of semiotics here to respond to reviewers, and to demonstrate the general relationship between semiotics and the information definition presented here. However, the linkage, though interesting and promising, is not necessary to the argument made here. One can think and talk about Information 1 and 2 and knowledge as described herein, without any reference to semiotic theory, and without loss to the model presented here.

Applications and implications of information forms in information science

There are numerous potential applications and implications of these fundamental information forms for information science. Three broad areas will be addressed here, as examples of the still broader possible uses for this approach to the concept of information.

Research in information seeking behavior

In this section, I will illustrate the value and applicability of these concepts of information by applying them to the sub-field of information-seeking behavior. Ultimately, it is desirable to apply these terms throughout information science.

An illustrative example can be found in the work of Edwin Hutchins in his book, *Cognition in the Wild* (1995). Hutchins studied a group of people working together who were relatively isolated physically from the rest of the world; they were the crew of a U.S. Navy vessel. His orientation was that of the book's title—How does cognition work in the real world, as opposed to the laboratory? However, we can study his work from the standpoint of information seeking with equal benefit. Hutchins studied the way the crew handled navigation, including an emergency incident, on the ship. He found that their collective production of successful navigation arose out of a wide range of sources, not just the training of the crew, though that was part of it. The very layout of the ship and the design of the bridge promoted the smooth flow of information from the exterior of the ship to the crew and among the crewmembers. Each crewmember took on a distinct, but coordinated, role. Critical information was posted at just the right locations for use. Likewise, navigation practices required that not one, but two crewmembers have certain crucial pieces of information at the same time to reduce the likelihood of error. Posting of critical information at the right locations for use, redundancy of information in the crew's knowledge and in the materials they worked with, and a variety of other information factors in the situation promoted the effective coordinated response. Even the format design of the forms the crew filled out made it easier for them to complete their work successfully.

Though Hutchins was studying collective cognition, he also produced groundbreaking work relevant to information studies, because he incorporated *all* the forms of information that were supporting the crew's performance, not just the forms that we conventionally label "information" (see, for example, Hutchins, 1995, pp. 263–285). People get information not just from paper sources, not just from other people, but also from the physical layout of their workspaces, from the design, not just the content, of informational genres, and above all, from the interaction of these various factors in a real situation. All the patterns of organization of matter and energy—cognitive, physical, architectural, social, linguistic—are informative. Therefore, to understand fully this information seeking and use situation requires the identification of the roles of all these forms of information.

The terms defined in this article can be seen as an initial effort to identify the various information forms needed for the study of people in their information milieus. In the example of the ship's crew, it would be possible to distinguish the nature and role of *experienced information*—what the crew was perceiving and thinking about, *enacted information*—what they did, and observed others doing, at each step of the process, *expressed information*—what they said and understood from their own and others' verbal and body language, *embedded information*—how the architectural layout and design of instruments and documents affected their information use, and *recorded information*—the documentary resources used.

Information genres

Distinguishing genres of information has been a long-standing necessity for library catalogers, who wish to include information type within their descriptions of various kinds of documents (Wilson & Robinson, 1990). However, while much work has gone into the making of definitions for practical cataloging, there has been no larger established theory of genre types within library and information science. Recently, the advent of new informational forms on the World Wide Web has provoked another round of interest in genre in information studies (Bates & Lu, 1997; Crowston &

Williams, 2000; Vaughan & Dillon, 1998, among others). In a completely different realm, genre has also been a focus of interest in literary studies, the arts, and other fields, where a given genre can be seen to be an expression of, and a vehicle for, a particular kind of communication (Ingarden, 1989; Trosborg, 2000).

Perhaps the information types described in this article can provide the basis for a more theoretically grounded understanding of genre. For example, within the humanities we can see the performing arts (dance, theater, music) as the disciplines of expressed and enacted information, the plastic arts (painting, sculpture) as the disciplines of embedded information, literary studies as the disciplines of recorded information, and so on. Starting with a consideration of these fundamental differences and distinctions in the *object of study* by the practitioners of these disciplines, perhaps we can develop a taxonomy of the material culture of the arts and humanities that has a novel basis. This taxonomy may also be useful for grounding the development of a suitable classification of genres for library work as well.

The information and curatorial sciences

Of late, there has been much interest in the relationship between library and information science, archives, and museum studies. In the digital era, all of these disciplines are involved in digitizing parts of their collections, and the challenges facing these fields appear to be converging. Museum collections management databases are being revamped and made available to the public online, just as library catalogs have been for some years now.

I believe that the distinctions made in this article among types of information can help clarify just what the relationship is and should be among these disciplines. They are all what might be called *collections disciplines*, as their primary purpose is to create collections of objects, that is, to bring together objects of social interest for research, learning, and entertainment, and make them available to an audience. All these disciplines create some sort of organized access to their collections, and house those collections in institutions of a certain type and organizational design.

Those institutions, the library, the archive, and the museum, arose for different purposes, however, and have different traditions. The different purposes arose around the collecting of distinct types of objects. In other words, *collections sciences are distinguished one from another by the kinds of objects of social interest that they collect.* Though digitization makes for some similarities in the challenges these disciplines face, it should also be kept in mind that each discipline arose around, and is designed to meet the needs of, the underlying social objects collected.

Libraries house published recorded information, archives house unpublished recorded information, and museums house various kinds of embedded information, from works of art to archeological artifacts. Natural history museums house embodied genetic information, that is, partial or whole (and no longer living) phenotypes. Still other museum-like institutions, specifically, zoos, aquariums, arboretums, and gardens, collect and house living phenotypes, also embodied genetic information.

The distinctions among types of information detailed here may be used as a basis for analyzing and distinguishing various sorts of collections sciences. Those that collect recorded information of various types are among the information sciences, and those that collect embedded objects and embodied phenotypes are part of the curatorial sciences.

Summary and conclusions

Information is the pattern of organization of matter and energy. All information is natural information, in that it exists in the material world of matter and energy. Represented information is natural information that is encoded or embodied. Encoded information is information that has symbolic, linguistic, or signal-based patterns of organization. Embodied information is the corporeal expression or manifestation of information previously in encoded form.

Goonatilake's model of three broad streams of information transmission over the history of life on the planet is utilized; he calls the streams "information flow lineages." These are the genetic, neural-cultural, and exosomatic flow lines (Goonatilake, 1991). I have proposed several fundamental forms of information and have assigned them to the Goonatilake flow lines. These are genetic information in the genetic line; experienced, enacted, and expressed information in the neural-cultural line; and embedded and recorded information in the exosomatic line. Genetic and neural-cultural information are *encoded*, respectively, as the genotype and as nervous system structures and action potentials. Genetic and neural-cultural information are *embodied*, respectively, as the phenotype and as experienced information (experience, consciousness), enacted information (actions), and expressed information (communication).

Exosomatic information, that is, information stored externally to the body, has been developed in complex ways by human beings, and has been differentiated here as embedded information (the pattern of organization of the enduring effects of the presence of animals on the earth) and recorded information (communicatory or memorial information preserved in a durable medium). Recorded information is the chief focus of the information professions, and embedded objects and embodied phenotypes are the chief foci of the curatorial professions, including museum studies and zoo management. Examples have been provided to illustrate the relevance of these terms to two further areas of information studies, namely, information seeking behavior and information genres. It should be possible to develop comparable examples for other areas of information studies.

Information deteriorating from its former relation to the living back into inert natural information is called trace information. Trace information, seen as ignored detritus in most other disciplines, takes on immense significance in information studies, archival studies, history, archeology, and the curatorial disciplines.

Finally, a distinction has been made between Information 1, the pattern of organization of matter and energy, and Information 2, some pattern of organization of matter and energy given meaning by a living being (or a component thereof). Knowledge has been defined as information given meaning and integrated with other contents of understanding.

This perspective and corresponding terminology have been developed to provide a fresh conceptualization of some of the classical issues in information science/studies. The discussion of Information 1 enables us to start at the root of all information description and information seeking—the patterns of organization in the universe, including those generated by other living beings—that animals and humans respond to and use in interpreting and giving meaning to their experience.

Having begun at that foundational level, we found that we could also identify several distinct forms of information of value to our and related collections disciplines. These terms provide a larger vocabulary than we had previously to describe the various kinds of information that are important to information science/studies. Perhaps we are now just a little closer to being able to build a more principled understanding of our and related disciplines.

ACKNOWLEDGMENTS

My thanks to Suresh Bhavnani, Terry Brooks, Michael Buckland, Michele Cloonan, William Cooper, Laura Gould, Jenna Hartel, Robert M. Hayes, P. Bryan Heidorn, Anders Hektor, Birger Hjørland, Peter Ingwersen, Jarkko Kari, Jaana Kekäläinen, Kimmo Kettunen, Erkki Korpimäki, Poul S. Larsen,

Mary N. Maack, Marianne L. Nielsen, Pamela Sandstrom, Reijo Savolainen, Dagobert Soergel, Sanna Talja, Pertti Vakkari, and the anonymous reviewers for their thoughtful and insightful comments on this and the companion paper (Bates, 2005).

REFERENCES

American Heritage Dictionary of the English Language (4th ed.). (2000). Boston, MA: Houghton Mifflin.

Bates, M.J. (1999). The invisible substrate of information science. *Journal of the American Society for Information Science, 50*(12), 1043–1050.

Bates, M.J. (2005). Information and knowledge: An evolutionary framework for information science. *Information Research, 10*(4), paper 239. Retrieved from http://InformationR.net/ir/10-4/paper239.html.

Bates, M.J., & Lu, S. (1997). Exploratory profile of personal home pages: Content, design, metaphors. *Online & CDROM Review, 21*(6), 331–340.

Bateson, G. (1972). *Steps to an ecology of mind.* New York: Ballantine.

Baumard, P. (1999). *Tacit knowledge in organizations.* London: Sage.

Blackler, F. (1995). Knowledge, knowledge work and organizations: An overview and interpretation. *Organization Studies, 16*(6), 1021–1046.

Blumentritt, R., & Johnston, R. (1999). Towards a strategy for knowledge management. *Technology Analysis & Strategic Management, 11*(3), 287–300.

Brier, S. (1996). Cybersemiotics: A new interdisciplinary development applied to the problems of knowledge organisation and document retrieval in information science. *Journal of Documentation, 52*(3), 296–344.

Brookes, B.C. (1975). The fundamental problem of information science. In V. Horsnell (Ed.), *Informatics 2.* Proceedings of a Conference Held by the ASLIB Coordinate Indexing Group (pp. 42–49). London: ASLIB.

Brunk, B. (2001). Exoinformation and interface design. *Bulletin of the American Society for Information Science and Technology, 27*(6), 11–13.

Chalmers, D.J. (1996). *The conscious mind: In search of a fundamental theory.* New York: Oxford University Press.

Chandler, D. (2004). *Semiotics for beginners.* Retrieved January 6, 2005, from http://www.aber.ac.ukmedia/Documents/S4B/sem02.html.

Collins, H.M. (1996). Embedded or embodied? [Review of the book *What computers still can't do* by H. Dreyfus]. *Artificial Intelligence, 80*(1), 99–117.

Crowston, K., & Williams, M. (2000). Reproduced and emergent genres of communication on the World Wide Web. *The Information Society, 16*(3), 201–215.

Damasio, A. (1999). *The feeling of what happens: Body and emotion in the making of consciousness.* San Diego: Harcourt.

Davenport, T.H., & Prusak, L. (1998). *Working knowledge: How organizations manage what they know.* Boston, MA: Harvard Business School Press.

Dawkins, R. (1976). *The selfish gene.* Oxford, England: Oxford University Press.

Dawkins, R. (1982). *The extended phenotype: The long reach of the gene.* Oxford, England: Oxford University Press.

Dennett, D.C. (1991). *Consciousness explained.* Boston, MA: Little, Brown.

Encyclopædia Britannica. (n.d.). Semiotics. *Encyclopædia Britannica online.* Retrieved November 4, 2004, from http://www.britannica.com.

Goonatilake, S. (1991). *The evolution of information: Lineages in gene, culture and artefact.* London: Pinter.

Havelock, E.A. (1980). The coming of literate communication to western culture. *Journal of Communication, 30*(1), 90–98.

Heilprin, L.B., & Goodman, F.L. (1965). Analogy between information retrieval and education. In L.B. Heilprin, B.E. Markuson, & F.L. Goodman (Eds.), *Proceedings of the Symposium on Education for Information Science* (pp. 13–21). London: Macmillan.

Hjørland, B. (2002). Principia informatica: Foundational theory of information and principles of information services. In H. Bruce, R. Fidel, P. Ingwersen, & P. Vakkari (Eds.), *Emerging frameworks and methods: Proceedings of the Fourth International Conference on Conceptions of Library and Information Science* (CoLIS4), (pp. 109–121). Greenwood Village, CO: Libraries Unlimited.

Hoffmeyer, J. (1997). The global semiosphere. In I. Rauch, & G.F. Carr (Eds.), *Semiotics around the world.* Proceedings of the Fifth Congress of the International Association for Semiotic Studies (pp. 933–936). Berlin: Mouton de Gruyter. Retrieved August 13, 2004, from http://www.molbio.ku.dk/molbiopages/abk/personalpages/jesper /Semiosphere.html.

Hoopes, J. (Ed.). (1991). *Peirce on signs: Writings on semiotic by Charles Sanders Peirce.* Chapel Hill: University of North Carolina Press.

Hutchins, E. (1995). *Cognition in the wild.* Cambridge, MA: MIT Press.

Ingarden, R. (1989). *Ontology of the work of art: The musical work, the picture, the architectural work, the film.* Athens, OH: Ohio University Press.

Lidov, D. (1998). Semiosis. In P. Bouissac (Ed.), *Encyclopedia of semiotics* (pp. 561–563). New York: Oxford University Press.

McCauley, R.N. (Ed.). (1996). *The Churchlands and their critics.* Cambridge, MA: Blackwell.

Ong, W.J. (1982). *Orality and literacy: The technologizing of the word.* London: Methuen.

Parker, E.B. (1974). Information and society. In C.A. Cuadra & M.J. Bates (Eds.), *Library and information service needs of the nation: Proceedings of a conference on the needs of occupational, ethnic and other groups in the United States* (pp. 9–50). Washington, DC: U.S.G.P.O. (ERIC #ED 101 716).

Random House Unabridged Dictionary (2nd ed.) (1993). New York: Random House.

Swanson, C.P. (1983). Ever-expanding horizons: The dual informational sources of human evolution. Amherst, MA: University of Massachusetts Press.

Taborsky, E. (n.d.). *Semiosis, semiosis/evolution/energy terminology.* Retrieved May 12, 2005, from http://www.library.utoronto.ca/see/pages/semiosisdef.html.

Trosborg, A. (Ed.). (2000). *Analysing professional genres* (Pragmatics and Beyond: New Series, Vol. 74). Amsterdam: John Benjamins Publishing.

Varela, F.J., Thompson, E., & Rosch, E. (1993). *The embodied mind: Cognitive science and human experience.* Cambridge, MA: MIT Press.

Vaughan, M.W., & Dillon, A. (1998). The role of genre in shaping our understanding of digital documents. *Proceedings of the ASIS Annual Meeting, 35,* 559–566.

White, H.D. (1992). External memory. In H.D. White, M.J. Bates, & P. Wilson (Eds.), *For information specialists: Interpretations of reference and bibliographic work* (pp. 249–294). Norwood, NJ: Ablex.

Wiener, N. (1961). *Cybernetics: Or control and communication in the animal and the machine* (2nd ed.). Cambridge, MA: MIT Press.

Wilkinson, R.H. (1994). *Symbol & magic in Egyptian art.* New York: Thames and Hudson.

Wilson, P., & Robinson, N. (1990). Form subdivisions and genre. *Library Resources & Technical Services, 34*(1), 36–43.

3

Information
(Encyclopedia entry)

ABSTRACT

A selection of representative definitions of information is drawn from information science and related disciplines, and discussed and compared. Defining information remains such a contested project that any claim to present a unified, singular vision of the topic would be disingenuous. Seven categories of definitions are described: Communicatory or semiotic; activity-based (i.e., information as event); propositional; structural; social; multitype; and deconstructionist. The impact of Norbert Wiener and Claude Shannon is discussed, as well as the widespread influence of Karl Popper's ideas. The Data-Information-Knowledge-Wisdom (DIKW) continuum is also addressed. Work of these authors is reviewed: Marcia J. Bates, Gregory Bateson, B.C. Brookes, Michael Buckland, Ian Cornelius, Ronald Day, Richard Derr, Brenda Dervin, Fred Dretske, Jason Farradane, Christopher Fox, Bernd Frohmann, Jonathan Furner, J.A. Goguen, Robert Losee, A.D. Madden, D.M. MacKay, Doede Nauta, A.D. Pratt, Frederick Thompson.

Introduction

The concept "information" is of signal importance to all the information disciplines. Perhaps for that reason, it is a term that has been defined in countless ways, over many decades. It would be fair to say that there is no

First published as Bates, M.J. (2010). Information. In M. J. Bates & M. N. Maack (Eds.), *Encyclopedia of Library and Information Sciences* (3rd ed., Vol. 3, pp. 2347–2360). Boca Raton, FL: CRC Press.

widely agreed-upon definition or theoretical conception of this term. The meaning of this term is still highly contested. In this regard, the status of the term is similar to that of "communication" in the communication sciences.

In light of the lack of agreement about the definition of the term "information," the main objective of this entry will be to lay out some of the major classes of definitions and theoretical constructions of the term that are currently or recently in play. No effort will be made to capture and discuss every definition that has been provided in the literature; rather major types will be presented, as well as popular ideas that are recurrent in the literature. The discussion draws from writings over the last 60 years; the approach is by category rather than by chronology.

The effort to define information is active in other disciplines besides those explicitly concerned with the topic; philosophy, cognitive science, electrical engineering, computer science, and systems theory, among others, have been active players on this scene as well. The objective of this entry, however, is to concentrate on the ideas about information that have been either developed within the information disciplines or, in a few cases, which have come from other fields but have also been influential in the information disciplines. For coverage of other approaches to the concept, the reader is directed to reviews by Aspray (1985), Belkin (1978), Capurro and Hjørland (2003), Cornelius (2002), Meadow and Yuan (1997), and Wersig and Neveling (1975).

Some authors embed a discussion of information within a much larger philosophical or theoretical program. In other words, exposition of the meaning of the term "information" is not a primary goal, but only incidental to much larger projects. It is beyond the focus of this entry to attempt a review of these larger intellectual programs. Prime examples include: Søren Brier's "cybersemiotics" (Brier, 1998), Benny Karpatschof's dissertation on "Human Activity" (Karpatschof, 2000), Howard Resnikoff's analysis of information within a mathematical, physical, and signal detection framework (Resnikoff, 1989), Jan Kåhre's "mathematical theory of information" (Kåhre, 2002), and Stonier's 3-volume disquisition on biology, physics, and information (Stonier 1990, 1992, 1997). Another author in this category is Luciano Floridi, a philosopher, who is developing and promoting an area of philosophy to be known as "Philosophy of information" (Floridi, 2002). He develops his own view of information as a philosopher, with attention to the issues of concern to that discipline in the entry "Information" (Floridi, 2004).

Because Norbert Wiener's and Claude Shannon's ideas of information were so influential at the dawn of the "Information Age," their influence is discussed in a preamble below. Shannon's actual "information theory," however, is reviewed elsewhere in an entry by that name in this encyclopedia.

After the preamble, conceptions of information of the following types will be reviewed, in succession:

- Communicatory or semiotic

- Activity-based (i.e., information as event)

- Propositional

- Structural

- Social

- Multitype

- Deconstructionist

Between these categories, we will take two interludes, one to discuss the "Problem of Popper's Worlds," regarding the philosopher Karl Popper, and the other to address "DIKW" or the commonly discussed sequence known as "Data-Information-Knowledge-Wisdom." Both of these, Popper's "three worlds" concept, and the DIKW sequence, have motivated so much discussion in the information sciences around information, that they merit separate discussion.

In the process, the work of the following people will be addressed: Marcia J. Bates, Gregory Bateson, B.C. Brookes, Michael Buckland, Ian Cornelius, Ronald Day, Richard Derr, Brenda Dervin, Fred Dretske, Jason Farradane, Christopher Fox, Bernd Frohmann, Jonathan Furner, J.A. Goguen, Robert Losee, A.D. Madden, D.M. MacKay, Doede Nauta, A.D. Pratt, and Frederick Thompson.

In this entry, no summary conclusion is made about "the best" or "the truest" understanding of the concept of information. Rather, the purpose is to present the array of ideas flowing around this core concept in the information disciplines, so that the reader may become acquainted with the issues.

Preamble: The Roles of Wiener and Shannon

It is almost impossible to overestimate the impact of Claude Shannon's ideas about information on the (American) intellectual culture of the

1950s and 1960s. In that era there was a tremendous amount of attention directed to the technical revolution(s) that had become possible with the development of computers, television, new communication technologies, and a new way of thinking about information. This new way of thinking percolated out of the academic world into the society at large, and imbued at a subconscious level the thinking of people who had no understanding of Shannon's ideas per se.

Today, many scholars write dismissively about the concept of information (see the last portion of this entry), and reject the earlier excitement around the "Information Age" and the "Information Society" as a love affair with a cold, technical, even militaristic conception of the technology-driven society (Day, 2001). Indeed, Frank Webster, writing in this encyclopedia about the "Information Society," analyzes the term's many weaknesses and confusions, without seeming to recognize the positive value originally gained from the ideas carried by the concept. Shannon appears in Webster's discussion only in terms of the "deracinated" definition of information that arises out of Shannon's writing. Ironically, as we shall see, it was, in fact, the very fecund power of that deracinated definition of information—i.e., a concept of information as independent of meaning—that allowed an explosion of scientific and social development around information and its social and technical role.

The new conception of information that came with Shannon was so fundamental, so pervasive in science and engineering, that today's critics do not actually see it *as it was to people then*. The several streams of new thinking on information were startling, different, and stimulating, compared to prior understanding. I believe that the impact was so fundamental that an earlier generation can be forgiven for inventing ideas like the "Information Age," the "Information Society," and "information explosion." The consciousness of information was so new, and changed so many established ideas, that it really felt to the participants like a new age marked by the new awareness of information.

It is fashionable now to deride that earlier absorption with the new concepts, but we are *able* to deride these concepts only because we have so thoroughly absorbed the learning from that time, that it feels easy to dismiss it in favor of newer ideas. The ever-present fact is that people both build on and react to what was present earlier in their lives. Today's critics are, of course, doing that too. This author is old enough to remember that earlier time, and I choose to present that era as I understood it, as a bit of a counter to the somewhat dismissive attitude toward it that is popular nowadays.

Boulding (1961) wrote about three levels of organization in life: 1) static structures; 2) clockwork, i.e., the world of mechanics; and 3) thermostats, that is, control mechanisms that maintain a stable condition by responding to feedback from their environment (Boulding, 1961, p. 20). These three levels have some parallel in the development of science in the Western world—the medieval belief in a static world created by God, followed by the Newtonian discovery and analysis of dynamic processes, followed by the cybernetic understanding of the role of information in life processes.

The world of Newton and his epigones was one in which the theory of forces and impacts of recognizable regular, measurable change was developed. The quintessential model of the mechanical universe is that of billiard balls being hit and rolling into other balls and making them move in a certain direction with a certain force.

The movement of the planets was closely measured, the mathematics of change in the form of the calculus was developed, and a deep threat to the medieval concept of the static universe arose. The long history of religious controversy, with Galileo as a prime example, and proceeding through the inquisition, the Reformation, and the Counter-Reformation, was in no small part due to the fundamental challenge offered by this new dynamic idea of how the universe worked.

In the twentieth century, information began to become important in the thinking of science and society. Problems of observation and the impact of observation emerged in early twentieth-century physics. Finally, during the 1940s and 1950s the role of information was made theoretically explicit in the theory of cybernetics. The term came from a Greek root word meaning to govern or steer. Norbert Wiener's conception of cybernetics involved the governing of action through the feedback of information (Wiener, 1961).

"Feedback" is a tediously trite term nowadays, usually employed in the context of customer relations or group therapy. But the idea behind it was revolutionary in Wiener's day. Wiener illustrated just how significant the idea was in his description of research that he conducted with a physician on physiological processes (Wiener, 1961, Chapter 4). Again, grossly simplifying, the thinking in that day in physiology was that when I reach out to pick up a pencil, this process is achieved by my brain sending a signal to muscles and tendons along the lines of "go get it," and the machinery of my arm goes into action and picks up the pencil. This was a classically mechanical concept of my actions. A pulse goes out to my arm to do a certain thing, I act, then the pulse diminishes.

Wiener and his colleague demonstrated that the process did not work that way. Instead, when I start to pick up the pencil, I extend my arm in the direction of the pencil, and then, *utilizing constant kinesthetic and visual feedback*, I microadjust the position of my arm repeatedly and successively until it successfully lands on the pencil, grasps it, and picks it up. Picking up the pencil is not a single, mechanical, act, but rather an extended behavior utilizing continuous information feedback telling me whether my hand is on or off course, and if off course, enabling me to adjust the tension in muscles and the direction of my reach so that I can successfully touch and pick up the pencil (Wiener, 1961, p. 8).

Thus, in cybernetic situations, two processes are going on continuously in parallel—the physical forces, and the detection and utilization of information about the physical forces, which information is used to affect the physical actions. While the billiard ball model was the one commonly used for the mechanical understanding of the universe, with the impact of cybernetic thinking, the household thermostat became the standard model of cybernetics and feedback. In the summer heat, I set the thermostat for a certain temperature. When the heat in the room affects a sensing mechanism in the thermostat beyond a certain point, the air conditioning starts, and cools the room down to where the sensing mechanism again achieves its desired temperature, and the air conditioning shuts off. The sensing mechanism provides continuous information, and the design of the thermostat is such that when the information indicates that the room temperature is outside a desired range, the air conditioning comes on. Governing, or steering, is about utilizing information feedback, to direct the ship of action.

In the larger history of scientific thinking, the development of cybernetics drew attention to the *distinct role of information* in physical and social processes. Previously, the kinesthetic and visual feedback I get while picking up the pencil—as well as in countless other information-based processes—had been almost entirely invisible in the thinking of science. This may be one reason why the 1950s and 1960s were so obsessed with information—the role of information had at last emerged as a focal topic of interest in science; its role in influencing physical and social processes at last came to the fore, and once seen, was studied with fascination in many domains of science.

So what was Claude Shannon's role in all this scientific development? Shannon and Wiener worked on some of the same ideas during this fertile period. Shannon, working at the Bell Laboratories, however, developed the mathematical and engineering theory to put an understanding of information on a firm basis (Shannon & Weaver, 1975). Strictly speaking, Shannon did not define information at all, at least in any conventionally understandable way. Shannon found a way to measure the *amount* of information going over a transmission channel. As Wiener puts it, "...we had to develop a statistical theory of the *amount of information*, in which the unit amount of information was that transmitted as a single decision between equally probably alternatives" (Wiener, 1961, p. 10).

Since the alternative messages, letters, words, or other units of communication are not always sent with equal probability, the formula Shannon developed measured the amount of information as a function of two things—the number of alternatives out of which a message might be selected for sending, and the probabilities of the various messages. The more possible messages from which the sent message is selected, and the more equiprobable the messages, the greater the amount of information transmitted.

Shannon's analysis was revolutionary in several senses. Before him, engineers really did not have a means of computing the maximum amount of information that could be transmitted through a channel of a given size or configuration. It was assumed that it would be possible to go on improving channels to carry more and more information. Shannon's formulas enable the calculation of the maximum possible information transmission for a given physical configuration.

Once Shannon developed a firm model of the amount of information, actual and possible, in a channel, he could clarify the role of redundancy, of error rates, and noise in a channel. For example, since the letters of the alphabet do not appear in written text with equal probabilities, the amount of information conveyed with English text is well less than 100% of the amount of information that could be conveyed if each letter were equiprobable. Further, Shannon mathematically analyzed the role of noise in a channel, and the ways in which redundancy could compensate for the noise. These discoveries were immensely important for all sorts of communication engineering situations. Go to the right section of an engineering library, and one can find textbooks full of hundreds of mathematical formulas developed out of these crucial insights by Shannon. Shannon's work revolutionized communication engineering. His key paper has been cited over 7000 times in the Institute for Scientific Information (ISI) "Web of Science" database (Shannon, 1948).

His impact went well beyond engineering, however. It was as if for the first time people saw the informational regularities beneath the surface variety of the text sent over a wire or the words spoken on the telephone or written in a book. Shannon's model of information is dismissed today because he separated information from meaning. What is currently forgotten, however, is that this separation was in fact an *achievement*. People had not been able to make that differentiation before. Now, with far greater clarity and understanding, the handling of information in quantitative terms at last came on its own. In 1951, the psychologist and linguist George Miller wrote the book *Language and Communication*, which essentially consisted of working out the implications of Shannon's work for those disciplines (Miller, 1951). In the process, Miller educated a generation of social scientists on this way of thinking.

The fundamental clarification of the relationship between messages and the amount of information they convey (Shannon), and the concomitant recognition of the important role of information throughout life processes (Wiener) led to an enormous surge of research and theorizing throughout science about information. (Other researchers, such as John von Neumann and Oskar Morgenstern, had important roles as well.) Just as those hundreds of mathematical formulas had to be worked out in the engineering world, so also did the social sciences need to work with these same ideas and transform some parts of those disciplines.

The application of Shannon's approach to the social and behavioral sciences was not straightforward, however. There were many insights gained, but many problems encountered as well. After a while, the initial enthusiasm in the social sciences waned—to the point where it is now fashionable to deride these post-War ideas. But we are shaped by those ideas so thoroughly nonetheless, that we can only attempt to throw off their influence.

Communicatory or semiotic definitions of information

We begin with definitions/conceptions of information that are framed within a communicatory or signaling context. Some authors even take the approach to the point of identifying meaning with information.

A.D. Madden

As a recent example, in 2000 Madden defined information as "a stimulus originating in one system that affects the interpretation by another system of either the second system's relationship to the first or of the relationship the two systems share with a given environment. . ." (Madden, 2000, p. 348). In 2004, he simplified it to "a stimulus which expands or amends the World View of the informed" (Madden, 2004, p. 9). The latter definition is reminiscent of Boulding's concept of the "image," which is the grand total of my (or any individual's) subjective knowledge, my mental image of the world, and my place in it (Boulding, 1961). Boulding argued that behavior depends on this image. So in Madden's approach, information is, in effect, something that alters the image.

Gregory Bateson

Gregory Bateson (1972) applied concepts of information and feedback to the psychodynamics of human relations, most famously writing of the pathological, feedback-based, "Double Bind" relationship (Bateson, 1972, pp. 271–278). He also wrote about information science. Expressing the semiotic approach still more generally, Bateson said that information is a difference that makes a difference (Bateson, 1972, p. 453). This approach has its roots in the idea of the single difference being the elementary unit of amount of information, the single bit, the zero or one. The difference that makes a difference, presumably, makes that difference to a sensing being.

B.C. Brookes

Brookes (1980), one of the grand old men of British information science, took a similar tack. The following is his "fundamental equation" for the relationship between information and knowledge.

$K[S] + \Delta I = K[S + \Delta S]$, which states in its very general way that the knowledge structure K[S] is changed to the new modified structure $K[S + \Delta S]$ by the information ΔI, the ΔS indicating the effect of the modification (Brookes, 1980, p. 131).

Thus the (human) knowledge structure in the mind is changed in some way with the input of information.

Doede Nauta

Nauta (1972) takes information quite explicitly to be meaning, but in a particular sense. "Information is that which is common to all representations that are synonymous to the interpreter (synonymity is identity of meaning)" (Nauta, 1972, p. 201). Thus, information is the meaning that is common to all the different ways of expressing that meaning. Here, it would seem that only representations can contain redundancies, because information is the common meaning core to all the different possible representations.

In this approach Nauta drew on both semiotic and (Shannon) information theoretic approaches.

Robert Losee

Losee (1997) developed what he calls a "discipline-independent definition" of information. "Information is produced by all processes and it is the values in the characteristics of the processes' output that are information" (Losee, 1997, p. 256).

> Information is always informative about something, being a component of the output or result of the process. This 'about-ness' or representation is the result of a process or function producing the representation of the input, which might, in turn, be the output of another function and represent its input, and so forth (Losee, 1997, p. 258).

Losee takes as his central example the baking of a cake.

> Examining the cake provides information about both the process and the original ingredients. . . . The choice of high quality ingredients . . . will affect the outcome. . . . Varying the process, such as the amount of time in the oven . . . also changes the final product. . . (Losee, 1997, p. 258).

This definition raises a couple of questions. First, what about situations that are not processes? Can there be no information there? Or is process so universal that everything is a result of it?

Second, it is with cake-baking, as with many physical processes, that the information produced from the process is often quite incomplete, and can be misleading. Losee says that examining the cake provides information both about process and original ingredients. But, in fact, the person who had never seen a cake before almost certainly would not know enough to be able to figure out from the output alone all that went into making it, both in ingredients and processes. This is true with many processes. Often, there is an emergent result—yeast does things in the cake not in evidence when the baking begins. The baking causes chemical processes that lead to the final result being very different in qualities from the starting dough.

Thus, in the experience of human observers, information can be limited and distorted coming out of a process, and it can also be quite unambiguous and complete, as when we know of the process "add +1 to the output of prior process." In the latter case, the information we derive from the result is presumably complete and correct. Thus, it would appear that Losee's general formulation of information includes many situations where the information is limited or distorted, as well as situations where the information resulting from a process is unambiguously clear.

Information as event

Allan Pratt

Pratt (1977) provided probably the most developed version of a conception of information in which information is an *event*, that is, he looks at the process of being informed and derives the term "information" from that. Pratt, too, draws on Boulding's concept of "the image." He says:

> My Image of the world, and my relationships in it, which includes my perception of cause and effect, of time, of space, of values, of everything which impinges on my consciousness, is different from yours, and from that of every other person in the world (Pratt, 1977, p. 208).
>
> After a person has received and understood the content of a message, in ordinary speech we say that he has become informed about the matter at hand. This is a surprisingly precise and accurate statement. He has been 'in-formed.'... He has been inwardly shaped or formed; his Image has been altered or affected.
>
> In-formation is the alteration of the Image which occurs when it receives a message. Information is thus an event; an event which occurs at some unique point in time and space, to some particular individual (Pratt, 1977, p. 215).

He compares the concept "information" to the concept "explosion." "Every explosion is unique; no two are identical." "Further, explosions cannot be stored or retrieved. One may, of course, store and retrieve potentially explosive substances. . . ." Later, he admits: "informative event" may be a more felicitous term than "an 'information'" (Pratt, 1977, p. 215). Here he comes upon the practical problem of the mass-noun usage of the word "information"; "an information," as a count noun, just does not fit into English usage.

Propositional definitions of information

Propositional definitions are ones in which a piece of information is considered to be a claim about the world, a proposition.

Richard Derr

Perhaps the most accessible such definition is that of Derr (1985): "... information is an abstract, meaningful representation of determinations which have been made of objects." "A determination is a judgment of what is the case" (Derr, 1985, p. 491). Using the example of an ordinary sentence, he elaborates as follows:

Five necessary conditions of the truth of the first sentence have been identified: In order for Sentence 1 to be true, it is necessary that:

1. Information be a representation,

2. The representation be abstract,

3. The representation be meaningful,

4. The representation consists of determinations which have been made,

5. The determinations have been made of certain objects (Derr, 1985, p. 491).

Derr states that "[n]one of these conditions by itself is sufficient to insure the truth of this sentence; however, jointly, they constitute a set of sufficient conditions" (Derr, 1985, p. 491). He goes on to argue that, based on these five essential properties, four derivative properties of information can be identified as well: Information is *communicable, informing, empowering*, i.e., one can take action based on having the information, and *quantitative*, i.e., information varies in amount (Derr, 1985, pp. 493–494).

Fred Dretske

Dretske (1981), a philosopher, has been widely influential in his conceptualization of information. He draws heavily on the Shannon conception of information. Much of his book is taken up with working through the logical and philosophical implications of his position, and no attempt will be made to expatiate that here. Instead, let us take his simpler, introductory description of information.

Roughly speaking, information is that commodity capable of yielding knowledge, and what information a signal carries is what we can learn from it. If everything I say to you is false, then I have given you no information. At least I have given you no information of the kind I purported to be giving (Dretske, 1981, p. 44).

Later: "Information is what is capable of yielding knowledge, and since knowledge requires truth, information requires it also" (Dretske, 1981, p. 45). Thus, to be information, a proposition must be true. Dretske takes his analysis through many complex arguments, but at the heart of every argument is the core statement "s is F." The latter is the generic statement of a proposition, a claim.

He distinguishes information from meaning in the following way. A sentence, such as "Joe is at the office," has a meaning that arises straightforwardly from interpreting the sentence for anyone literate in English. A sentence may, however, carry much more information as well, beyond the meaning of the sentence itself.

> The acoustic signal that tells us someone is at our door carries not only the information that someone is at the door, but also the information that the door button is depressed, that electricity is flowing through the doorbell circuit, that the clapper on the doorbell is vibrating, and much else besides (Dretske, 1981, p. 72).

Thus, Dretske's sense of information includes all the demonstrable implications of a proposition, not only the (more limited) meaning of the proposition itself.

Christopher Fox

Fox (1983) takes his propositional view of information through many transformations as well. Again, using the simplest formulation for our discussion here, Fox is essentially claiming that the information (if any) contained in a set of sentences in a particular context is the proposition *p* such that, first, *p* is the conglomerate proposition expressed by the set of sentences in that particular context, and second, that the agent of that context is in a position to know that *p* (Fox, 1983, p. 203). Put still more colloquially, information is the collective propositional claim of a set of statements in a given context, provided the agent in that context is in a position to know that *p*.

Fox titles his book *Information and Misinformation*, and spends some time analyzing the latter concept as well. He disagrees with Dretske in that he does not feel that the act of "informing" necessarily means informing someone of a *true* proposition. He concludes, through a series of arguments based on the logical and linguistic character of true and false statements, that *informing* does not require truth, that is, the claim "that p" need not be true for someone to be informed of it. He argues, however, that we use "misinform" in a stronger sense—that to misinform someone necessarily means that the person is being informed of something that is not true. So we *inform* people of things that may or may not be true, but when we *misinform* someone, it necessarily involves telling the person something that is not true (Fox, 1983, p. 154ff).

We are now in a position to return to Losee's conception of information. Losee's position admits of misinformation too. He says, "The value of a variable is information about the input; when the information is only partial and is tainted by error, it is better understood as misinformation. Essentially, this is information that is partly or wholly false" (Losee, 1997, p. 267). So in Losee's conception, the value produced by a process may indeed be inadequate or inaccurate, and should then be known as misinformation.

Structural definitions

In addition to Frederick Thompson, discussed below, we will review two other authors whose definitions are largely structural, D.M. MacKay and Marcia Bates. Their definitions are "multitype," however, and will be reviewed in the section on multitype definitions.

Frederick Thompson

Thompson (1968) makes a discursive argument about information as a kind of structure. He recognizes the ways in which the structuring and organizing of information contains its own information, and is therefore likewise informative. He describes information as "a product that results from applying the processes of organization to the raw material of experience, much like steel is obtained from iron ore" (Thompson, 1968, p. 305). He also compares the scientist to the artist, in that "[d]ata are to the scientist like the colors on the palette of the painter. It is by the artistry of his theories that we are informed. It is the organization that is the information" (Thompson, 1968, p. 306).

Commonly, in daily life, the forms of organization of information are seen to be neutral, content-free. Indeed, they are often taken for granted, and not even noticed. Thompson brings out the ways in which these organizing activities are themselves content, influencing, to a greater or lesser extent, the overall meaning of the text or other body of information. This is an intellectual position—highlighting the impacts of the form and organization of information—that has been taken up by many more recent scholars, for example, Geoffrey Bowker and Leigh Star (1999).

Social definitions

The definitions considered to this point are drawn largely from logical and scientific points of view. This is sometimes known as a *nomothetic* approach; the fundamental effort in the sciences is to discover causes, effects, patterns, and tendencies that underlie the surface variety and particularity of life. The humanities perspective of valuing and studying the unique, specific characteristics of a situation, social group, or event, known as an *idiographic* approach, has not yet been represented here. In fact, probably due to the influence of the post–World War II scientific interest in information, many people coming from a humanities point of view have dismissed "information" as a specifically technological and heartless concept, to be perennially contrasted with the rich detail of specific institutions and historical moments, such as a study of the library in nineteenth century Illinois. However, the authors discussed in this section do embrace the concept of information, but argue that it must be seen as embedded in a social context.

Ian Cornelius

In 1996, Ian Cornelius (1996) wrote about an "interpretive viewpoint" to information. Taking the practice of law for his example, he argued that information should be seen as socially constructed within a set of practices. A practice is a "coherent set of actions and beliefs which we conform to along with the other people in our practice (whatever it may be, profession or game), and it has its own internal logic and ethic" (Cornelius, 1996, p. 15).

> My claim is that information is properly seen not as an objective independent entity as part of a 'real world,' but that it is a human artefact, constructed and reconstructed within social

situations. As in law, every bit of information is only information when understood within its own cultural packaging
which allows us to interpret it (Cornelius, 1996, p. 19).

Further,

> ... [T]here is no separate entity of information to discover
> independent of our practices. Up to the point that it is sought
> by a practitioner within a practice it is not information and
> cannot be interpreted.
> When a practice is seeking to impose meaning on some-
> thing, that thing will already have come within the interpre-
> tive range of that practice and will already be at an early stage
> in a process of interpretation (Cornelius, 1996, p. 20).

Thus, information must be seen within the dense context of social
relations, negotiations, and understandings operative in a particular social
context. Cornelius compares the embeddedness of information received
and interpreted with the embeddedness of most interpretations of the law.

Joseph Goguen

Writing a year later, Joseph Goguen (1997) developed a more detailed concept
of socially embedded information. His more idiographic approach may have
succeeded in humanizing and de-technologizing the concept for those of a
more social science or humanities bent. He defines information as follows:
"An item of information is an interpretation of a configuration of signs for
which members of some social group are accountable" (Goguen, 1997, p. 31).
He argues that all information is situated within a context, and can only
be fully understood within that context. He then addresses the question of
whether any information is ever context-free; is some information more
bound within a specific context than other information is? He describes a
continuum of the character of information, from "wet" to "dry." He states
that processes of abstraction and formalization are attempts to take information out of contexts and make it as generally applicable as possible. The
more decontextualized, the "drier" the information (Goguen, 1997, p. 32).
Certainly, many scientific results fit this description. We do not need
to know the specific circumstances in which Robert Boyle discovered
Boyle's Law (physics), in order to gain full value from the use of Boyle's
Law. On the other hand, an enormous amount of contextual knowledge is

needed in order properly to interpret the reasoning and strategy involved in directing the prosecution of World War II, or, in the information system design context, the hostile attitudes toward use of an information system in a particular government agency.

Goguen:

> In general, information cannot be fully context sensitive (for then it could only be understood when and where it is produced) nor fully context insensitive (for then it could be understood by anyone in any time and place).
>
> According to our social theory of information, meaning is an ongoing achievement of some social group; it takes *work* to interpret configurations of signs, and this work necessarily occurs in some particular context, including a particular time, place and group. The meaning of an item of information consists of the relations of accountability that are attached to it in that context, and . . . the narratives in which it is embedded (Goguen, 1997, p. 34).

He draws on ethnomethodology for the *principle of accountability:* *"Members are held accountable for certain actions by their social groups; exactly those actions are the ones constructed as socially significant by those groups"* (Goguen, 1997, p. 40).

It should be understood that Goguen is developing this concept of information in the context of information system design and usability testing. He refers to the numerous instances where massively expensive information systems were abandoned because they were dysfunctional for the organization for which they were designed. He argues that the close observation of people as called for in ethnomethodology is the gold standard for learning how a group organizes the work of their institution—what systems of categories, what contrasts, what relationships matter to the group, how the group divides up the processes and objects in their world in order to carry out their ongoing activities. Goguen:

> In particular, ethnomethodology looks at the *categories* and *methods* members use to render their actions intelligible to one another; this contrasts with presupposing that the categories and methods of the analyst are necessarily superior to those of members (Goguen, 1997, pp. 40–41).

This is a powerful conceptualization of information that incorporates its social role. Goguen argues that information, as understood through his construction of the term, is *situated, local, emergent, contingent, embodied, vague,* and *open* (Goguen, 1997, pp. 34–35). Recall Derr's characterization of information and compare it to Goguen's view: "abstract, meaningful representation of determinations which have been made of objects" (Derr, 1985, p. 491). The idiographic and nomothetic world views are on display in this contrast. For my part, they both carry a lot of truth. I do not see that one or the other approach has to "win" in the culture wars. By incorporating both approaches in our thinking, we may end up with the richest possible understanding.

First interlude: the problem of Popper's Worlds

We interrupt this recital of different approaches to the concept of information in order to consider the role of the work of Karl Popper (1979), a well-known twentieth century philosopher of science, in discussions of the nature of information. Popper was not himself interested in the concept of information, and used the word very little. However, he developed another idea, of three worlds in the scientist's life, that has had enormous impact in information science, and recurs again and again through the literature on information. Popper wrote over a number of years about what he called three "worlds": "We can call the physical world 'world 1,' the world of our conscious experiences 'world 2,' and the world of the logical *contents* of books, libraries, computer memories, and suchlike 'world 3'" (his italics) (Popper, 1979, p. 74).

It was important to Popper to distinguish one person's subjective understanding of a scientific topic from a more objective existence of knowledge of science that is independent of individual people. Once Robert Boyle died, the knowledge of "Boyle's Law" did not die with him. It was expressed and recorded in the literature of science, and was in the minds of many people who studied science. In short, scientific learning has a public, independent, and thus, reasonably objective existence apart from any one person's subjective understanding of a scientific finding.

It is the daily bread of the work of people in the information disciplines to address the recorded, "exosomatic" (outside the body) forms of information. When a philosopher writes about objective knowledge, argues for its importance, and is at pains to distinguish it from other "worlds," it may seem that at last someone from outside the information disciplines has paid recorded knowledge the attention that it deserves. At last, the very heart of the work of information managers has been recognized and validated as a key part of science!

Unfortunately, however, a close reading of Popper indicates that he did not view information resources in the way that information professionals do. Though he would sometimes refer to objective knowledge in books and journals, thus making it easy to conflate the "thingness" of books in with the objective nature of scientific findings, in fact, it is evident again and again in Popper's writing that when he is speaking of the objective nature of world 3, he means, as his definition above indicates, the *logical contents* of the books and journals, not the objects themselves. When he pays attention to the books themselves, he always calls them a part of world 1. Indeed, the scientific examples he uses are generally of the very "driest" sort, in Goguen's terms—generally mathematical and physical examples. To make his point about objective science, he is determined to select examples that can live independently in the most unambiguous way from individual knowers.

Popper was a philosopher of science, and was concerned about various debates he was having with his philosopher colleagues. It was important to him to argue for the idea that the body of scientific learning can be thought of as independent from any one individual's (world 2) experience. He called this "epistemology without a knowing subject" (Popper, 1979, pp. 106–152). A prime argument of his was that we can learn more about epistemological issues associated with science if we attend to the world 3 form of science, rather than to questions of how an individual person comes to know and believe something, the traditional subject matter of philosophical epistemology.

But it appears that he was not particularly conscious of the contents of libraries as a domain of study or professional practice. Indeed, Popper appears to have been unaware of John Ziman's book: *Public Knowledge: An Essay Concerning the Social Dimension of Science*, published in 1968, which took a more socially sophisticated look at the senses in which science has a public existence (Ziman, 1968). Popper was not a social scientist, and he rather ignored distinctions that would be important in the social sciences. He saw libraries as the abode of the logical content, but it was only the logical content that interested him—not the vast infrastructure of information institutions, professional associations, and laboratories supporting these exosomatic forms of information storage. In fairness to Popper, it was the philosophical, not the social or professional dimensions of the material aspects of science that interested him.

Nonetheless, Popper has held a continuing fascination for information studies, and discussion of his three worlds appears in the work of many, including Brookes (1980), Rudd (1983), Neill (1987, 1982), Capurro and Hjørland (2003), and other information scientists' work. It is not my objective here to analyze these appearances of Popper in information studies, except to say that Popper is often misunderstood, or the subtle distinctions he made are ignored. To take just one example: Capurro and Hjørland define Popper's world 3 as follows: "intellectual contents such as books, documents, scientific theories, etc." (Capurro & Hjørland, 2003, p. 393). Treating scientific theories as being in the same category as books and documents, thus conflating world 1 with world 3, risks perpetuating the confusions.

If the reader finds tackling the large body of writings by Popper to be daunting, J.W. Grove (1980) provides a good, knowledgeable discussion of world 3. The best places to find Popper's own discussions of world 3 are in his book *Objective Knowledge* (Popper, 1979) and in chapter P2 of Popper and Eccles, *The Self and Its Brain* (Popper & Eccles, 1985).

Multitype information definitions

Donald MacKay

Donald MacKay (1969), writing at about the same time as Thompson, who was reviewed above, also saw information structurally, specifically, "Information: that which determines form" (MacKay, 1969, p. 160). MacKay took a more rigorous scientific approach to conceptualizing information than Thompson did. Elsewhere, he defines it as "that which adds to a representation" (MacKay, 1969, p. 163). Within this general formulation, three senses of information are distinguished:

1. Structural information content, measured by "logons."

2. Metrical information content, measured by "metrons."

3. Selective information content, measured by "bits."

MacKay divided the above three measures into two types, constructive and selective. The first two measures are labeled "constructive." The third measure, "bits," corresponds to Shannon-type information, and is labeled "selective." Since the Shannon approach measures amount of information, in part, by the number of options from which a message is selected—a message chosen from 100 possible equiprobable messages is more informative than a message selected from 10 possible equiprobable messages—the amount is measured in an indirect way, in terms of the number of messages *not* sent, hence "selective."

It is interesting, therefore, to contemplate MacKay's effort to measure information in a more direct, or positive way, with his first two "constructive" measures, logons and metrons. The logons he sees as "[t]he number of *distinguishable groups or clusters* in a representation . . . its *dimensionality or number of degrees of freedom*" (MacKay, 1969, p. 165). The metron, on the other hand, is "that which supplies one element for a pattern." Later: "Thus the amount of metrical information in a pattern measures the *weight of evidence* to which it is equivalent" (MacKay, 1969, p. 166). This approach is redolent of inferential statistics, though it appears to have been intended to be still more general than that. Whatever the case, it does not appear to have caught on as a general means of measuring information.

Brenda Dervin

Brenda Dervin (1977) developed an Information$_{1,2,3}$ formulation of information, drawing heavily on the work of Richard Carter, though the distinctions she makes bear some resemblance to Popper's. (Carter's works were notoriously fugitive, seldom appearing in conventional publication venues.)

Dervin states: "In the most general sense, Information$_1$ refers to external reality; Information$_2$ refers to internal reality." This distinction "forces our attention to the notion that information can be whatever an individual finds 'informing.' It moves our attention away from 'objective' information, toward assessing the 'cognitive maps or pictures' of an individual" (Dervin, 1977, p. 22). She then asks how the individual moves between these two realities. "How does he impose sense on reality when he finds none there?" (Dervin, 1977, p. 23).

She argues that "[b]oth an individual's selection and use of Information$_1$, *and* his creation of Information$_2$ result from some kind of behavior. It is suggested here that these behaviors are in themselves legitimate informational inputs: Information$_3$" (Dervin, 1977, p. 23). In other words, Dervin sees Information$_3$ as all the ways in which people make sense of their worlds, the set of techniques they have to make this reconciliation between internal and external world. She then provides examples of different ways that people make sense of their experiences: by "decisioning," by using a "liking-disliking procedure," by a "relating to others" strategy, i.e., getting advice from others. In sum, we can say that in Dervin's terms, Information$_1$

is external world, Information$_2$ is internal world, and Information$_3$ is the procedure used to reconcile these two worlds.

Michael Buckland

Buckland (1991) argues for three types of information:

1. Information-as-process: "When someone is informed, what they know is changed." This definition is similar to Pratt's concept of "in-formation" (see above).

2. Information-as-knowledge: "'Information' is also used to denote that which is perceived in 'information-as-process.'"

3. Information-as-thing: "The term 'information' is also used attributively for objects, such as data and documents that are referred to as 'information' because they are regarded as being informative . . ." (Buckland, 1991, p. 351).

Buckland titled his article "Information as thing" and addresses this sense of information the most in the article's discussion. He sees the information disciplines as primarily being concerned with information in this sense.

"[T]he means provided, what is handled and operated upon, what is stored and retrieved, is physical information (information-as-thing)" (Buckland, 1991, p. 352). In this regard, his thinking most resembles that of Jason Farradane (1979), who defined information as a "physical surrogate of knowledge" (Farradane, 1979, p. 17).

Farradane argued that such a tangible concept was needed in order to develop a proper science out of information science.

Citing examples of documents, statistical data, statutes, photographs, etc. Buckland says: "In each case it is reasonable to view information-as-thing as *evidence*, though without implying that what was read, viewed, listened to, or otherwise perceived or observed was necessarily accurate, useful, or even pertinent to the user's purposes." Later: "If something cannot be viewed as having the characteristics of evidence, then it is difficult to see how it could regarded as information" (Buckland, 1991, p. 353). This character-ization does leave the reader wondering if Buckland always distinguishes information-as-knowledge from information-as-thing. Suppose a legal precedent is considered evidence. Is the physical book where the precedent is published the evidence, or the "information-as-knowledge" within the book that is the evidence? It would seem to be important to distinguish the physical form of the record from its logical content.

Buckland goes on to say that most anything can be informative, therefore information is everything. "We conclude that *we are unable to say confidently of anything that it could not be information*" (Buckland, 1991, p. 356).

Buckland concludes by suggesting areas in information science where distinguishing "information-as-thing" could be helpful. He notes that historical bibliography, the study of books as physical objects, and statistical analysis of information objects (now known as "informetrics") are areas where information-as-thing has primacy. Second, in the professional activity of the information disciplines, an understanding of the physical handling of different kinds of objects and materials is vital to effective management of those resources.

Marcia J. Bates

Marcia Bates (Bates, 2005, 2006), sets out self-consciously to build a conception of information that is suitable and productive for the information disciplines to use in theory and practice.

Her basic definition is a structural one—arguably the most suitable approach for disciplines concerned with the material storage and access to information. "Information is the pattern of organization of matter and energy" (Bates, 2006, p. 1033).

She aligns with Wiener on the nature of information: Wiener has said, "Information is information, not matter or energy" (Wiener, 1961, p. 132). That is, she argues that information is not identical to the physical material that composes it; rather information is the *pattern of organization* of that material, not the material itself. (This contrasts with Buckland's conclusion that information is everything.)

On the other hand, though the material world is indisputably materially "there" and not a figment of solipsistic imagination, how living beings perceive and conceive of that pattern of organization is immensely variable across species and across individuals and across the same individual at different times. The bug crawling across a checkerboard may not be aware of the alternating red and black squares at all, whereas the game-playing human looking at the board may be so focused on the squares that he or she does not even notice what material the board is composed of. Thus, nature has a physical reality independent of human beings, in which we can posit the existence of patterns of organization, even if we perceive them poorly or not at all, while living beings experience patterns of organization in their lives that may be uniquely their own, either as a species or an individual.

At the same time, she conceptualizes information in the larger context of the development of life on earth, drawing on a conception developed by Susantha Goonatilake of the evolution of information transfer over the history of life on earth. Goonatilake's view of the broad fundamental types of information is well suited to the needs of the information disciplines (Goonatilake, 1991).

He argues that over the history of life on earth, information has been stored and transmitted in three fundamental ways: genetically, neural-culturally, and exosomatically, that is, through the genome, through cultural transmission, and through storage devices external to the body. Humans, who began by memorizing and telling stories to pass on culture neural-culturally, developed an explosively growing capacity for exosomatic storage once they began to carve, draw, and write on more or less durable surfaces.

In developing her conception of information, Bates argues that in living organisms, information tends to move back and forth between encoded and embodied forms. Genetic information is encoded in the genome and embodied in the living animal, the phenotype. The phenotype, in turn, encodes the genome in reproductive material to the next generation.

Neural-cultural information is encoded in the brain of an animal, and embodied in the experiences, actions, and communications of the animal. Exosomatic information is encoded in the external materials in which it is stored, whence it is again embodied in human experience during reading, touching, or observing.

All such life-associated information she calls "represented" information, because embodied information can be thought to represent encoded information, and encoded information can be thought to represent embodied information.

To Goonatilake's three channels of information transmission she adds a fourth channel, "Residue." Residue is the flow line of extinction, where previously encoded or embodied information degrades, as in the Biblical "dust to dust." This flow line is important in the information disciplines, because relatively durable information storage does not endure forever, and can degrade to the point where no socially meaningful evidence remains.

Thus, in parallel to Goonatilake's three flow lines, Bates defined the following forms of information:

- The genetic flow line is associated with Genetic information.

- The neural-cultural flow line is associated with Experienced information (in the mind), Enacted information (actions in the world), and Expressed information (nondurable communications).

- The exosomatic flow line is associated with Embedded information and Recorded information. Embedded information is associated with the enduring effects of animals on the earth—from a path through the woods to deliberately fashioned tools, homes, and other objects. Recorded information is communicatory or memorial information preserved in a durable medium. This type of information is the prime domain of most of the information disciplines—libraries, archives, knowledge management, etc.

- Finally, residue is association with Trace information—the pattern of organization of the residue that is incidental to living processes or which remains after living processes are finished with it. The importance of Trace information to the archaeological and museum worlds is self-evident.

Bates argues for the value of such an approach by describing several example applications of these multitype terms for information in information studies. For example, the study of information seeking behavior needs to incorporate an awareness of *all* the forms of information a person takes in.

People get information not just from paper sources, not just from other people, but also from the physical layout of their workspaces, from the design, not just the content, of informational genres, and above all, from the interaction of these various factors in a real situation. All the patterns of organization of matter and energy—cognitive, physical, architectural, social, linguistic—are informative (Bates, 2006, p. 1043).

Thus, all types and forms of information need to be incorporated in our thinking in doing information behavior research, and having these multiple types of information in mind can promote that awareness and use in research.

Second interlude: DIKW

DIKW stands for Data–Information–Knowledge–Wisdom. Just as Popper's three worlds frequently show up in considering information, so also does this sequence. Discussion of this sequence is based on the assumption that the terms go from the least to most processed or integrated, with data the rawest, and wisdom the most rarefied.

Arguably, this view of these terms comes ultimately from their popular usage. Intuitively, we see each term in the sequence as more developed, "cooked" (Hammarberg, 1981), or worked through than the term to its left. Discussions of this sequence, in more or less formally worked through thinking, show up many places, including, at least, Houston and Harmon (2002), Hammarberg (1981), Meadow and Yuan (1997), and Ferris (2005). Sharma traced the hierarchy through the knowledge management and information science literature in the 1980s back through to the poet T.S. Eliot in 1934 (Sharma, 2009). Thompson suggests that "signal" should be a fifth term in the sequence, preceding data (See the entry on "Telecommunications," p. 5156).

Though this sequence may feel intuitively right, it is difficult to take it from its popular meaning and develop it into something sufficiently refined to be useful for research. Most discussions of DIKW really, at base, elaborate the intuitive understanding, and do not take theory much further.

Deconstructing "information"

"Information" has had so much importance in the thinking of many disciplines, including library and information science, over the last 60 or more years, that it is not surprising that the time has come where authors set out to debunk or "deflate" the importance of the concept of information. They choose various ways of doing this; three are illustrated below.

Ronald Day

In Ronald Day's deconstruction of *The Modern Invention of Information*, he states that

> I attempt to show how professional and authoritative texts
> about the social importance of information tried to use
> language (particularly through books) to construct a social,
> utopian value for information and helped to raise information
> and its connotations of factuality and quantitative measure to
> a privileged, even totalitarian, form of knowledge and dis-
> course (Day, 2001, p. 2).

He says his objective is not to say what information "really is" nor to conclude that information is good or bad, but "rather that certain connotations of information, and the social and cultural privileging of certain technologies and techniques associated with it, are cultural and social productions that elevate certain values over other values . . ." (Day, 2001, p. 117). Thus, he is not defining so much as critiquing the uses of the term "information."

Bernd Frohmann

Bernd Frohmann's objective in his book *Deflating Information* is to replace the centrality of "information" as the focus of much of information studies and science studies with the centrality of practices surrounding documents (Frohmann, 2004). His objective is thus not so much defining or theorizing information as it is, as his title says, "deflating information."

> One of the aims of this book is to show how rich and varied the practices with scientific documents can be, especially compared to the simplistic idea that there is no more to the informativeness of a document than what happens in the mind of someone who understands it (Frohmann, 2004, p. 16).

Some would complain, however, that good research on information seeking has, from the beginning, also examined scientific practices around documents, and linked those practices to the subjective points of view of the people being studied. Most of the articles on information seeking in the "ancient" 1958 International Conference on Scientific Information (National Academy of Sciences, National Research Council, 1959), or the 1960s era American Psychological Association's Project on Scientific Information Exchange in Psychology (American Psychological Association, 1968) could be shown to have done that. What we can say of recent decades, though, is that consciousness of all the practices, situations, and cultural beliefs surrounding information seeking and use has grown substantially, and Frohmann's book-length treatment demonstrates that.

Jonathan Furner

Furner suggests that rather than relating information to epistemology, we should relate it to the philosophy of language. Further, "we shall find that philosophers of language have modeled the phenomena fundamental to human communication in ways that do not require us to commit to a separate concept of 'information.'" And, "[o]nce the concepts of interest have been labeled with conventional names such as 'data,' 'meaning,' 'communication,' 'relevance,' etc., nothing is left (so it may be argued) to which to apply the term 'information.'" In fact, he claims that the entire field may be misnamed and "that its subject matter should more appropriately be treated as a branch of communication studies, semiotics, or library studies" (Furner, 2004, p. 428). Furner might not call himself a Post-Modernist, but the effort to design an approach that does without *the* central concept in a discipline, has an objective very much in a postmodernist spirit of deconstructing hitherto core values and ideas.

He identifies what he considers to be the several ways in which "information" is used and discussed in information studies, and categorizes them in three broad groupings as follows: "Information-as-particular," namely "Utterances," "Thoughts," and "Situations;" "Information-as-action," namely "Communication;" and "Information-as-universal," namely, "Informativeness," and "Relevance" (Furner, 2004, p. 438). He then argues that other fields address most of these topics with deeper understanding and research results than information studies does, and that even in those topic areas where information studies has done much work, such as relevance, we can draw from work in pragmatics and the philosophy of language (Furner, 2004, p. 444).

One can certainly agree that we have not advanced as far as desirable in understanding information, and, as well, that we often lack an adequate theoretical basis out of which to develop our ideas. However, one can see this situation as a problem of failing to date to develop our unique intellectual substrate, rather than that we are re-inventing lumpy versions of other disciplines' well-wrought wheels.

The universe of study of the physicist is the physical processes and dynamics that govern our universe; the universe of study of the biologist is the world of living things. The primary domain of study in the information disciplines is the world of exosomatic information, and human beings' relationship to that world as creators, designers, and users. Heretofore, society has seen that world, if it noticed it at all, as merely an epiphenomenon of the real world of things that matter. Increasingly, as the twenty-first century develops, we are more and more often grasping the reality that the world of information has become its own universe of study, one of immense importance to human beings, full of intriguing phenomena to observe and understand. That study should surely include identification of key terms, along with the provision of our own disciplinary definitions.

Conclusions

The understanding of the core concept of "information" in information science remains a highly contested area. Information is seen as

- A proposition, a structure, a message, or an event

- As requiring truth or indifferent to truth

- As socially embedded and under perpetual re-interpretation, or as measurable in bits

- As a worn-out idea deserving of dispatch, or as an exciting conception understandable in terms of evolutionary forces.

The much-debated concept of information remains at the lively heart of information science.

REFERENCES

American Psychological Association. (1968). *Project on Scientific Information Exchange in Psychology, 1963-1968* (21 reports). Washington, DC: American Psychological Association.

Aspray, W. (1985) . The scientific conceptualization of information: A survey. *Annals of the History of Computing, 7*(2), 117-140.

Bates, M.J. (2005). Information and knowledge: An evolutionary framework for information science. *Information Research, 10*(4), paper 239. Retrieved from http://InformationR.net/ir/10-4/paper239.html.

Bates, M.J. (2006). Fundamental forms of information. *Journal of the American Society for Information Science and Technology, 57*(8), 1033-1045.

Bateson, G. (1972). *Steps to an ecology of mind.* New York: Ballantine.

Belkin, N.J. (1978). Information concepts for information science. *Journal of Documentation, 34*(1), 55-85.

Boulding, K.E. (1961). *The image: Knowledge of life and society.* Ann Arbor, MI: University of Michigan Press.

Bowker, G.C., & Star, S.L. (1999). *Sorting things out: Classification and its consequences.* Cambridge, MA: MIT Press.

Brier, S. (1998). Cybersemiotics: A transdisciplinary framework for information studies. *BioSystems, 46*(1-2), 185-191.

Brookes, B.C. (1980). The foundations of information science; Part 1: Philosophical aspects. *Journal of Information Science, 2*(3-4), 125-133.

Buckland, M.K. (1991). Information as thing. *Journal of the American Society for Information Science, 42*(5), 351-360.

Capurro, R., & Hjørland, B. (2003). The concept of information. *Annual Review of Information Science and Technology 37,* 343-411.

Cornelius, I. (1996). Information and interpretation. In *Integration in Perspective, Proceedings of CoLIS 2: Second International Conference on Conceptions of Library and*

Information Science, Copenhagen, Denmark, (pp. 11–21). Copenhagen, Denmark: The Royal School of Librarianship.

Cornelius, I. (2002). Theorizing information for information science. *Annual Review of Information Science and Technology, 36*, 393–425.

Day, R.E. (2001). *The modern invention of information.* Carbondale, IL: Southern Illinois University Press.

Derr, R.L. (1985). The concept of information in ordinary discourse. *Information Processing & Management, 21*(6), 489–499.

Dervin, B. (1977). Useful theory for librarianship: Communication not information. *Drexel Library Quarterly, 13*(3), 16–32.

Dretske, F.I. (1981). *Knowledge and the flow of information.* Cambridge, MA: MIT Press.

Farradane, J. (1979). The nature of information. *Journal of Information Science, 1*(1), 13–17.

Ferris, T.L.J. (2005). Characteristics of data, information, knowledge, and wisdom. In P.H. Sydenham, & R. Thorn (Eds.), *Handbook of Measuring System Design,* (Vol. 1, pp. 231–234). Chichester, England: Wiley.

Floridi, L. (2002). What is the philosophy of information? *Metaphilosophy, 33*(1–2), 123–145.

Floridi, L. (2004). Information. In L. Floridi (Ed.), *Blackwell guide to the philosophy of computing and information,* (pp. 40–61). Oxford, England: Blackwell.

Fox, C.J. (1983). *Information and misinformation: An investigation of the notions of information, misinformation, informing and misinforming.* Westport, CT: Greenwood Press.

Frohmann, B. (2004). *Deflating information: From science studies to documentation.* Toronto, Canada: University of Toronto Press.

Furner, J. (2004). Information studies without information. *Library Trends, 52*(3), 427–446.

Goguen, J.A. (1997). Towards a social, ethical theory of information. In G. Bowker, L. Gasser, L. Star, & W. Turner (Eds.), *Social science research, technical systems, and cooperative work: Beyond the great divide,* (pp. 27–56). Mahwah, NJ: Lawrence Erlbaum Associates.

Goonatilake, S. (1991). *The evolution of information: Lineages in gene, culture, and artefact.* London: Pinter.

Grove, J.W. (1980). Popper "demystified": The curious ideas of Bloor (and some others) about World 3. *Philosophy of the Social Sciences, 10*(2), 173–180.

Hammarberg, R. (1981). The cooked and the raw. *Journal of Information Science, 3*(6), 261–267.

Houston, R.D., & Harmon, E.G. (2002). Re-envisioning the information concept: Systematic definitions. In H. Bruce, R. Fidel, P. Ingwersen, & P. Vakkari (Eds.), *Emerging frameworks and methods: Proceedings of the Fourth International Conference on Conceptions of Library and Information Science* (CoLIS4) (pp. 305–308). Greenwood Village, CO: Libraries Unlimited.

Kåhre, J. (2002). *The mathematical theory of information.* Boston, MA: Kluwer Academic.

Karpatschof, B. (2000). *Human activity: Contributions to the anthropological sciences from a perspective of activity theory.* Copenhagen, Denmark: Dansk Psykologisk Forlag.

Losee, R.M. (1997). A discipline independent definition of information. *Journal of the American Society for Information Science, 48*(3), 254–269.

MacKay, D.M. (1969). *Information, mechanism, and meaning.* Cambridge, MA: MIT Press.

Madden, A.D. (2000). A definition of information. *Aslib Proceedings, 52*(9), 343–349.

Madden, A.D. (2004). Evolution and information. *Journal of Documentation, 60*(1), 9–23.

Meadow, C.T., & Yuan, W. (1997). Measuring the impact of information: Defining the concepts. *Information Processing & Management, 33*(6), 697–714.

Miller, G.A. (1951). *Language and communication.* New York: McGraw-Hill.

Nauta, D. (1972). *The meaning of information.* The Hague, the Netherlands: Mouton.

Neill, S.D. (1982). Brookes, Popper, and objective knowledge. *Journal of Information Science, 4*(1), 33–39.

Neill, S.D. (1987). The dilemma of the subjective in information organisation and retrieval. *Journal of Documentation, 43*(3), 193–211.

Popper, K.R. (1979). *Objective knowledge: An evolutionary approach*. Oxford, England: Clarendon Press.

Popper, K.R., & Eccles, J.C. (1985). *The self and its brain*. Berlin, Germany: Springer International.

Pratt, A.D. (1977). The information of the image: A model of the communications process. *Libri, 27*(3), 204–220.

Proceedings of the International Conference on Scientific Information. (1959). Washington, DC (2 Vols.). Washington, DC: National Academy of Sciences, National Research Council.

Resnikoff, H.L. (1989). *The illusion of reality*. New York: Springer-Verlag.

Rudd, D. (1983). Do we really need World 3? Information science with or without Popper. *Journal of Information Science, 7*(3), 99–105.

Shannon, C.E. (1948). A mathematical theory of communication. *Bell System Technical Journal, 27*(3), 379–423.

Shannon, C.E., & Weaver, W. (1975). *The mathematical theory of communication*. Urbana, IL: University of Illinois Press.

Sharma, N. (2008). The origin of the "data information knowledge wisdom" hierarchy. http://www.personal.si.umich.edu/~nsharma/dikw_origin.htm. (Accessed January 2009.)

Stonier, T. (1990). *Information and the internal structure of the universe: An exploration into information physics*. London: Springer-Verlag.

Stonier, T. (1992). *Beyond information: The natural history of intelligence*. London: Springer-Verlag.

Stonier, T. (1997). *Information and meaning: An evolutionary perspective*. Berlin: Springer.

Thompson, F.B. (1968). The organization is the information. *American Documentation, 19*(3), 305–308.

Wersig, G., & Neveling, U. (1975). The phenomena of interest to information science. *The Information Scientist, 9*(4), 127–140.

Wiener, N. (1961). *Cybernetics: Or control and communication in the animal and the machine* (2nd ed.). Cambridge, MA: MIT Press.

Ziman, J.M. (1968). *Public knowledge: An essay concerning the social dimension of science*. London: Cambridge University Press.

4

Information: The last variable

ABSTRACT

In research in the behaviorally-oriented domains of information science, such as information seeking, search strategy, and human factors in online searching, we have borrowed the variables, appropriately enough, that are typically studied in behavioral science research—specifically, social, demographic, and cognitive variables. All these are valuable, but there is one last variable that is seldom studied—the information itself, specifically, the structure and organization of the information. In the occasional published behavioral studies that do examine this variable, highly significant results are found associated with it. Several areas of needed research on behavioral aspects of information and methods for studying this variable are discussed.

Introduction

In research in the behaviorally-oriented domains of information science, we have used, appropriately enough, the language and methodology of the social/behavioral sciences. Those fields typically study the effects of various independent variables on dependent variables of interest. Applied to our field, this approach has led us to look at questions like the relationship between

First published as Bates, M.J. (1987). Information: The last variable. *Proceedings of the 50th ASIS Annual Meeting, 24*, 6-10.

income (independent variable) and propensity to use libraries (dependent variable), or between training and performance in online searching.

In information science we have thus studied behavioral topics such as information seeking behavior, information use, online search behavior, and search strategy in terms of a wide range of social, demographic, cognitive and other types of social science variables. In the area of information seeking behavior, variables studied in our field have included race, ethnicity, age, occupation, education level, environmental factors, and state of mind. (See, for example, Berelson, 1949; Dervin, 1976; Garvey, 1979; Mick, 1980; Warner, 1973). With regard to online searching, variables studied have included degree of training, experience with search system and database, and various cognitive factors (e.g., Bellardo, 1985; Borgman, 1986; Brindle, 1981; Fenichel, 1981; Fidel, 1984; Woelfl, 1984).

All these variables are valuable, but there is one last variable that is left out—*the information itself*, specifically, the organization of the information. It is ironic that in borrowing the useful techniques of the social sciences we have largely overlooked the one variable which is most distinctively and characteristically our own, the one area where we might make the most powerfully unique contribution to the broader domain of social science research. We do, of course, study information itself extensively in information science, but mostly in bibliometric research. We do not much examine the features I will be referring to, however, in behavioral research.

In the area of online searching, Fidel and Soergel (1983), in their massive review of factors involved in database searching, refer to the information variable through mention of the database as a factor in search performance. They do not formulate the issue in the manner that will be done here, however; furthermore, little research has been done on this element of searching, no matter how formulated.

In this article I will deal with two areas of behavioral information science research, general information seeking behavior, and online searching. First, research in each will be discussed which indicates that information is a powerful and relatively unexamined variable, and then suggestions will be made for ways in which information may be profitably studied.

Behavioral research on information

Let us now examine two research articles in the area of information seeking that are exceptional in their focus on the information itself. These articles first gave me the germ of the idea being presented here. (There are probably other such articles of this sort as well—no claim is being made to a universal review of the literature. However, few such have been found.)

In 1962 L.J.B. Mote published a study that contained some provocative results, the significance of which, I believe, has not been appreciated even to this day. Mote divided the scientific users of the Shell Thornton Research Centre Library (U.K.) into three groups according to whether their fields of research were low, medium, or high scatter. Low scatter fields were defined as those in "which the underlying principles are well developed, the literature is well organized, and the width of the subject area is fairly well defined" (Mote, 1962, p. 170). In high scatter fields the number of different subjects is great and the organization of the literature is almost non-existent. The medium group fell in between in degree of structure.

With a sample totaling 178 people, Mote found that the average number of inquiries requiring 30 or more minutes to answer per person during a three-year period was, for the low to high scatter group, 1.4, 3.6, and 20 (yes, twenty!) respectively. No one in the first group made more than six inquiries and no one in the third group made fewer than ten (Mote, 1962, p. 172). Mote also found the same pattern with under-30-minute requests studied in a smaller sampling over a three-month period. Most information use studies—those done on the usual social/demographic variables—get results that vary from one group to another by a few percentage points. The variation in the Mote study is by an order of magnitude, an astonishing range.

More recently, Packer and Soergel (1979) examined the relationship between scatter of information and need for various forms of current awareness. Their results: Scientists in high scatter fields had their efficiency (success per time spent) increased by using selective dissemination of information (SDI), while those in low scatter fields had their efficiency decreased by using SDI. Here, the difference in the nature of the available information led to two diametrically opposed strategies for current awareness—again a very strong difference due to the information variable.

These two articles were dealing with broad patterns of information seeking behavior. We shall return to that area with suggestions for further research. But first, let us look at a second area of behavioral research in information science, viz., online searching. On the whole, studies of online searching have been somewhat disappointing in explaining variations in search success. While some differences in performance have been found to be associated with the variables studied—for example, Borgman (1986) found that humanities and social science students tended to have more difficulty with searching than science students—in general, many of the cognitive and social variables studied in a spate of recent dissertations have explained little of the variance in search performance. Even training and experience—two of the variables which are presumably most closely and directly associated with searching, show little effect (Fenichel, 1981; Wanger et al., 1980).

On the other hand, in a study that examined a number of variables in the searching of nearly two hundred searchers, Wanger et al. found that the search request itself "showed a consistent, significant effect on recall and precision scores" (Wanger et al., 1980, p. vi–5). They noted: "Although the online user community has speculated for some time that searching is a highly individualized process and that it defies any rigid or precise modeling, the study findings help to dramatize the influence of the search request on search results" (Wanger et al., 1980, p. vi–5). Recall scores averaged across searchers on 18 different search questions (each searched by about 30 searchers) ranged between 7.5% and 70.2%. In other words, the recall, averaged across all searchers on a given query, was as low as 7.5% on one query and as high as 70.2% on another query. A similar wide range of averages of 10.5% to 94.3% was found for precision (Wanger et al., 1980, p. vi–6).

Ordinarily, averaging many individuals leads to central tendencies that are fairly similar. That these recall and precision scores retained extreme differences even across many searchers suggests that the information itself—the query and its associated documents—is a huge source of variation in search results. Here, as with the Mote and Packer studies, variations in results associated with the information itself is much greater than the few percentage points ordinarily found with other variables.

Implications for research

If we accept that looking at the information itself is at least a promising area, and possibly even one of major significance in behavioral information science studies, how might we go about studying this variable in information seeking and online searching?

There may be several distinct senses of the term "high scatter" operating in information seeking. For a researcher's given area of interest:

1. There may be few or poor indexing and abstracting services for that area.

2. The research area may cut across the common method of organizing that topic, as in research in an interdisciplinary field. Since most bibliographic resources are

discipline-oriented, an interdisciplinary topic would likely lead to difficulties in accessing all the relevant material for it.

3. The topic may cut across sub-disciplines within the field, or have subtler mismatches with the terminology of the field.

These problems are conceptually distinct at one level, but at another level may be viewed as all the same problem. Namely, for any given area of interest there may be bibliographic control which is well organized to suit searches in that area—or not.

This fact may be expected to have ramifications all along the chain of activities from need to satisfaction of need. The Mote research suggests that the routine searching done by end user professionals for themselves is harder for them in some fields, so leads to higher use of professional information specialists. Such results have clear staffing implications for special libraries. If we could develop a measure of searchability of information in various fields, then we might be able to say whether the clients for a specific library were likely to need more help than the average or not. Such a measure would be a measure of the organization and accessibility of the information, i.e., of the quality of the bibliographic control of the field.

The second area where the Mote and Packer results have implications is search strategy. Some very interesting studies wait to be performed in this area. We may hypothesize that in doing research on a topic that is central to a field with low scatter (e.g., a field having one major abstracting and indexing service covering the whole field, as with *BIOSIS* for biology, or *Chemical Abstracts* for chemistry), then the best search strategy is to search exhaustively online in just one or two databases. As Packer found, use of SDI services would be of little use. On the other hand, in a topic area with poor bibliographic control in one of the ways listed above, the searcher might need to search the database of several disciplines at a minimum, and/or may search multiple-field databases such as *SCISearch* or *NTIS*. In this circumstance it may also pay the searcher to use those databases that search a topic across many databases in order to identify the best ones, e.g., BRS's *CROS* database. In the case of a well-controlled topic, use of *CROS* may be wasteful and unnecessary. Thus we can see that quite strikingly different online search strategies may be appropriate to queries of these different types.

Finally, a closer study of the information variable may shed light on the sources of variation in studies of online searching, and may at last enable us to diagnose what leads to good search performance and what to poor performance. Researchers such as Saracevic (1983) and Jahoda and

Braunagel (1980) have been analyzing the nature of questions of late on philosophical and pragmatic grounds.

I am suggesting here a somewhat different approach, however. When a searcher interviews a client, analyzes the query, and formulates it in terms understandable by the information system, her problem is to represent the query in a way that will pull the documents that the requester will ultimately view as relevant. Countless information retrieval studies have identified the relevant documents in the database and then calculated the recall and precision of the searches, using just this way of thinking about the problem.

What I am suggesting, however, is a closer analysis, a micro-analysis of the query and the answering relevant documents. In each case, we would ask, "What, specifically, would the searcher have had to do to retrieve this particular, document?" If the searcher's query did not contain good matching terms, how would the search have had to be altered to retrieve the item? Were there cross-references in the thesaurus (on or offline) that could have led the searcher to the document? Or did the document use obscure terms outside the common vocabulary of the field?

We may find that the searcher failed to make use of available capabilities of the system. For example, suppose the searcher ANDed two terms and retrieved a huge output and so rejected that pair of terms and moved on to other less useful terms. And suppose further that the searcher could have used a proximity operator on the two terms and gotten a smaller, manageable set containing given relevant documents. In this case we would say that the retrieval failure lay with the searcher, who did not use a standard capability of the system where it could have helped. In the case of the obscure vocabulary, we would say the failure lay with the indexing practices or the thesaurus itself. Such an analysis would finally point to the specific factors in the query and document information that are associated with recall and precision variations.

Some work along this line has been done already. For example, Markey et al (1980), examined online searches in detail to determine the relative contribution of controlled and free text vocabulary. Markey (1984) has also examined terms in studying online catalog retrieval and Belkin (1986) has analyzed user's natural language expressions of information need for purposes of designing an end-user retrieval interface. Much could be gained, I believe, by using this technique on standard intermediary online search studies as well. Behavioral studies to date that examine many behavioral variables have not explained much of the variance in online searching success. A detailed micro-analysis of each query against each relevant bibliographic record, i.e., a hard look at the information itself, may prove more fruitful.

REFERENCES

Belkin, N.J., & Kwasnik, B.H. (1986). *Using structural representations of anomalous states of knowledge for choosing document retrieval strategies.* Paper delivered at the Association for Computing Machinery, Information Retrieval Conference, Pisa, Italy.

Bellardo, T. (1985). An investigation of online searcher traits and their relationship to search outcome. *Journal of the American Society for Information Science, 36*(4), 241–250.

Berelson, B. (1949). *The library's public.* New York: Columbia University Press.

Borgman, C. (1986). The user's mental model of an information retrieval system: An experiment on a prototype online catalog. *International Journal of Man–Machine Studies, 24*(1), 47–64.

Brindle, E.A. (1981). *The relationship between characteristics of searchers and their behaviors while using an online interactive retrieval system* (Doctoral dissertation). Syracuse University, NY.

Dervin, B., et al. (1976). *The development of strategies for dealing with the information needs of urban residents: Phase I; Citizen study. Final report.* (ERIC ED 125 640). Seattle: University of Washington.

Fenichel, C.H. (1981). Online searching: Measures that discriminate among users with different types of experiences. *Journal of the American Society for Information Science, 32*(1), 23–32.

Fidel, R. (1984). Online searching styles: A case-study–based model of searching behavior. *Journal of the American Society for Information Science, 35*(4), 211–221.

Fidel, R., & Soergel, D. (1983). Factors affecting online bibliographic retrieval: A conceptual framework for research. *Journal of the American Society for Information Science, 34*(3), 163–180.

Garvey, W.D. (1979). *Communication: The essence of science: Facilitating information exchange among librarians, scientists, engineers, and students.* New York: Pergamon Press.

Jahoda, G., & Braunagel, J.S. (1980). *The librarian and reference queries.* New York: Academic Press.

Markey, K. (1984). *Subject searching in library catalogs: Before and after the introduction of online catalogs.* Dublin, OH: OCLC Online Computer Library Center.

Markey, K., et al. (1980). An analysis of controlled vocabulary and free text search statements in online searches. *Online Review, 4,* 225-238.

Mick, C.K., et al. (1980). Toward usable user studies. *Journal of the American Society for information Science, 31*(5), 347–356.

Mote, L. J. B. (1962). Reasons for the variations in the information needs of scientists. *Journal of Documentation. 18*(4), 169–175.

Packer, K.H., & Soergel, D. (1979). The importance of SDI for current awareness in fields with severe scatter of information. *Journal of the American Society for Information Science, 30*(3), 125–135.

Saracevic, T. (1983). On a method for studying the structure and nature of requests in information retrieval. *Proceedings of the 46ᵗʰ American Society for Information Science Annual Meeting, 20,* 22–25.

Wanger, J., et al. (1980). *Evaluation of the online search process: A final report.* Santa Monica, CA: Cuadra Associates.

Warner, E.S., et al. (1973). *Information needs of urban residents,* (ERIC ED 088 464). Baltimore and Rockville, MD: Regional Planning Council and Westat Research, Inc.

Woelfl, N.N. (1984). *Individual differences in online bibliographic searching: The effect of learning styles and cognitive abilities on process and outcome* (Doctoral dissertation). Case Western Reserve University, Cleveland, OH.

TWO

The information professions

The invisible substrate of information science

ABSTRACT

The explicit, above-the-water-line paradigm of information science is well known and widely discussed. Every disciplinary paradigm, however, contains elements that are less conscious and explicit in the thinking of its practitioners. The purpose of this article is to elucidate key elements of the below-the-water-line portion of the information science paradigm. Particular emphasis is given to information science's role as a meta-science— conducting research and developing theory around the documentary products of other disciplines and activities. The mental activities of the professional practice of the field are seen to center around *representation and organization* of information rather than *knowing* information. It is argued that such representation engages fundamentally different talents and skills from those required in other professions and intellectual disciplines. Methodological approaches and values of information science are also considered.

Introduction

Recently, digital information and new forms of information technology have become the focus of tremendous amounts of attention and energy in our society. Money is pouring into the development of all manner of

First published as Bates, M.J. (1999). The invisible substrate of information science. *Journal of the American Society for Information Science, 50*(12), 1043–1050.

technologies and information systems. The excitement penetrates not only the business world and the general society, but also academia, where computer scientists, cognitive scientists, and social scientists are thinking about information and the social impacts of information technology in new ways.

This new context poses a challenge to information science. Currently, the wheel is being reinvented every day on the information superhighway. Our expertise is ignored while newcomers to information questions stumble through tens of millions of dollars of research and startup money to rediscover what information science knew in 1960. We in the field need to make our research and theory better known and more understandable to the newcomers flooding in—or be washed away in the flood.

To do that effectively, however, we need to become more fully conscious of the research and practice paradigm from which we operate. A field's paradigm, in Thomas Kuhn's (1970) sense, consists of the core body of theory and methodology of a field, along with an associated world view regarding the phenomena of interest to the field. In the sciences, there is generally an organizing theoretical model of great scope, which generates research questions for decades. (As Kuhn described for the field of physics, first, Newton's Laws, then Einstein's theory of relativity were successive defining theoretical models for that field.)

As will be noted shortly, the explicit paradigm of information science has been very well described before. However, a field's paradigm is much more than the explicit theoretical model it works from. Certain methodological approaches and a world view are generally integrally linked with the questions studied. A field tends to draw people with certain cognitive styles, who produce research of a certain character. The field has a history, great names, stories, customs, mores, and values. One does not work long in information science without knowing the names Wilf Lancaster, Gerard Salton, and Llewellyn C. Puppybreath III, and what they are known for.

Much of the paradigm of any field lurks below the water line, largely unconscious and unarticulated, even by its practitioners. Researchers often soak up the paradigm, understanding the subject matter and what it means in every way to be, say, a physicist or engineer, without being able to articulate it well.

Today, in information science, many newcomers without a background in the field are coming in. At this historical juncture, information scientists need to become more conscious of the thought world we are operating out of, so that we can communicate it more rapidly and effectively to large numbers of new people, and so that we can continue to influence the future of information in the 21st century.

The purpose of this article is to attempt to articulate at least some of those parts of our field that are largely unarticulated ordinarily—to describe the invisible substrate of information science, the part of the field that is below the water line.

Paradigm above the water line

Information science does have an explicit, above-the-water-line, paradigmatic definition, and an understanding of that explicit expression is important to an understanding of the intrinsic unity of the whole paradigm. Information science is the study of the gathering, organizing, storing, retrieving, and dissemination of information. That definition has been quite stable and unvarying over at least the last 30 years. In fact, in a January 1968 article in this journal, Harold Borko wrote the following:

> Information science is that discipline that investigates the properties and behavior of information, the forces governing the flow of information, and the means of processing information for optimum accessibility and usability. It is concerned with that body of knowledge relating to the origination, collection, organization, storage, retrieval, interpretation, transmission, transformation, and utilization of information. It has both a pure science component, which inquires into the subject without regard to its application, and an applied science component, which develops services and products (Borko, 1968, p. 3).

Paradigm below the water line

The meta-field of information science

It is first of all important to recognize that information science, like education and journalism, among others, is a field that cuts across, or is orthogonal to, the conventional academic disciplines. All three of the above-named fields deal with distinct parts of the transmission of human knowledge—information science with the storage and retrieval of it in recorded form, education with the teaching and learning of it, and journalism with the discovery and transmission of news. Under these circumstances, such fields cut across all of what we might call "content" disciplines. Art historians focus on the study of art; information scientists, on the other hand, take art information as but one slice of the full range of information content with which we deal. Likewise, art education is but one part of education, etc.

Paisley has made a similar distinction by contrasting behavioral science "level fields" and "variable fields." He defined the former as disciplines that study human behavior at different levels of organization—psychology at the individual level, sociology at the group level, and anthropology at the culture level (Paisley, 1972, p. i). Paisley states that variable fields, on the other hand, look at one variable across all the conventional levels. For example, political science looks at political behavior across the several levels.

Here, however, the distinction being made is between conventional disciplines and meta-disciplines. The meta-disciplines are distinguished by the fact that they are interested in the subject matter of all the conventional disciplines, they do something with that subject matter that is of value for society (see next paragraph), and that something is unique to each meta-discipline. (Though they are research disciplines, the three examples of information science, education, and communication/journalism also have distinct professional cores, which are vital to their natures.)

Information science organizes that subject content for retrieval, education uses teaching skills to convey that knowledge to learners, journalism uses reporting and writing skills to convey news events to others, etc. In each case, it is not a variable that is being studied, but rather the content of all the conventional disciplines is being shaped and molded for a societal objective through different types of professional activities involving the manipulation and transmission of knowledge.

Research in these various meta-fields analyzes the *processes and domains* associated with the professional activities being carried out in each case. Information seeking and searching, teaching and learning, and communication are (some of) the *processes* being studied in the research of the three example fields of information science, education, and communication research/journalism. These processes are identifiable in human interactions with virtually all kinds of knowledge or information, and contain many research-worthy questions.

Although each field covers all kinds of knowledge or information, each, nonetheless, has particular domains it studies, which cut across all the conventional subject disciplines. These domains are distinguished not by their subject content, which can be highly various, but rather by their rhetorical character in the broadest sense, that is, by their selection, design, and objectives. The *domain* of information science is the universe of recorded information that is selected and retained for later access; the domain of education is curricula; and the domain of journalism is the journalistic product of all the newsworthy areas of life (science reporting, political reporting, etc.).

Note that in each case, the intellectual product of research or of social or cultural activities is selected, designed, and shaped for some social purpose. Different talents, training, and experience are required to select and index documents (information science), to select and organize information in a curriculum to optimize learning (teaching), or to sleuth for information and shape a news story (journalism).

What then, is the nature of that shaping for a purpose in information science, particularly? This can most readily be understood by looking first at the practice of the field. After that, research and theory of information science will be examined in this perspective as well.

The content of form

In applied information science, we find ourselves primarily concerned with the form and organization of information, its underlying structure, and only secondarily with its content. In the sciences and humanities, it is the content that is of dominating concern. In fact, the organization of the information they are using is usually virtually or entirely invisible to the practitioners of those disciplines; they have simply never thought of it, never realized that extensive and intellectually demanding work is needed to develop index and database standards, to select and catalog resources, etc.

Most people outside our field do not realize that there is a content to the study of form and organization. I believe that this is one of the chief reasons our field is commonly thought to have no "there" there. The average person, whether Ph.D. scholar or high school graduate, never notices the structure that organizes their information, because they are so caught up in absorbing and relating to the content. And, in fairness to them, they are not interested in the structure. *We* are interested in the structure.

As a practical matter, when one does the work to gather, store, organize, retrieve, and disseminate information—the classic elements of the formal, above-the-water-line paradigm definition of information science—one *necessarily* gets involved with understanding and manipulating its form, structure, and organization. One's attention is drawn, again and again, to these features of the information, simply to get the job done.

People who come into this field, whether formally educated in it or who drift in through a job, sooner or later go through a transformation, wherein they shift their primary focus of attention from the information content to the information form, organization, and structure. The Ph.D. art historian who gets a job working with art history information out of a love of the subject matter eventually finds him or herself working with the core questions of information science, not of art history.

Being and representing

The transformation involves even more than a shift in the subject content of one's actions or thoughts, however. Work in the meta-discipline of information science, both at the practical and theoretical levels, draws upon different cognitive talents than most of the work within conventional subject disciplines.

Perhaps the best parallel for illustrating the difference in the mental processing of information scientists and conventional disciplinary practitioners is to use an analogy of the relationship between actors and physicians. We take it for granted that when we see a film or television program like "ER" ("Emergency Room"), that it is actors who portray the physicians, because that is the way it has always been done.

On reflection, however, is it not strange that the people who have the most experience and knowledge about being doctors—the doctors themselves—are not the ones who portray doctors in drama? Why not? Why do not the competitive demands of the television marketplace pressure the heads of networks to hire real physicians—the true experts, after all—to portray physicians in a medical drama?

The answer here is that although the physicians know the most about medicine, *portraying* a physician is different from *being* a physician. Portraying a physician requires a different body of talents than being a physician does. Occasionally, some people have both types of talent, but usually not. Actors, with little or no medical knowledge, but with experience portraying a variety of characters, do a better job at portraying physicians than even physicians themselves do. Is that not remarkable, when we reflect on it?

In like manner, *representing* information—whether you are indexing or formulating a search strategy or helping someone articulate what they want to find—is different from *knowing* the information. After all, the physicians *know* much more about hospitals, medicine, and treating patients. They know so much, it is in their bones, a part of their every action. Yet when called upon to portray a medical situation instead of simply live it, physicians generally make poor actors. A talented actor, without a day's experience in medical school, can do a much better job.

Above, it was said that the rhetorical character of each of the domains of information science, education, and journalism differed from each other and from the conventional disciplines. For information science, a particular kind of representation is at the heart of the rhetorical stance of the field toward its domain, the universe of recorded information. (Here it is left to the other two fields to define their rhetorical stances.) Creating databases and catalogs involves creating representations of forms of information. The skill a reference librarian or information specialist develops also involves representation—figuring out how to conceptualize and represent a user's query, then in turn, translating the query (representing it) into a form an information system uses, which in turn arises from the representations of documents in the information system.

Subject expertise

A perennial issue in the information field revolves around how much subject knowledge an information specialist needs. Surely, it is said, one must be an expert in molecular biology to be a good information specialist at a biotechnology firm. I am among the many, however, who contest this assumption. I would argue that what one mainly needs is information expertise and talent, not content expertise. The latter is a nice bonus, if it is present, but is not essential.

When taking on a new part, actors sometimes research the context or the role they will play. Such research can enable them to do a better job of acting, but researching hospital life or medicine does not entail getting a medical degree! Actors, in fact, may be looking at entirely different things when they research the role than would ever occur to a doctor living that life. The actor needs different knowledge and different talents to do a good job at portraying physicians. This author was once told by a manager at the National Library of Medicine that some years earlier they had tried using physicians to index the *Index Medicus* database, but had given it up, because the physicians did not do as good a job as trained indexers did.

When an information specialist takes a position at a biotechnology firm, it would be a good idea to read some popular books about molecular biology, and learn who the dominant individuals are and what the major research issues and approaches are. But what will be most important for the information specialist will be to use information-related talents. People who are attracted to this field generally have somewhere within them, wide subject interests, good skills with language, with getting the big picture about a subject matter, and a knack for operating at the level of *representation* of the subject matter, rather than just working *in* the subject matter.

Over many years of teaching, I have observed that master's students in information science programs complete the mental transformation to thinking like information specialists within a few months. Often they have considerable difficulties during the first few weeks of the program, because *at first it feels alien to think about a resource in terms of the features that matter to the organization and retrieval of it, rather than in terms of mastering its content.* In a job, without formal information training, this transformation process may take longer.

It is argued here, however, that unless that transformation occurs, one cannot do a fully effective job as an information professional. And this claim is made unequivocally. If you want to portray a doctor, you have to be a good actor, not a doctor; if you want to work with information organization and retrieval, you have to be a good information person, not a subject specialist without information training. All the subject expertise in the world is not enough, if you do not possess the mental framework and skills in information work.

If drama and acting were being invented today, instead of deep in the mists of history, people might well be making the primitive assumption today that doctors must play doctors in drama. The beginning assumption would be that to produce good acting, we must use people who are in exactly the same situation found in the play—only pregnant women can portray pregnant women, only physicians play physicians, etc. However, humankind long ago learned that actors, not specialists in the circumstances of the story, are best at acting.

Description of information and retrieval from large bodies of information are quite recent phenomena, however. Collections did not become large enough to require extensive and systematic organizational approaches until the 19th century. Information expertise developed within the library, and later the documentation, fields over many decades, but the numbers of people pursuing these activities were still very small, and marginal to the larger society.

Now at the end of the 20th century, however, the society at large is discovering information and problems of information description and organization. Members of the broader society are consistently committing what we might call the First Fallacy of Information Work; they are thinking that organizing information requires deep subject expertise and no information expertise. In just the last couple of years I have heard of three very well-funded projects of this sort. In one, dozens of historians, exclusively, untrained in cataloging, were hired to catalog historical material, with consequent enormous waste of time and resources as they fumbled their way, finally, to creating a usable indexing vocabulary. In two other projects, educators, also untrained in information work, were enlisted to index educational materials without any guidance from anyone with the least background in information work. In my view, doing this is analogous to hiring physicians to portray physicians in the theater. The sooner society moves beyond this misapprehension of the nature of information work, the better for all.

(Sometimes highly gifted people come into the information field with no training and do well. Likewise, some actors are so naturally talented that training and experience are virtually unneeded for them. But most people in both fields benefit substantially from training and experience.)

My litmus test for the newcomers who are now interested in information work is whether I can observe evidence that they have gone through the transformation of becoming an information expert. That perspective is the *lens* through which information scientists see their world—both at the theoretical and practical levels. It is the single most defining framework element for their world view. There are many other elements to the part of information science that is below the water line, that will be addressed shortly—but this is the single most important, in my view. It is what the newcomer or outsider does not understand; it is what the insider takes for granted (and, therefore, seldom notices or articulates). This perspective drenches the thought and actions of the information person.

Librarianship and information science

Both people who call themselves librarians and people who call themselves information scientists share this information perspective. Other fields with which information science might have been thought originally to have much in common, such as computer science, cognitive science, computational linguistics, or artificial intelligence, did not, in fact, prove to be good matches. Both librarianship and practical information science, however, have the information perspective in common, and the phrase "library and information science," or "LIS" has become very common. I believe that this coming together arose out of deep commonalities in the way of thinking and doing necessary to achieve information work objectives. Although librarianship and information science have very different histories, and, in particular, different methodological and values perspectives, they have in common this core relationship to the material of their work.

Information science theory

The distinctive perspective discussed above in information work carries over into and is integral to the theory of the information science field as well. In 1970, this journal changed its name from *American Documentation* to the current one. That year is as good as any to mark the formal recognition of the then new field. The roots of information science lay in the theory and thinking in several related fields, particularly in the years 1930–1970. What those disparate theories and elements had in common, what enabled them to be a reasonably coherent intellectual discipline when brought together in information science, was their interest in form and structure, in particular, in information form and structure.

General systems theory (Bertalanffy, 1968; also see historical discussion in Checkland, 1981), which developed in the 1930s, and later, drew attention to the underlying structure or pattern in social and technical institutions and devices. Once the concept of the system was developed and elaborated, systems could be recognized as underlying countless disparate social, technical, and physical phenomena. Operations research and systems analysis during and after World War II developed these ideas further into a variety of applied realms.

John von Neumann and Oskar Morgenstern (1967) developed game theory—which is a way of seeing an underlying common structure of trade-offs, benefits, and disadvantages within a variety of social and economic situations. Perhaps the best-known game is "Prisoner's Dilemma." In this situation, the prosecution's evidence on a burglary charge against two men is weak. The prosecutor interrogates each man separately. Each is told that if he confesses and his friend does not, he will receive a light sentence, and that if he does not confess and his companion implicates *him,* he will receive a very heavy sentence. If both confess, they will receive a middling sentence, and if neither confesses they will both get a smaller charge of carrying a concealed weapon (Jones, 1980, pp. 77–78). This game structure is so robust that it underlies countless cops and robbers programs on television, as the television writers imaginatively play out one variation or another of it.

This recognition of underlying structures arose throughout the social and engineering sciences in the decades after the War. If we define information as "the pattern of organization of matter and energy" (Parker, 1974, p. 10), then structure of all kinds itself constitutes a kind of information.

Within that context, Norbert Wiener (1961) identified the role of information in natural and human systems in a way that had never been recognized before. He developed the field of cybernetics, which deals with the guiding or governing of systems. Wiener demonstrated that many systems are driven not primarily or only by mechanical forces, but rather are determined by the feedback of information to a governing element of the system. We are so used to hearing the word "feedback" in its common everyday overuse, that the iconoclastic newness of Wiener's ideas has been lost nowadays. Wiener demonstrated, for example, that when a person reaches for an object, it is done with continual visual and kinesthetic feedback of information, which is then used to guide the hand further. The hand does not just respond to a single impulse from the brain to "grab."

The early work that had perhaps the single most electrifying impact of all was Claude Shannon's information theory (Shannon & Weaver, 1949). Shannon measured the amount of information going through a telephone wire. Such a development does not on the face of it sound revolutionary, but it was, because his theory was abstract, and seemingly applicable to many environments, including not only the technical but also human language and psychology. The limits of Shannon's theory for the human sciences ultimately became evident, but the legacy of a new, abstract sense of information as reducing uncertainty by measurable amounts, remained.

Similarly, Noam Chomsky's theory of syntactic structures in language (1971)—common patterns underlying all different languages—had an explosive impact on several fields, and was the engine that drove the field of psycholinguistics. (In my 1980 study of citations in information science, Chomsky was the single most cited individual [Bates, 1980, p. 278].) Miller, Galanter, and Pribram, three well-known psychologists, wrote *Plans and the Structure of Behavior* (1960), which posited a common underlying structure to all, or virtually all, human behaviors.

Gregory Bateson identified common underlying structures in learning, as well as metastructures in communication that reference other communications. He dealt, thus, in many different ways with representations of representations. It is no accident that the cover of the 1972 paperback of his *Steps to An Ecology of Mind* states: "The new information sciences can lead to a new understanding of man" (Bateson, 1972). He is best known for his "double-bind" theory of schizophrenia, but his theories referred to all communication and learning among humans, not just schizophrenia. When biochemical explanations of schizophrenia largely displaced psychological ones, Bateson's work fell into disrepute—a most unfortunate failure of our intellectual world to recognize that he was a theorist of all communication, not just the unhealthy communications that are sometimes associated with dysfunctional families.

Finally, the recognition of form and structure found the purest expression of all in G. Spencer-Brown's *Laws of Form* (1972), which analyzed the mathematical rudiments, the absolute essence, of form. Spencer-Brown started with the irreducibly smallest distinction, a single difference, which is the first step in creating form out of the formless void.

All of these thinkers had in common the recognition of underlying structure beneath the surface variety of life. They contributed theories and ideas of great power, and were the theoretical driving force behind information science. Thus, when the theory and practice of information science are described in the way they have been so far in this article, the close relationship between theory and practice—through their common attention to form and structure, becomes evident.

Information science's universe

So both from a theoretical and a practical standpoint, information scientists are interested in the structure of their object of study—information. But as the examples above indicate, many social and behavioral scientists are interested in underlying structures also. Many engineers, based on Shannon's and Wiener's work, among others, are interested in information. What, then, is distinctive about information science's theory?

We are interested in information as a social and psychological phenomenon. The information we study generally originates from human agency in some way, whether it is the data beamed down from a satellite or the text of a book on Immanuel Kant's philosophy. Our primary, but not sole focus, is on *recorded* information and people's relationship to it.

All the academic disciplines can be seen as studying different universes of phenomena. The natural sciences study the natural world, the social sciences study the social worlds produced by humans, and the arts and humanities study the content and context of the creative works of human beings, from philosophy to literature to the arts.

Information science has a distinct universe that it studies also—the world of recorded information produced by human agency. We can imagine all the human activities in studying the above natural, social, and artistic universes themselves producing information entities—books, articles, databases, data files, etc.—thus creating a fourth universe, that of recorded information.

The recorded information universe contains many other kinds of information besides research results—popular literature, business records, personal archives, music, film, etc., and, of course, all of these in electronic form as well. In short, the documentary products of human activity themselves form a universe deserving of study, and study of that universe—and how human beings produce it, seek it, retrieve it, and use it—is the intellectual domain of information science.

I argued in an earlier paper (Bates, 1987) that one of the primary concerns of our field should, therefore, be to define the parameters and variables associated with our universe—information produced by human agency. We have been slow to do that, however, and still tend to borrow the variables used in the other social sciences, rather than develop those unique to our field.

This study of the information universe finds its purest expression in bibliometrics, or the study of the statistical properties of recorded information. However, the field's interest is in human-produced information, and therefore, how human beings relate to this information—how they seek it, use it, ignore it, retrieve it—is of central research importance also.

In comparison to other social and behavioral science fields, we are always looking for the red thread of information in the social texture of people's lives. When we study people, we do so with the purpose of understanding information creation, seeking, and use. We do not just study people in general. The rest of the social sciences do various forms of that. Sometimes this can be a very fine distinction; other times it is very easy to see. In communications research, a cousin to our field, the emphasis is on the communication process and its effects on people; in information science we study that process *in service of information transfer.*

For another example, there are social scientists today who are observing people doing collaborative work through new types of networked systems in the field of computer-supported cooperative work (CSCW). The sociologist or social psychologist identifies and describes the network of relationships and the social hierarchy that develops under these circumstances. They may examine the impact of technology on those social relationships and on the work of the individuals involved.

The information scientist, on the other hand, follows the information the way Woodward and Bernstein "followed the money" in their Watergate investigations. That's the red thread in the social tapestry. When we look at that social hierarchy, we are not interested in the hierarchy per se, but, rather, we ask how it impedes or promotes the transfer of information. We ask what kinds of information people prefer to communicate through this or that new channel of information technology. We always follow the information.

Information science's big questions

Three Big Questions can be identified within the above framework: (1) the physical question: What are the features and laws of the recorded-information universe? (2) The social question: How do people relate to, seek, and use information? (3) The design question: How can access to recorded information be made most rapid and effective?

The second question above deals with all kinds of information, but the other two have "recorded information" at their heart. We need to understand how people relate to and use all kinds of information, and in their social contexts—the second question—to contribute to our understanding of the first question and to do the best job possible answering the third question. But one of the defining characteristics of our field—and another feature that unites us with librarianship—is that we deal principally with *recorded* information.

There's a reason for that—it is not just an historical happenstance. A fundamental difference between recorded information and more evanescent or ephemeral forms is that recorded information generally lasts a long time. It can hang around for months, years, or centuries. And that, in turn, means that it can pile up. One of the fundamental challenges for both librarianship and information science is to find a way to contend with ever-larger piles or stacks or sets of information.

Because of the linguistic, psychological, cognitive, social, and technical complexities of information retrieval, each increase in size of the information source or database requires different solutions; scalability is a fundamental problem in this field. I believe that at some point a historian will show that the information explosion (with us since the invention of printing) has driven most of the major innovations in information organization and access. Each time the average collection grew to a new level, a new access method had to be devised. This has been particularly noticeable in the last century—from the development of subject headings to the development of hyperlinks.

The development of each new medium or technological device also requires a sophisticated blend of our technological and sociopsychological understanding to produce the best information retrieval system result. Each new thing we learn about the universe of information can be used in answering the design question also. Thus, although these three questions are posed as distinct questions, it can be seen that the research in response to each of them is also mutually supportive and valuable for answering the other questions.

Methodological substrate

Let us turn now to the other elements of the substrate of our field. The fundamental methodological stance of information science can be described as sociotechnical. The two most important methodological traditions we draw on are the social sciences and the engineering sciences. Some individuals in the field are more interested in, and more capable in, either the social or the technical side of this equation. But to function effectively in information science, one must be at least comfortable with both sides of this dual tradition. Thus, only those computer scientists endure in the field who do not dismiss the linguistic or psychological complexities of information retrieval as squishy nonsense, and only those social scientists survive who are interested in the technology and do not hand it an unreflective Luddite rejection.

Consequently, information science tends to draw multitalented people, people who enjoy the mixture of kinds of cognition that the nature of our field requires to solve its research problems. This is one of the reasons we have failed to coalesce as a field around one standard methodological paradigm. For one thing, we need this methodological variety to solve these problems, and for another thing, the people in our field, by our multitalented natures, enjoy and tolerate that variety much better than is the case in other fields. If a field can solve its research problems with a smaller range of cognitive and methodological approaches, it can then attract and maintain its activities using people who are more single-minded, or single-talented—but that is not the case in information science.

Methodologically, we may use bibliometric techniques, other statistical techniques, and philosophico-analytic approaches for studying the first question. Much qualitative research remains to be done analyzing the social significance of the design characteristics of various documentary forms (Hjørland, 1997, p. 127). The second question draws on the full range of social science techniques, from the quantitative to the qualitative. The third, design question draws on the above two approaches as well, but most distinctively uses engineering techniques. A fundamental approach in information retrieval system design is formative evaluation. First, a system is developed that embodies solutions to information retrieval for a designated context. That system is then tested for performance, which produces learning that is practical and immediate as well as more fundamental. These discoveries are then applied to a new design, and performance proceeds to improve. It can thus be seen that to solve the information science field's problems, a mix of methodologies are needed.

A final comment on methodology: regarding the great methodological shift sweeping through the social sciences, the shift to the qualitative, multiple-perspective, post-Modernist approaches—these new techniques simply add to and enrich the armamentarium of techniques available to the information scientist for studying the subject matter of our field. For reasons that have already been argued, this field *requires* multiple methodological approaches to conduct its research. In mid-20th century social science we have had a series of waves of methodological fashion—each wave declaring the prior approach to be hopelessly bankrupt and inadequate. It is to be hoped that it is finally recognized that all of these methodological approaches can be powerful and useful—especially in information science.

Values

The values of information science have tended to follow the "value neutral" science or engineering model. The emphasis at the applied level is on getting the job done well—the engineering approach—without political or explicitly value-laden objectives. At the pure science level, the effort is to find out the truth (or multiple truths) of the matter studied, regardless of personal agenda. The advantage of removing political issues from scientific endeavors has been that communication in information science (as with other sciences) has been able to proceed across political boundaries and among people with very different political philosophies.

Librarianship, in contrast, follows a more service-oriented and empowerment-oriented value system. The library is there to produce a certain desirable social result, and, as a consequence, many of the activities of the library field are organized and directed to meet that values-laden goal. The mix of values driving library work varies from country to country, and, appropriately, is suited to the particular circumstances of each nation.

There is one more characteristic—we might even call it a value—of the field of information science that merits recognition: a sense of humor. As another article in these anniversary issues documents, the American Society for Information Science has long had a Special Interest Group devoted to presenting spoofs of papers (SIG-CON), and idolizing its founding member and "information racketeer," the never-seen Llewellyn C. Puppybreath III. I took this element of our field for granted until I visited one year a social science academic association (which shall remain nameless), and came upon a most deadly serious group of people whose sense of humor was well hidden and whose anxiety at each article presentation was palpable. I was happy to return to a group of people able to laugh at themselves, yet still deeply value their own work.

Conclusions

Information science does not consist only of the explicit paradigm of the study of the selecting, gathering, organizing, accessing, and retrieving of information that is the usual description of the field. As with most intellectual domains, the field of information science has many unarticulated, but important, elements "below the water line."

It has been the purpose of this article to bring some of those elements to the surface, so that we may better understand our own work and communicate it to the many people from the broader society who are now excited by information questions and problems.

REFERENCES

Bates, M.J. (1980). A criterion citation rate for information scientists. *Proceedings of the American Society for Information Science Annual Meeting, 17,* 276–278.

Bates, M.J. (1987). Information: The last variable. *Proceedings of the American Society for Information Science Annual Meeting, 24,* 6–10.

Bateson, G. (1972). Steps to an ecology of mind. New York: Ballantine.

Bertalanffy, L. (1968). General system theory: Foundations, development, applications. New York: George Braziller.

Borko, H. (1968). Information science: What is it? *Journal of the American Society for Information Science, 19*(1), 3–5.

Checkland, P. (1981). *Systems thinking, systems practice.* New York: Wiley.

Chomsky, N. (1971). *Syntactic structures.* The Hague: Mouton.

Hjørland, B. (1997). *Information seeking and subject representation: An activity-theoretical approach to information science.* Westport, CT: Greenwood Press.

Jones, A.J. (1980). *Game theory: Mathematical models of conflict.* New York: Wiley.

Kuhn, T.S. (1970). *The structure of scientific revolutions* (2nd ed.) Chicago, IL: University of Chicago Press.

Miller, G.A., Galanter, I., & Pribram, K.H. (1960). *Plans and the structure of behavior.* New York: Holt.

Paisley, W. (1972). *Communication research as a behavioral discipline* (Draft for criticism). Palo Alto, CA: Stanford University Institute for Communication Research.

Parker, E.B. (1974). Information and society. In C.A. Cuadra & M.J. Bates (Eds.), *Library and information service needs of the nation: Proceedings of a conference on the needs of occupational, ethnic and other groups in the United States* (pp. 9–50). Washington, DC: U.S.G.P.O. (ERIC #ED 101 716).

Shannon, C.E., & Weaver, W. (1949). *The mathematical theory of communication.* Urbana, IL: University of Illinois Press.

Spencer-Brown, G. (1972). Laws of form. New York: Julian Press.

von Neumann, J., & Morgenstern, O. (1967). *Theory of games and economic behavior* (3rd ed.). New York: Wiley.

Wiener, N. (1961). *Cybernetics: Or control and communication in the animal and the machine* (2nd ed.). Cambridge, MA: MIT Press.

2

The information professions: Knowledge, memory, heritage

ABSTRACT

Introduction Along with leading to growth in the numbers of people doing information work, the increasing role of information in our contemporary society has led to an explosion of new information professions as well. The labels for these fields can be confusing and overlapping, and what does and does not constitute an information profession has become unclear.

Method We have available a body of theory and analysis that can form the basis of a review of this new professional landscape, and enable us to clarify and rationalize the array of emerging information professions.

Analysis Work by the author and others on the philosophical nature of, and core elements of, the information professions is drawn upon and applied to the current professional scene. The nature of information itself and of information-related activities are defined and closely analysed to produce models of the disciplines and the work elements within the disciplines of information.

Results The analysis makes possible the incorporation of popular new information disciplines into an overarching framework that includes pre-existing fields as well.

First published as Bates, M.J. (2015). The information professions: Knowledge, memory, heritage. *Information Research*, 20(1), paper 655. http://www.informationr.net/ir.

Conclusions The analysis provides a perspective that clarifies the relationships among the information disciplines as well as their relationship to other professional activities in society.

Introduction

At this particular historical moment, we are taking part in an extraordinary sea change in how information science, libraries, archives, and all the information-related disciplines are viewed. Each week, it seems, we learn of a new information-related field. Knowledge management: Is that the same as information management? Bioinformatics—*not* the same as biomedical informatics. Digital humanities—or was that humanities informatics? And so on and so on. There is a lot of uncertainty and confusion now—as well as a lot of creative ferment going into the creation of new information professions.

We need to be formulating a conception of the information professions that makes sense out of the ferment, one that rationalizes and clarifies just what these fields are, including where the *existing* information professions play a role in this new landscape. It is important to be proactive, lest others, with far less understanding of the requirements of information collection, organization, storage, and retrieval, set an agenda for these professions that is founded on ignorance of the actual requirements of information management.

Here I want to present a framework for thinking about these many information professions. It is to be hoped that this framework will make it easier to assess new future claims about information professions, and constitute one step toward making sense out of the blooming, buzzing confusion of the new professional landscape.

The nature of the information disciplines relative to other disciplines

In a 1999 article, "The Invisible Substrate of Information Science" (Bates, 1999), I argued that information science needed to be seen as a different type of discipline, in comparison to the usual array of disciplines. Normally, we think of the academic disciplines as being on a spectrum, from the study of the arts at one end, through the humanities to the social sciences to

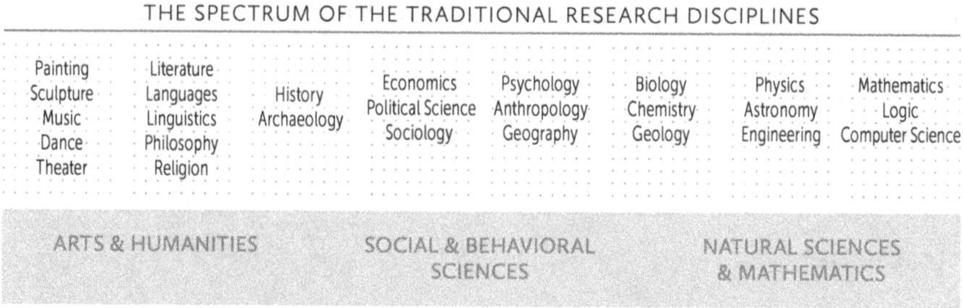

Painting Sculpture Music Dance Theater	Literature Languages Linguistics Philosophy Religion	History Archaeology	Economics Political Science Sociology	Psychology Anthropology Geography	Biology Chemistry Geology	Physics Astronomy Engineering	Mathematics Logic Computer Science
ARTS & HUMANITIES			SOCIAL & BEHAVIORAL SCIENCES		NATURAL SCIENCES & MATHEMATICS		

FIG. 1. *Spectrum of traditional academic disciplines*

the biological, earth, and physical sciences and mathematics at the other end. See Figure 1: Spectrum of traditional academic disciplines. (Figures in this article are reproduced or updated versions of illustrations appearing in Bates, 2007.)

There are some fields, however, that cut all the way across this spectrum; they deal with every traditional subject matter, but do so from a particular perspective. These fields organize themselves around some particular social purpose or interest, which then becomes the lens through which the subject fields, such as literature, geology, etc., are regarded. There are both theoretical and research questions to study, looking through that lens, and practical, professional matters to address. I call these fields "meta-disciplines." Three prime examples of these meta-fields are the information disciplines, communication/journalism, and education. (See also Bates, 2007.) See Figure 2.

Each meta-discipline deals with knowledge in all the conventional fields on the academic spectrum, but does so from a particular orientation or position that is needed to accomplish the work and the theorizing of its area. Educators work on the theory and practice of teaching and learning—how learning is best achieved, across all subject domains. Communication researchers study the transmission of messages and their impacts in various contexts, and communication practitioners, namely, journalists, learn to identify topics of interest, sleuth for news and shape and present a story. The information disciplines all deal with the collection, organization, retrieval, and presentation of information in various contexts and on various subject matters. That social purpose of collecting, organizing, and disseminating information shapes all the activities of the information disciplines; it is the lens through which all the subject content of the traditional disciplines is viewed, and the framework for the work in

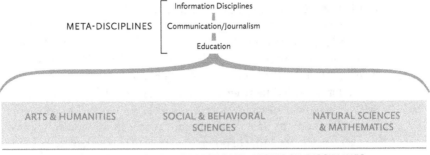

META-DISCIPLINES
- Information Disciplines
- Communication/Journalism
- Education

ARTS & HUMANITIES   SOCIAL & BEHAVIORAL   NATURAL SCIENCES
                           SCIENCES   & MATHEMATICS

THE SPECTRUM OF THE TRADITIONAL RESEARCH DISCIPLINES

FIG. 2. *The "orthogonal disciplines," or "meta-disciplines," shape the subject matter of all the traditional disciplines according to the social purpose of the meta-discipline.*

that area. In all three of these areas, "...the content of all the conventional disciplines is being shaped and moulded for a societal objective through different types of professional activities involving the manipulation and transmission of knowledge" (Bates, 1999, p. 1044).

It is important to see these fields in this way, because the meta-disciplines do not obviously fit into the conventional spectrum, in the way that biochemistry fits between biology and chemistry, for example. People do not know where to place us, and tend to be dismissive. They do not see the shaping character of our professional objectives; they see only the traditional subject content. So, for example, we have all heard the scathing dismissals of education courses at universities as being "content-free"; some of us may have heard that about the information disciplines as well.

So we start this discussion with the information fields being rather seriously misunderstood by many in society, including highly educated people. Some of the confusion comes out in discussions about interdisciplinarity. If you are studying art information, it does not mean that your work is interdisciplinary in the conventional sense, as in a *blend* between biology and chemistry to create biochemical research. Art is the subject matter being managed, but art information work does not blend art research and information research in any conventional sense of interdisciplinarity. The information research is always at a "meta" level. For instance, we study how art historians go about their work in order to make better information systems for them; we do not research historical questions in art.

These examples are raised, because all these disciplines are misunderstood due to their being meta-disciplines, that is, outside the worldview of the content disciplines per se. They address all the traditional subject matter from a particular perspective needed to achieve their social purposes. For the most part, the content of education courses does not consist of the subject matter the teachers teach per se; rather it concerns the meta questions of how to structure and organize the material in order to teach it effectively. Likewise, you don't enter a library science program and just read a lot of books. In other words, you don't become a good librarian solely by having read a lot, valuable though that may be. You also have to know a lot about the "meta" subject matter of the information disciplines. You get your LIS training in order to learn how to select, organize, store, retrieve information, etc.

The origins of information professions

From where, then, do new information disciplines arise? The fundamental engine of development is need. Human beings want to retain informational resources, and, after a very short time, these resources collect at such a rate that some principles of selection, organization, etc., need to be brought to bear, in order for the resources to continue to be available for effective use. As resources collect, interested individuals recognize the problems and then resolve them through theoretical and professional development of ideas and practices. Those individuals either draw upon earlier information disciplines or invent or re-invent solutions to their problems.

In almost all these cases, however, the interested individuals come out of one or more of the traditional academic or professional disciplines. Thus, for example, the need to organize historical archives was first tackled, usually, by historians. The need to store and retrieve radiological records first became known to medical personnel, and so it was the medical professions that first tackled the problem of radiology informatics. As a consequence, the writing and thinking in archival theory is strongly humanities-oriented in character, while a more technical approach, arising from the needs of medicine, drives radiology informatics.

So, the disciplines of origin often have a marked impact on the character of the information fields that arise from them. For that reason, I have arrayed the information disciplines across the traditional academic spectrum of fields—*with the understanding that in most cases the information discipline is, in fact, applicable to a much broader range of information solutions than its origins indicate.* As a rule, all information disciplines are in the process of becoming more generally applicable, as the discipline gains sophistication and breadth of understanding. Figure 3 portrays the information disciplines selected for the *Encyclopedia of Library and Information Sciences, 3rd ed.* (Bates & Maack, 2010; Bates, 2007), arrayed across the spectrum of traditional

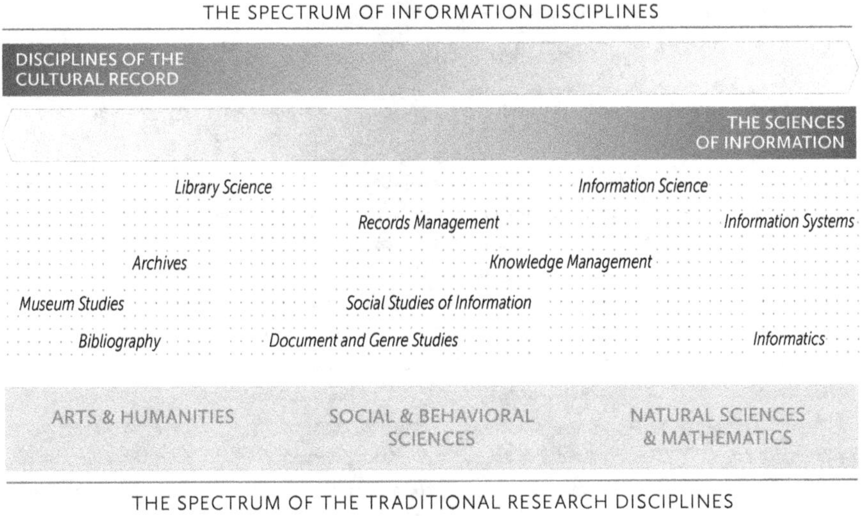

FIG. 3. *The information disciplines*

disciplines, and with the understanding that their applicability is generally broader than their original outlook.

A better name for the encyclopaedia would have been *The Encyclopedia of the Information Disciplines*. We defined the content much more broadly than library and information science, and I am not proposing here that all these disciplines be *subsumed* under LIS. Because the publisher required that we retain the old name for the encyclopaedia, this is a misunderstanding that can easily occur, and I want to correct it. On the contrary, LIS is one of the information disciplines; it is a sibling of the other disciplines, not their master.

In considering and evaluating the potential areas to cover within the encyclopaedia, we soon became aware that one can approach many of these questions from either of the two classic approaches of scholarship—the famous "two cultures" of the humanities and the sciences. The more humanities-oriented fields I call the "disciplines of the cultural record" and the more science-oriented fields, I call the "information sciences." Some of the information fields with their roots in the social sciences may draw upon either or both orientations, so the lines in Figure 3 are seen to overlap in the centre of the diagram.

In the encyclopaedia, we endeavoured to include entries on all these information disciplines, as well as the major sub-disciplines. Figure 4 portrays the sub-disciplines, again arrayed from left to right by traditional

DISCIPLINES OF THE CULTURAL RECORD				
				THE SCIENCES OF INFORMATION
Reading interests	Information policy/law	Information behavior	Library automation	Information retrieval
Art librarianship	Library special collections & manuscripts		Management information systems	Database management systems
Theological librarianship		Business records management		Information architecture
Information arts	Semiotics	Government archives	Business archives	Science/engineering librarianship
Publishing studies	Diplomatics	Government records management		Electronic publishing
Digital asset management	College and academic libraries		Special libraries	Digital libraries
Social epistemology	Genealogical archives	Social science data archives		Technical writing
Film and broadcast archives	Business process management		Sound and audio archives	Data mining
Art museums	Historical and archaeological museums	Natural history museums		Science museums
Analytical bibliography	Site museums and monuments	Museum visitor studies		Biblio-/Webo-/Sciento-metrics
Cultural informatics	Social informatics	HCI for info systems		Grid storage of data
Museum informatics	Legal informatics	Business informatics	Bioinformatics	Geoinformatics
Digital humanities	Security informatics	Medical/health informatics		Chemical informatics
ARTS & HUMANITIES	SOCIAL & BEHAVIORAL SCIENCES		NATURAL SCIENCES & MATHEMATICS	

THE SPECTRUM OF THE TRADITIONAL RESEARCH DISCIPLINES

FIG. 4. *The Information sub-disciplines*

academic disciplinary origins (Bates, 2007). Again, I argue that most of these fields now have broader applicability than the disciplines of their origin, as their internal theories and processes have become better understood and more generalized in theory.

So the information disciplines may have roots in particular traditional disciplines, but through time, as they work to meet the needs of information collection, storage, dissemination, and use, they become more and more "meta," that is, they operate on the subject content from the particular perspective of the needs of the information disciplines, and less so from the research perspectives of the content disciplines.

A more fundamental look at the character of the information professions

The discussion of the information disciplines to this point has assumed a prior understanding of and familiarity with them. But with the growing profusion of information fields recently, we need a closer and deeper look at what is distinctive about these fields relative to the rest of social activity in modern societies. So in this section I address the fundamental types of information and who manages those types in our society.

To date, I have written extensively on what information is (Bates, 2005, 2006), and have not been able to fully develop those ideas, in order to bring them closer to the day-to-day practice of the information professions. Because it would take far too long in the context of this paper to be fully persuasive regarding the approach I have taken to the topic, the reader is referred to those papers. Here, I assume a general understanding of what information is, with the caveat that I believe that information exists both subjectively and objectively—subjectively as our human experience of novelty, learning, emotion, perception, etc., and objectively, as the pattern of organization of matter and energy, the marks that take up the pages of books, or the electronic ones and zeroes that exist in digitized information stores.

A man by the name of Susantha Goonatilake has written about what he calls three information flow lineages through the history of life on earth—genetic, neural-cultural, and exosomatic (Goonatilake, 1991). He has emphasized the sense in which each of these channels both stores and communicates to a later time the information they contain.

Since the beginning of life on earth, the genetic line, that is, the DNA in the genome of plants and animals, has carried its generation's information from one generation to the next. That is, information reflecting the entire history of life on earth is stored in the DNA, including the so-called "junk DNA" that was once useful and now may or may not play a role in the creation of a new living being. So encoded genetic material is a form of information storage—not only of information needed currently for the maintenance of the animal, but also of the genetic history of species.

So the first of Goonatilake's lineages is genetic information. His second is what he calls the neural-cultural lineage. Through history, the more developed the brains and memories of animals became, the more possibilities there were for information to be stored, used and passed on during the lifetime of the animals. Mother tigers teach their young how to hunt. Drawing on her neural information stores in her brain, the mother tiger hunts with the cub, and the cub learns, and stores in its brain, the methods of hunting, thus carrying this knowledge forward another generation.

With the coming of modern human beings, and especially with the development of language, it became possible to store neurologically, and to pass on culturally, large amounts of very specific information to the next generation, and so on down through many generations. Some scientists believe that the Biblical story of Noah and the flood dates to an actual flood that occurred in Eurasia when the last Ice Age receded. One generation told the next generation, over hundreds and thousands of years, and the story still comes to us, so long after the event. That is information flow through the neural-cultural channel.

Goonatilake's third channel of transmission is a more recent one, and begins when human beings figured out how to store information exosomatically, that is outside the body. He calls this the exosomatic flow line. This channel began with drawings and carvings in rocks, and once writing was developed, exploded in many forms, from cuneiform writing in clay to Chinese characters in ink. In the prior, oral, age, humanity's total store of knowledge necessarily consisted of that which could be held in one or a few human brains. Memorization was the chief means of retaining knowledge, and knowledge could be passed on from one person to another only in the actual presence of that other person—living being to living being. But once human beings found a way to record information in more or less permanent form outside the body, then that information could be retained for indefinitely many future generations. The person sending the information and the person receiving it did not have to be in each other's living presence. Furthermore, that exosomatic information was no longer limited in quantity to what a single individual could learn and memorize. Stores of that exosomatic information could build up, so that human beings could consult and use the expertise of countless other human beings. The storage and management of exosomatic information was one of the major contributors to the exponential growth of human knowledge and power over nature.

For each of Goonatilake's three channels I have identified, described, and labelled major types of information of interest to the information disciplines (Fig. 5). Three types of neural-cultural information are defined there—experienced, enacted, and expressed information—but will not be further discussed here. With regard to exosomatic information, I distinguish between recorded and embedded information. *Recorded information* is communicatory or memorial information preserved in a durable medium. Recorded information is the chief focus of libraries and archives. *Embedded information* refers to the enduring effects of the presence of animals on the earth; it may be incidental, as a path worn through the woods, or deliberate, as a fashioned tool or structure (Bates, 2006). Embedded information, such as is found in the artefacts from earlier cultures that are uncovered by archaeologists, may be quite informative but it generally lacks the deliberate communicative intent that recorded information has. It is informative only as an incidental consequence of the activities and skills of the people leaving the artefacts. For example, an archaeologist can identify a characteristic style of ceramics left by an earlier civilization, and

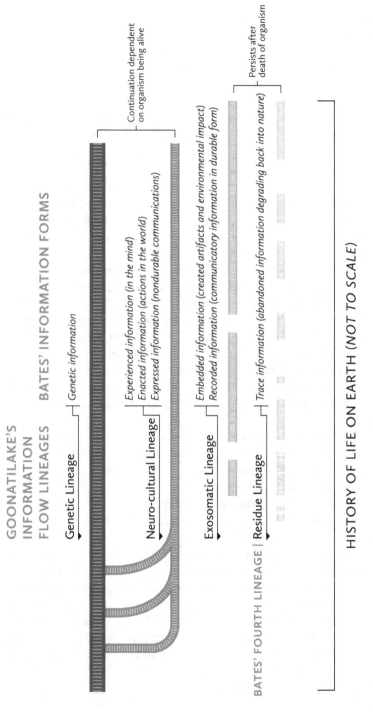

GOONATILAKE'S
INFORMATION
FLOW LINEAGES   BATES' INFORMATION FORMS

Genetic Lineage   Genetic information

Neuro-cultural Lineage   Experienced information (in the mind)
                                Enacted information (actions in the world)
                                Expressed information (nondurable communications)

Continuation dependent
on organism being alive

Exosomatic Lineage   Embedded information (created artifacts and environmental impact)
                            Recorded information (communicatory information in durable form)

BATES' FOURTH LINEAGE | Residue Lineage   Trace information (abandoned information degrading back into nature)

Persists after
death of organism

HISTORY OF LIFE ON EARTH (NOT TO SCALE)

FIG. 5. *Information flow lineages (Goonatilake) and information forms (Bates) (Bates, 2006)*

distinguish one set of pots from another set produced by a different culture. Embedded information is a major focus of museum studies.

I suggest that, in addition to Goonatilake's three information lineages, there is a fourth one, which I call "residue." The type of information carried by the residue lineage is "trace information" (Bates, 2006). So we write books or build memorial statues, or develop tools and objects to serve our life purposes, such as to communicate or to enable us to do things that would be otherwise impossible to do without our cultural artefacts. But there comes a time when people—or civilizations—are done with their books or objects. These then are thrown away, buried, lost. They deteriorate, ultimately, to the point where they have blended back with the earth. Dust to dust, as the saying goes. But during that time when they are lost to current use but still have not disintegrated totally, they contain trace information. This form of information is of obvious importance to museum and other information professions.

So what have we so far? Genetic information is encoded in DNA and passed on through reproduction. Neural-cultural information exists in living animals and is passed on in the presence of other animals. Exosomatic information comes in the forms of recorded and embedded information. Embedded information is to be seen in the environment impacted by animals and in artefacts touched or created by animals. Recorded information has existed since human beings have marked external objects with drawings, symbols, or other communicatory information. Finally, information in the channel "residue," which is in the process of disintegrating, is trace information.

As information professionals, we are interested in all these forms of information, because we need to understand the role of information throughout human life. I would argue, however, that at the heart of the information disciplines is exosomatic information, the external stores of information, that is, Goonatilake's third stream of information transmission. What unites all the information professions is that they manage the record of our culture for all its uses, from entertainment and education to preservation for future generations. Even the bioinformatics databases that store the DNA information of species of animals and plants are a part of our culture—in this broad sense—just as are books, papers, audio clips, videos, and digital resources of all types. The term "culture" is used in the broad sense of all that we have created as a species, as many peoples and many individuals. It is our entire social heritage as a species. Compare this with what J.M. Balkin called "cultural software" (Balkin, 1998). In the information disciplines, we manage the physical (including digital) form of the cultural software of humanity.

The universe of documentation

Figure 6 (overleaf) demonstrates how I am thinking about the universe of recorded information (Bates, 2007).

In business, government, medicine, we see people going about their activities. In that process, they may engage with letters, books, images, or countless other forms of recorded information. For example, in her office with a patient, the physician reviews an X-ray image and talks with the patient. She draws on her memory to identify a drug to prescribe, then she pulls down a medical reference book and determines the amount of medication to prescribe. She writes more in the patient's record.

Now for the physician and the patient (Figure 7, p. 137), what is important is the medical situation and whether they will succeed in solving it. Will the prescribed drug cure the patient's problem? The physician and the patient are concerned with the diagnostic situation. We in the information disciplines, on the other hand, are concerned with the information and documentation in this situation. How shall we store and retrieve the patient's record? How will the results of the diagnostic tests be kept? How best to design the information system interface to enable the physician to use her electronic medical records software easily? Does the physician have the best reference book for her purposes? Even when we study the interpersonal situation between the two people, we are interested in its impact on the *success of information transfer.* We are not sociologists studying the role of medical professionals in society, nor are we physicians concerned with the quality of her diagnosis. Rather, we are information people, and we observe the information transfer in this situation, and work to store safely the private medical information and retrieve the relevant factual information that pertains to it. So all these worlds at the heart of Figure 6 are where most people live their lives most of the time. The universe of documentation around the edges of the image, however, is the universe of study for the information disciplines.

A note about Figure 6: I have deliberately extracted out the many forms of information into a separate "universe" to draw attention to their importance for the information disciplines. Indeed, much of research and practice in the i-fields is about that universe. However, as the above physician example illustrates, in our daily lives information forms are fully integrated into our lives—often to the point that we fail to notice their role. Therefore, information researchers and practitioners must also attend to information in the full stream of life and work, in order to fully understand human use of information, as is done in information behavior research.

This universe of documentation is the focus of the several information-oriented disciplines such as library and information science, archives,

UNIVERSE OF LIVING

Self-expression
Aesthetic/cultural life
Political/social activities
Sports
Hobbies/clubs
Family life
Entertainment
Religion
Education
Government
Research and development
Business
Law/Crime/Judicial system
Military
Medicine
Social welfare
Industry
Energy
Agriculture
Mining
Construction

UNIVERSE OF DOCUMENTATION

UNIVERSE OF LIVING

In the process of living, people produce recorded information, which then constitutes the universe of documentation

The domain of research and practice in the information professions

UNIVERSE OF DOCUMENTATION
(Mixing information genres, channels, technologies)

WWW
Tweets
Broadcasts
Newspapers
Emails
Journals
Monographs
Databases
Simulations
Images
Documents
Video
Film
Recordings
Music
Audio files
Bibliographies
Cable
Digital libraries
Internet
Grid computing
Server farms
Cloud storage

FIG. 6. The universe of documentation and the universe of living

FIG. 7. *Physician with patient*

records management, informatics, knowledge management, bibliography, etc. However, we can see a very close kinship with museum studies as well. In libraries and other information institutions we store the record of our civilizations. We emphasize the *recorded information* record, while the museums emphasize the *embedded information* record (Bates, 2007). All these disciplines are engaged in preserving and carrying forward the cultural heritage of civilizations. Indeed, the words in the sub-title of this paper, "Knowledge, Memory, Heritage" can be seen to reflect the stored information in the three classical information institutions of libraries, archives, and museums, respectively, although all three institutions store all these cultural forms. See Figure 8 (overleaf): Knowledge, memory, heritage disciplines.

Sorting out the disciplinary confusion

Now, how do we take this discussion to the next step, and build a map of the information disciplines? First, the relationship between the terms "discipline" and "profession" needs to be clarified. All professions—from medicine to accounting to clinical psychology to horticulture—are mixtures of theory and practice. If a job is so simple that it consists of a series of steps that you carry out one after the other; in other words, if a job is

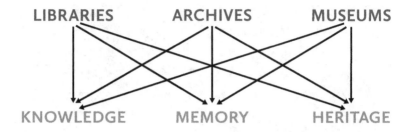

FIG. 8. *Knowledge, memory, heritage disciplines*

algorithmic, then it is not a profession. All professions require the mastery of a body of general theory and understanding, which the practitioner then applies selectively and creatively, as needed, to a series of real-world problems. The application of the general knowledge requires judgment and experience to do well.

Thus, every profession necessarily has that body of general knowledge, which consists of theory, research, practice-based principles, and the long experience and reflection of the senior practitioners. Some of that theory and research tackles questions of general academic interest related to the important issues of the profession. For example, in information studies and information science we have a lot of research on information seeking behavior. Studying how people go about finding information has taught us some surprising and counter-intuitive things about people and information. Applying what we learn from this research to practice enables reference librarians to provide better service to clients in libraries. At the same time, the growth in understanding about people and information that is produced by the information seeking research also contributes to the social sciences generally, and other academic fields such as sociology, psychology, and science studies can benefit from our research. So every profession is substantial enough in its nature that it produces knowledge of general academic value, as well as supporting the solving of real-world problems in the area of its professional expertise.

Thus every profession has both academic disciplinary aspects and professional practice aspects. So, to some extent, but not entirely, "discipline" and "profession" overlap in meaning. Here, as a rule, when I emphasize the academic, theoretical aspects of the information fields, I use the term "discipline" and when I emphasize the professional practice aspects, I use the term "profession." Because theory and practice are so closely linked, we wanted to be sure to address both in developing the contents of the encyclopaedia, and considering how to handle the relationship between

INFORMATION DISCIPLINES

Library and Information Science
Archival Science
Museum Studies
Bibliography
Records Management
Knowledge Management
Informatics
Information Systems
Document and Genre Theory
Social Studies of Information

FIG. 9. *Information disciplines in the* Encyclopedia of Library and Information Sciences, 3rd ed. *(Bates & Maack, 2010). [Preferred title:* Encyclopedia of the Information Disciplines].

discipline and profession was one of the important issues in the encyclopaedia's design. See Figure 9: Information Disciplines.

After *much* sorting out, analysing, and re-analysing—it was harder than it looks to arrive at this final result—we see in Figure 10 (overleaf) the top-level categories used in the Topical Table of Contents of the encyclopaedia.

I found that we could arrange these categories by S.R. Ranganathan's classic five types of facet, personality, matter, energy, space, and time.

The sorting out of topics that we did for the encyclopaedia was appropriate to the encyclopaedia, but now we have a somewhat different task—providing a useful structure for discussing the various information disciplines that seem to be cropping up everyday. As the list displayed above implies, there are several key facets that characterize the information professions. Figure 11 (p. 141) displays the facets being discussed here, and Figure 12 (p. 142) lists some facet elements to be found within the respective facets in the corresponding locations within the circle.

At the heart of information work, there are the **services and functions** that professionals carry out in order to achieve the objectives of the fields. These include some form or other of appraising, collecting, organizing, storing, retrieving, disseminating, and working with users of the recorded or embedded information being managed. These have many names in the several information professions, but in every case there is some form or other of these functions. Services and functions represent the very essence of the work of the information professions. Over the centuries, the balance

TOP-LEVEL CATEGORIES	ORGANIZING FACET
Information Disciplines and Professions Concepts, Theories, Ideas Research Areas	PERSONALITY
Institutions Systems and Networks Literatures, Genres, and Documents	MATTER
Professional Services and Activities People Using Cultural Resources Organizations	ENERGY
National Cultural Institutions and Resources	SPACE
History	TIME

FIG. 10. *Key organizing categories for the* Encyclopedia of Library and Information Sciences, 3rd ed. *(Bates & Maack, 2010), with associated Ranganathan PMEST facets.*

among these activities has shifted; serious attention to understanding and working with the people who access the information has been the great contribution of the 20th and 21st centuries to the professional canon, and the research in information behavior done in recent decades has contributed to the social sciences.

One function may be called "appraisal" in archival science and "book selection" in librarianship. Another function may be called "cataloging" in librarianship and "registration" in museums. Another may be called "information behavior research" in information science and "visitor studies" in museums. Everywhere we look, this handful of functions crops up again and again. Every time there is something to *store,* some kind of information source to be preserved, if you want to be able use it later, then these functions have to be performed under one name or the other.

A second major facet is the **information** itself—that which is at the heart of all we do in the information professions. Nowadays, everyone is interested in information, and in searching for information. Google has

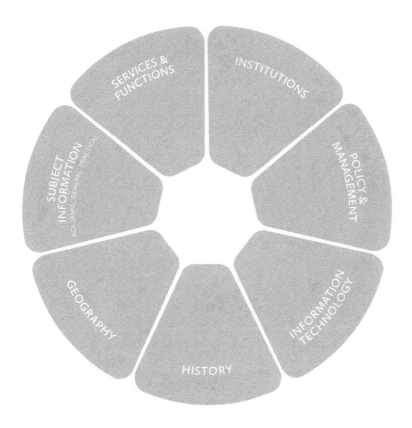

FIG. 11. *Information discipline facets*

unquestionably made that apparently so easy that everyone thinks they are an expert searcher. But in fact, *the study of information in and of itself* is the research area we should *own*. We do, in some respects, as with our long history of research in informetrics and bibliometrics. But in other respects we have done too little to date. In the encyclopaedia, we included several articles from earlier editions that discussed the history and development of the *literatures* of medicine, economics, and so on. In other words, the body of all texts produced in a field is itself a desirable focus for research. How scientific literatures develop, for example, can provide enlightenment on the nature of science and science communication. We need more such articles.

A third major facet is clearly the **institution**. The institutions that store the resources—the libraries, archives, and museums—have become large and important physical, social, and administrative presences in society. As we shall see, though institutions are of signal importance for

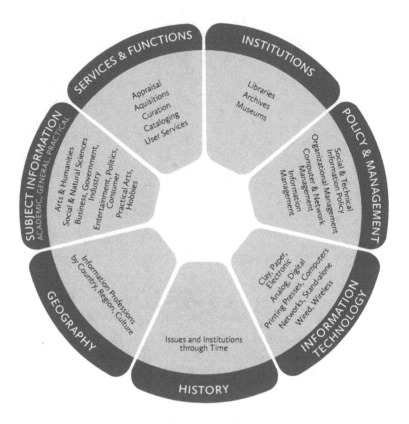

FIG. 12. *Example elements of information discipline facets*

some information professions, they are not so important for others. Some information professions, such as records management, or knowledge management, emphasize functions or other aspects of their work, and think of themselves as being mostly independent of specific institutions.

A fourth major facet is the **information technology** for organizing, storing, and retrieving the information. Nowadays, we think entirely in terms of computers and electronic devices such as smartphones and tablets. But finding a good technology for storage has been at the heart of the information professions for millennia. The transition from the scroll to the codex book, for instance, was a pivotal technical improvement in access to information, as was the development of the vertical file (cabinet) to replace the prior flat boxes for storage in the nineteenth century. It was no accident that Melville Dewey, in establishing the field of librarianship in the United States, also established a company to provide library supplies.

Part of the professionalization of the information professions consisted in the development of improved physical and conceptual approaches to conducting the work of the professions through standardization and optimization of functions.

The fifth facet I will mention is **policy and management**. In any activity in which you have large physical plants, large bodies of information, and lots of people working, you must also develop policies, including ethical principles, at the societal and local level, and find ways to manage both the physical store and the people working with that store. Every information profession has management tucked into it somewhere.

Now, Ranganathan's space and time, i.e., **geography** and **history**, are also important—the variations in professional outlook and practices from country to country, or through time, but here I want to emphasize the present and the North American context, which I know best.

So, finally, here is how I think we can view the multiplicity of information professions that we see growing up around us: The information professions are complicated by their "meta" character; there is much confusion around the relationship between the content disciplines and the meta disciplines, and around interdisciplinarity. In addition, there is *confusion around all the facets that were described above.* People, in coming up with new information professions—usually assuming that they are the first people to have ever thought about these questions—pull out aspects of one or more of these information profession facets that have been mentioned, and make those particular facets the focus of their thinking. There is an added complexity due to the fact that these are professions and not solely bodies of academic knowledge. Professions are a complex mixture of knowledge, physical plant, and professional activities and organizations.

The older information professions grew up around the management of specific storage institutions. Libraries, archives, and museums were the original focus of the information professions. In the nineteenth century, when the growth of resources was becoming a major issue in the handling of information, these physical institutions quite reasonably became the focus of professionalization. All three major institutions needed to grow substantially in the nineteenth century, especially in the latter part of the century and into the early twentieth century, so the building of buildings, and physical housing of resources was a natural focus of these professions. That the work centred around these physical institutions led to the professions of *library* science, *archival* science, and *museum* studies.

In the middle to late twentieth century, with increasing dependence on automation and digital storage, the physical storage per se became less important than the institutional needs centring around the use of the

FIG. 13. *Three example information disciplines combining facets*

information. For example, in the case of records management, the *location* of storage was not as important as the sheer volume of records—often stored in many different offices and warehouses—that had to be managed in some way or other by companies and governments. The problem usually started with the question, "What the heck do we do with all these records?" as those files grew and could no longer be handled just within the offices that generated them. The issue was the institutional records themselves, and the problem was managing them; hence, records management. More recently, knowledge management has followed a similar trajectory, as organizational science has become more and more sophisticated in thinking about information as an organizational resource to be managed carefully and exploited for the purposes of the organization.

With medical informatics, we can see that medicine is a world unto itself. It constitutes an enormous part of our economy, and has a range of issues to deal with that crop up again and again in different medical contexts, from hospitals to doctors' offices—privacy, access, speed, and so on. Medicine, being so well funded, will assume that its needs are so large and important that it generates its own information discipline—hence, medical informatics, which emphasizes the subject area of the information, medicine, and the application of information technology to that area, namely, informatics.

Thus, I would argue that, in part because of the confusions around disciplines vs. meta-disciplines, and also because of the relative newness of the information perspective altogether in academia and society, there has been much confusion around naming information disciplines and professions. *A number of these fields, in naming themselves, have drawn on one, two, or more of the above five to seven key facets.* Earlier fields, developed during the dominance of physical storage of information, organized around the chief institution served by their professions. Some of the more recent fields, which may store lots of information, but not in the huge physical plants formerly needed, emphasize other aspects, such as use of information technology, as in digital asset management or digital humanities.

Returning to our facets diagram, we can see how different fields named themselves by drawing on one or more of these key information profession facets.

In Figure 13, for example, we see art libraries combine institution and subject matter, enterprise ontologies combine function and business knowledge, and medical informatics combine information technology and the subject area of medicine.

Matters of size

More needs to be said about the three information disciplines that focus on the institution: library science, archival science, and museum studies. These are the oldest and most venerable of the disciplines, and the fact that they have historically focused around physical institutions is no accident. Nowadays, when billions of documents are stored compactly and out of sight on computers in server farms and in homes and offices, it is easy to forget that in the pre-digital era, acquisition of more information automatically meant that much more physical storage would be required.

It has been well documented that, in modern times, the growth of information, ever since the development of movable type in the fifteenth century, has proceeded at an exponential rate. That has meant in practice that the amount of information of a given type, for example book titles, or journal titles, doubles every ten to twenty years, and has continued to do so over hundreds of years. The Library of Congress, for example, was launched in 1801, with a collection of 740 volumes (Cole, 2010, p. 3419). At a doubling rate of once every 14 to 15 years, the current number of catalogued books, as of 2013, of 23,592,066, is reached (Annual Report of the Librarian of Congress, 2014).

Yet in 1851, the Library of Congress had only 55,000 volumes (Cole, 2010, p. 3420). That is the size of a big-city branch public library nowadays. In 1868, the first very big public library in the United States, in Boston, second then only to the Library of Congress in size, announced a collection of 144,000 volumes. The next two public libraries in size in the U.S. were New Bedford (21,000 volumes) and Cincinnati (20,000 volumes) (Sessa, p. 2127).

(How volumes and/or titles are counted can be quite complicated at the detail level. Do you count each volume of a journal, or do you count the journal title only? Each copy of each book? There are many questions like that that must be decided when counting and comparing collections. So counts from one institution to another may be based on different grounds. In the case of the Library of Congress, the growth pattern was further complicated by the occurrence of two large fires during the nineteenth century. However, the important point to keep in mind is that the overall pattern of exponential growth is quite robust, including rebounding after fires, with relatively small variations due to the counting rules.)

We forget that even though the pattern of doubling is quite consistent (with some variation for fires, wars, and depressions), the absolute numbers of books changes dramatically over these periods. It is one thing to deal with a doubling over 15 years of 1,000 to 2,000 books, or even 10,000 to 20,000 books; and quite another to deal with a doubling of 1 million to two million in just 15 years. In the first case one acquires a few more bookcases; in the second case, the librarian hires one or two more workers, and finds a second room somewhere to house the collection. In the last case, however, such small adjustments are out of the question. During the mid-twentieth century, many academic, public, and government libraries experienced the latter sort of massive growth of their physical collections, and needed vastly more space, usually a whole new building. I entered the field in the 1960's, and it was still the case at that time that no university librarian could retire with satisfaction unless he or she had managed to have a whole new building built for their library. This kind of growth of course required great resources and political arrangements that were not always available.

This large growth of physical resources also led to an obsession with various forms of miniaturized storage of information, with experiments on a wide range of types of microforms and films. Indeed, in the 1930's–1950's, in the early days of the American Documentation Institute, the predecessor organization for the Association for Information Science and Technology, one of the main interests of the organization and its journal was with finding forms of cheap, reduced-size storage.

I suspect that it is also no accident that the field of librarianship began to be professionalized in the 1870's with the activities of Melville Dewey and the founding of the American Library Association. Earlier, a slowly growing 8,000-volume college library could be run as a side activity by one of the college's professors. But when libraries grew into the tens of thousands of volumes, one or two people's idiosyncratic ways of organizing the collection were no longer sustainable. Consistent principles and practices had to be developed, and the skills and philosophies of library organization had to be taught to prospective librarians. One could argue that the growth in collection size necessarily led to the growth of professionalization of librarianship. During that same period, the library as a physical institution requiring large resources, especially a building, became a natural cynosure of the profession.

We see a roughly similar pattern with the growth of archives and of museums over the nineteenth and twentieth centuries, as the growth in bureaucracies and corporations led to exponential growth in archival materials, and research materials collected from around the world led to growth in museums. One of the great advantages of the digital age has been that museums, which previously had nowhere near enough display space to present their huge collections, could now house much of the collections online, and thus enable viewers to see far more of their resources.

Viewed in this light, it makes sense that the older professions had an institutional focus, expressive of their need for vast storage capacity, while the more recently developed information disciplines downplay the physical storage side of their work.

Ever since the beginnings of the applications of computers to information storage, there has been an on-and-off again hostility between some information professionals, which has manifested itself in the "information vs. library" argument that comes up now and again. I think it is important to recognize that there is not an either/or here. All these different kinds of information professions are important and valuable. We need to recognize that each profession has arisen in a particular historical context. The institution-based professions arose in the nineteenth century because those institutions were needed to store the materials physically; thus the institutions became the organizing principle for each of the respective professions. In the twentieth and twenty-first centuries, because of the power of the various information technologies we have, information can be seen in a more unified way, as pervading our lives and society, and requiring a sophisticated understanding of information behavior and of the numerous technical options for storing information, relatively independently from their storage in specific locations. So the more recently developed fields do not focus so much on particular institutions. Again, each field reflects the operative circumstances at the time of its founding.

Conclusions

Clearly, the names and coverage of these various information professions have historical and practical reasons why they developed the way they did and chose the name that they chose. I would argue, however, that at the heart of all of them are the key services and functions, which may or may not be associated with specific information institutions, and which all manage a body of exosomatic information, in physical or digital form, using numerous information technologies, in order to make that exosomatic information available for humanity to use.

It is to the distinct advantage of *all* these information professions to see the commonalities, while respecting the differences among them, in order to instruct society at large about the fundamentally important purposes we serve in human society. Goonatilake wrote about genetic, neural-cultural, and exosomatic information. It has been the exosomatic information, the information stored outside our bodies and not dependent on individual human beings to memorize and pass on, that has enabled the explosion of learning and social sophistication that has developed since recorded information began. It is *our* job to manage that exosomatic information.

This is a huge task, and a very important one, one that has been largely invisible and undervalued historically. It is time that all the information professions unite to make clear their role in society to the larger society, so that our value becomes clear to all. The modern social sciences came into their own in the twentieth century, after having been marginalized and thought to be unimportant in the nineteenth century. Well, the information disciplines have been marginalized in the twentieth century, but have the capacity to come into their own in the twenty-first century—provided they recognize and express their true power and importance.

REFERENCES

Annual Report of the Librarian of Congress for the fiscal year ending September 30, 2013. (2014). Washington, DC: Library of Congress. (www.loc.gov).

Balkin, J.M. (1998). *Cultural software: A theory of ideology.* New Haven: Yale University Press.

Bates, M.J. (1999). The invisible substrate of information science. *Journal of the American Society for Information Science, 50*(12), 1043–1050.

Bates, M.J. (2005). Information and knowledge: An evolutionary framework for information science. *Information Research, 10*(4), paper 239. Retrieved from http://InformationR.net/ir/10-4/paper239.html.

Bates, M.J. (2006). Fundamental forms of information. *Journal of the American Society for Information Science and Technology, 57*(8), 1033–1045.

Bates, M.J. (2007). Defining the information disciplines in encyclopedia development. *Information Research, 12*(4), paper colis29. Retrieved from http://InformationR.net/ir/12-4/colis/colis29.html.

Bates, M.J., & Maack, M.N. (Eds.). (2010). *Encyclopedia of Library and Information Sciences* (3rd ed., 7 Vols). New York: CRC Press.

Cole, J.Y. (2010). Library of Congress: History. In M.J. Bates & M.N. Maack (Eds.), *Encyclopedia of Library and Information Sciences* (3rd ed., Vol. 5, pp. 3418–3430). New York: CRC Press.

Goonatilake, S. (1991). *The evolution of information: Lineages in gene, culture and artefact.* London: Pinter.

Sessa, F.B. (2010). History of public libraries. In M.J. Bates & M.N. Maack (Eds.), *Encyclopedia of Library and Information Sciences* (3rd ed., Vol. 3, pp. 2119–2132). New York: CRC Press.

A tour of information science through the pages of *JASIS (following) was part of a two-issue, 50ᵗʰ anniversary special edition of the* Journal of ASIS *in 1999, for which Marcia J. Bates served as guest editor. As part of the front matter of the edition she adapted photos from the journal in 1951 depicting the state of information retrieval at that time.*

As an illustration of the "dream," the left-hand page reproduced the paragraph below describing Vannevar Bush's hypothetical "Memex" machine. Facing it on the right-hand page was the "reality:" two illustrations (opposite) from an article in the journal depicting one of the earliest (mechanical) information retrieval devices ever developed.

In 1950, this was the dream. . .

Bush, V. "As We May Think." *Atlantic Monthly, 176,* 101–108, July 1945.

   Although the volume of recorded scientific information has been expanding rapidly in modern times, our methods of consulting the record have not experienced a parallel development. The need of a fresh approach to the old problems of indexing and classifying is emphasized and the possibility of developing new machines to activate new methods of storing and handling information is discussed. Looking into the future, Dr. Bush describes a scholar's desk machine, or "Memex," in which books, records, and other papers may be stored and linked in a manner similar to that in which a human memory operates. (pp. 27–28 in Norman T. Ball, "Committee on Organization of Information." *American Documentation 1*(1), 24–34.)

In 1950, this was the reality. . .

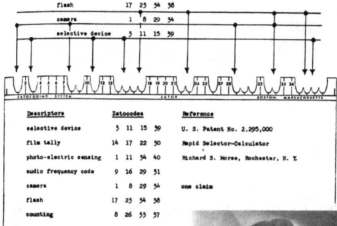

Simultaneously Selective Patterns

flash	17	23	34	38
camera	1	8	29	34
selective device	3	11	15	39

Descriptors	Zatocodes				Reference
selective device	3	11	15	39	U. S. Patent No. 2,295,000
film tally	14	17	22	30	Rapid Selector-Calculator
photo-electric sensing	1	11	34	40	Richard S. Morse, Rochester, N. Y.
audio frequency code	9	16	29	31	
camera	1	8	29	34	one claim
flash	17	23	34	38	
counting	8	26	33	37	

FIG. 1. ZATOCODING, illustrated with a 5 by 8 inch edge notched Zatocard for the tally. Note that the random Zatocode patterns in the field overlap and intermingle. Selection on the combination of three descriptors, "flash," "camera," and "selective device," is according to the inclusion of the pattern of arrows into the pattern of notches in the coding field. Zatocards are sorted by the selector shown in Figure 2.

Calvin N. Mooers. (1951). **Zatocoding applied to mechanical organization of knowledge.** *American Documentation* 2(1), 20-32.

FIG. 2. THE ZATOR SELECTOR, capable of sorting edge notched Zatocards at the rate of 800 cards per minute. The black box-like portion of the Zator Selector is vibrated by a motor in the base. At the bottom of the box, a row of holes allows pins to be inserted to form the selective grid pattern. For selection, a pack of Zatocards is placed notched edge down on the vibrating grid. The rejected cards are those whose edge notches do not fit the pattern of the grid; hence they are supported on top of the grid. The edge notches of the accepted cards fit the grid and the cards drop down through the grid to a distance equal to the depth of the notches. In this way, the accepted cards are offset slightly below the rejected cards. To separate the rejected cards from the accepted cards the operator spears the pack of rejected cards through the small hole on the edge opposite the notches and lifts the pack out. The accepted cards, being offset below the rejected cards, are not engaged by the spearing tool; they drop out from the pack onto the table.

3
A tour of information science through the pages of *JASIS*

Following are selected article titles and descriptive material drawn from the pages of the *Journal of the American Society for Information Science* (*JASIS*) and its precursor title, *American Documentation,* dating from the beginning of the journal in January 1950 until this issue went to press in spring 1999. All citations through the end of 1969 can be found in *American Documentation;* all later citations are from *JASIS.* The "tour" draws on regular and "brief communication" refereed articles. At the end of the tour are provided lists of the "Perspectives" and "Special Topic" issues that have appeared in the journal over the years. (See that section for further explanation of the special issues.)

Selections were made with the intent to illustrate the development of research, technology, and thinking in information science over the years. The pages of the journal have formed the crossroads for most, if not all, major researchers in the field. Famous researchers from outside information science proper, such as Noam Chomsky, Herbert Simon, and George Miller have also appeared in *JASIS.*

Furthermore, every significant area of research in information science is well represented in the pages of the journal. Information retrieval system design and evaluation, description of operational information systems and services, indexing theory and evaluation, search strategy, information seeking research, bibliometric analysis, information policy, and the

First published as Bates, M. J. (1999). A tour of information science through the pages of *JASIS. Journal of the American Society for Information Science, 50*(11), 975–993.

economics of information have all been addressed in *JASIS*. Despite the journal's reputation for an emphasis on the mathematical and technical, the first user needs study appears in *American Documentation* as early as 1951.

It is also striking how *early* many crucial ideas in information science appeared in these pages. H.P. Luhn proposes key word in context indexing in 1960, and selective dissemination of information in 1961—the latter now rediscovered in the age of the World Wide Web as "push" technology. William Goffman and associates propose the use of "meta-language" in information retrieval systems in 1964—and mean it in a way close to its current usage. In 1966, Lockheed staff describe a new system that enables a "dialogue" with an "on-line reference retrieval system." Murray Turoff and Starr Hiltz provide a "progress report" on electronic journals in 1982.

User-centered information system design did not start in the 1980's, as some would claim. Rather, it dates at least to Mooers' Law, published in 1960, and Edwin Parker's 1966 argument that "the system should adapt to the receiver or user, rather than the user to the system" (see below under Information Seeking and Needs). Victor Rosenberg's 1974 article calling for the introduction into information science research the "intuitive, the subjective, and the experiential" presages by decades the current wave of interest in qualitative methodology.

Just two more examples will be provided here of these early insights. In 1967, Burton Adkinson and Charles Stearns present a remarkably prescient description and forecast of the information scene. Their three stages of library automation (which actually address far more than just libraries themselves) are an excellent summary of what, in fact, has happened in the years since their article.

Finally, Watson Davis, the founder of this Society (under its original name, the American Documentation Institute), describes in a 1951 article the association's system of "auxiliary publication." This arrangement, first begun in 1936, enabled researchers to send in additional, supplemental, and otherwise unpublished materials to the American Documentation Institute. Other researchers who wished to use this material could then contact the Institute and have a copy made for their own use. The ability to self-publish such supplemental materials is one of the features most praised about the Internet for researchers today. The technology available today is much more supportive for auxiliary publication than was the case in 1936, but the idea was actually implemented back then.

In the early days, the technology for information retrieval was almost unbelievably primitive by current standards. Coordinate indexing was developed by Mortimer Taube and his associates by using a paper card for each index term. (This so-called "term-entry" system was the converse of the "item-entry" approach illustrated by Calvin Mooers' Zatocoding, displayed at the beginning of this anniversary section.) In indexing individual records, the document number of each record would be posted to all the cards bearing index terms appropriate to the document. The cards would then be filed. Upon need to fulfill a search query, the cards for the desired subject terms would be pulled out from their boxes and the document numbers on each card visually compared. A coordinate "AND" between two index terms would be achieved when all document numbers were identified that appeared on both index term cards. In general, in the early days, the ideas galloped well ahead of the available technology.

Whole research specialties have been spawned from the research presented in the journal's pages. Information retrieval research developed out of the work of Cyril Cleverdon, F.W. Lancaster, Lauren Doyle, and, above all, Gerard Salton, all of whom published in *JASIS*. (It is not being claimed that *JASIS* necessarily published the first or only article on the given topics, but articles in the journal have frequently been very influential in the development of information science research areas.) Bibliometric research reached its modern form out of the work of Burton and Kebler, M.M. Kessler, Eugene Garfield, Derek de Solla Price, Henry Small, and Howard White. Similar lists of names could be provided for research in both human and automatic indexing, search strategy, analysis of online database use and users, and information seeking.

We may expect to see some recent *JASIS* articles prove to be the productive sources of new research specialties and methodological approaches in the years to come—Birger Hjørland and Hanne Albrechtsen's domain analysis; new approaches to studying information seeking from Elfreda Chatman, Carol Kuhlthau, Barbara Wildemuth, Rob Kling, and Carole Palmer; new kinds of libraries from Richard Lucier, Joseph Janes, and Edward Fox. We may see a revolution in thinking about relevance arising out of Stephen Harter's 1992 article on the subject. Researchers studying new types of real-world operational information systems appearing today are too numerous to mention; it is anyone's guess what role multimedia, imagebases, search engines, intelligent agents, and, of course, the Internet, will take in the information retrieval world of the 21st century.

The above are just a few of the names and topics to be found in the pages of *American Documentation/Journal of the American Society for Information Science.* Indeed, even the lengthier "tour" of the pages of the journal that follows below is just a small part of the full history and contents of the journal. All 50 volumes of the journal were examined to select the items in the tour. I have endeavored to represent a wide range of topics, perspectives, and individuals in the selection of articles, but ignorance, biases rooted in personal interests, and general human failings will no doubt be responsible for the omission below of significant articles or the selection of lesser articles by researchers otherwise represented.

In the following, the descriptions are arranged by subject and chrono-logically within each subject. Article titles are in **boldface** to make it easy to scan down through the listings to see the topic development in each research area. An unconventional bibliographic format for the article information is used for ease of reading and follow-up.

To simplify, quotation marks are left out, but all text is taken directly from the articles; none is supplied by the editor. Text is usually from the abstracts, but sometimes comes from the text of the articles. (In the early years the journal did not require abstracts for articles.) Elisions in the selected texts are indicated by. . . . No text is provided where titles are self-explanatory.

Information and information science paradigm

1964   **Identifying Key Contributions to Information Science.** Carlos A. Cuadra *15*(4), 289–295.
       The present attempt to identify key contributions suggests that we are far from common agreement on the conceptual, methodological, or practical contributions to the information science field.

1968   **Information Science: What Is It?** H. Borko *19*(1), 3–5.
       Information science is that discipline that investigates the properties and behavior of information, the forces governing the flow of information, and the means of processing information for optimum accessibility and usability.

1974   **The Scientific Premises of Information Science.** Victor Rosenberg *25*(4), 263–269.
       I believe that the notion of man as a mechanical device and the notion that human behavior can be perfectly replicated by computers are fundamental wrong. . . . A more holistic approach is needed. . . . We must begin to look at the interrelationships between various parts of the information environment. . . . To deal effectively with the transcendent qualities of human communication we must admit as scientific evidence the intuitive, the subjective, and the experiential.

1976   **Information Science and the Phenomenon of Information.** Nicholas J. Belkin, & Stephen E. Robertson *27*(4), 197–204.
       . . . the fundamental phenomena of information science are deduced: the text and its structure, the structure of the recipient and changes in that structure, and the structure of the sender and the structuring of the text.

1977   Variety Generation—A Reinterpretation of Shannon's Mathematical
        Theory of Communication, and Its Implications for Information
        Science. Michael F. Lynch *28*(1), 19–25.
        The conventional interpretation of Shannon's mathematical theory
        of communication in relation to textual material is unduly restrictive
        and unhelpful. A reinterpretation which is based on the definition of
        new symbol sets, comprising approximately equally-frequent strings
        of characters, is presented.

1983   Entropy and Information: A Multidisciplinary Overview. Debora
        Shaw, & Charles H. Davis *34*(1), 67–74.
        The concept of entropy, from the second law of thermodynamics, has
        been used by numerous writers on information theory.

1991   Information as Thing. Michael K. Buckland *42*(5), 351–360.
        Three meanings of 'information' are distinguished: 'information-as-pro-
        cess'; 'information-as-knowledge'; and 'information-as-thing'. . . .

1992   The Communication–Information Relationship in System—Theoretic
        Perspective. Brent D. Ruben *43*(1), 15–27.
        There is a growing recognition that *information* and *communication*
        are interrelated in very fundamental ways.

1994   Operationalizing the Notion of Information as a Subjective Construct.
        Charles Cole *45*(7), 465–476.

1996   Information Calculus for Information Retrieval. C. J. van Rijsbergen,
        & M. Lalmas *47*(5), 385–398.
        The theory, which is based on Situation Theory, is expressed with a
        calculus defined on channels. The calculus was defined so that it satisfies
        properties that are attributed to information and its flows. This paper
        demonstrates the connection between this calculus and Information
        Retrieval, and proposes a model of an Information Retrieval system
        based on this calculus.

Theoretical information retrieval system design

1954   Machine Literature Searching I. A General Approach. J.W. Perry, Allen
        Kent, & M.M. Berry *5*(1), 18–22 (first of a series of 10 articles).
        The lengthening time required for searching larger and larger indexes
        and for consulting expanding collections of abstracts has become a
        source of mounting concern. Modern automatic equipment is able to
        scan and recognize index entries made in the form of holes on punched
        cards, magnetized spots on magnetic tape. . . .

1962   Indexing and Abstracting by Association. Lauren B. Doyle *13*(4), 378–390.
        This article discusses the possibility of exploiting the statistics of
        word co-occurrence in text for purposes of document retrieval. . . . It
        is shown that the most strongly co-occurring word pairs, which are
        therefore 'associated' in a statistical sense, can be represented in the
        form of an 'association map.'

1963   A Generalized Computer Method for Information Retrieval. Claire
        K. Schultz *14*(1), 39–48.
        A generalized method is given for performing information retrieval by
        computer. . . . A block process diagram, a detailed flow chart for appli-
        cation to a computer, and a description of the application are provided.

1965   The Evaluation of Automatic Retrieval Procedures—Selected Test
        Results Using the SMART System. Gerard Salton *16*(3), 209–222.

1969   Word-Word Associations in Document Retrieval Systems. M.E. Lesk
        20(1), 27–38.
        The SMART automatic document retrieval system is used to study
        association procedures for automatic content analysis. . . . There is
        little overlap between word relationships found through associations
        and those used in thesaurus construction, and the effects of word
        associations and a thesaurus in retrieval are independent. . . . A prop-
        erly constructed thesaurus, however, offers better performance than
        statistical association methods.

1974   A Clustering Algorithm Based on User Queries. Clement T. Yu *25*(4),
        218–226.

1978   On the Nature of Fuzz: A Diatribe. Stephen E. Robertson *29*(6), 304–307.
        The imprecision of some of the concepts which are used in formal
        models in information science has led to a spate of attempts to apply
        fuzzy set theory to aspects of information science. An analysis of the
        various kinds of imprecision that can occur indicates strongly that
        fuzzy set theory is not an appropriate formalism for these models.

1981   A Comparison of Two Systems of Weighted Boolean Retrieval. A.
        Bookstein *32*(4), 275–279.
        A major deficiency of traditional Boolean systems is their inability to
        represent the varying degrees to which a document may be written on
        a subject. In this article we isolate a number of criteria that should be
        met by any Boolean system generalized to have a weighting capability.

1987   Fuzzy Relational Databases: Representational Issues and Reduction
        Using Similarity Measures. Henri Prade, & Claudette Testemale *38*(2),
        118–126.
        The proposed similarity measure, based on a fuzzy Hausdorff distance,
        estimates the mismatch between two possibility distributions.

1987   I^{3}R: A New Approach to the Design of Document Retrieval Systems.
        W.B. Croft, & R.H. Thompson *38*(6), 389–404.
        The system uses a novel architecture to allow more than one system
        facility to be used at a given stage of a search session. Users influence
        the system actions by stating goals they wish to achieve, by evaluating
        system output, and by choosing particular facilities directly.

1990   Indexing by Latent Semantic Analysis. Scott Deerwester, Susan T.
        Dumais, George W. Furnas, & Thomas K. Landauer *41*(6), 391–407.
        A new method for automatic indexing and retrieval is described. . . .
        The particular technique used is singular-value decomposition, in

which a large term by document matrix is decomposed into a set of ca. 100 orthogonal factors from which the original matrix can be approximated by linear combination.

1991   User-Based Document Clustering by Redescribing Subject Descriptions with a Genetic Algorithm. Michael D. Gordon *42*(5), 311–322.
This article reports that clusters of co-relevant documents obtain increasingly similar descriptions when a genetic algorithm is used to adapt subject descriptions so that documents become more effective in matching relevant queries and failing to match nonrelevant queries.

1995   Machine Learning for Information Retrieval: Neural Networks, Symbolic Learning, and Genetic Algorithms. Hsinchun Chen *46*(3), 194–216.
These newer techniques, which are grounded on diverse paradigms, have provided great opportunities for researchers to enhance the information processing and retrieval capabilities of current information storage and retrieval systems.

1996   Inter-Record Linkage Structure in a Hypertext Bibliographic Retrieval System. Dietmar Wolfram *46*(10), 765–774.
The author explores inter-record linkage relationships of a bibliographic hypertext system through the use of descriptor term co-occurrences. . . . The observed distribution of term co-occurrences using term sizes follows a complex, but regular relationship.

1997   Clustering and Classification of Large Document Bases in a Parallel Environment. Anthony S. Ruocco, & Ophir Frieder *48*(10), 932–943.
Development of cluster-based search systems has been hampered by prohibitive times involved in clustering large document sets. . . . We propose the use of parallel computing systems to overcome the computationally intense clustering process.

1998   Application of Rough Sets to Information Retrieval. Sadaaki Miyamoto *49*(3), 195–205.
After a brief review of fuzzy sets, rough sets, and a fuzzy logical model for information retrieval, rough approximations for retrieved data are defined. The approximations are considered for both crisp and fuzzy cases.

1998   Coordinating Computer-Supported Cooperative Work: A Review of Research Issues and Strategies. Jonathan K. Kies, Robert C. Williges, & Mary Beth Rosson *49*(9), 776–791.

Information retrieval system evaluation

1964   Testing Indexes and Index Language Devices: The ASLIB Cranfield Project. F.W. Lancaster, & J. Mills *15*(1), 4–13.

1968   Expected Search Length: A Single Measure of Retrieval Effectiveness Based on the Weak Ordering Action of Retrieval Systems. William S. Cooper *19*(1), 30–41.

The new measure is based on calculations of the expected number of irrelevant documents in the collection which would have to be searched through before the desired number of relevant documents could be found.

1968   **The Browser's Retrieval Game.** Siegfried Treu *19*(4), 404–410.
The user starts with any single term(s) or term pair(s), 'browses' through the available index term vocabulary 'aisles' by means of a display console, and finally arrives at the game's objective: Either a query in terms of pair(s) of highly associated terms or the conclusion that the subject of interest is not represented.

1969   **Effectiveness of Information Retrieval Methods.** John A. Swets *20*(1), 72–89.
Results of some 50 different retrieval methods applied in three experimental retrieval systems were subjected to the analysis suggested by statistical decision theory. The analysis validates a previously proposed measure of effectiveness and demonstrates its several desirable properties.

1975   **Performing Evaluation Studies in Information Science.** Rowena Weiss Swanson *26*(3), 140–156.
The state-of-the-art of evaluation study in information science is analyzed with respect to 1. the scope of evaluation studies; 2. the use of laboratory-type environments; 3. the use of surrogate judges; 4. selection of variables; 5. frequency of study; and 6. comparability of study results.

1978   **Evaluation of Information Retrieval Systems: A Decision Theory Approach.** Donald H. Kraft, & Abraham Bookstein *29*(1), 31–40.
The Swets model of information retrieval, based on a decision theory approach, is discussed. . . . It is seen that when the variances are unequal, the Swets rule of retrieving a document if its Z value is large enough is not optimal.

1978   **The Application of Multiple-Criteria Utility Theory to the Evaluation of Information Systems.** Saul Herner, & Kurt J. Snapper *29*(6), 289–296.

1982   **Optimal Values of Recall and Precision.** Tore Olafsen, & Libena Vokac *33*(2), 92–96.
When the cost structure is known, the optimal value of recall and the optimal value of precision can be found.

1988   **Performance Measures of Information Retrieval Systems—An Experimental Approach.** John J. Regazzi *39*(4), 235–251.
The study finds that there is no operational difference between the relevance-theoretic and the utility-theoretic model of evaluation. It further suggests the need for performance measures based upon a complex set of factors including document and information attributes, the judge, and other environmental factors, particularly the effects of learning during the evaluation process.

1992   **Evaluation of Advanced Retrieval Techniques in an Experimental Online Catalog.** Ray R. Larson *43*(1), 34–53.

1995   Understanding Performance in Information Systems: Blending
         Relevance and Competence. Myke Gluck *46*(6), 446–460.
         The article . . . reports the results of an experiment that used a gen-
         eralized geographic information system to illustrate how collecting
         and analyzing data simultaneously from both a system and user views
         of performance can suggest improvements for information systems.

1996   Some Perspectives on the Evaluation of Information Retrieval Systems.
         Jean M. Tague-Sutcliffe *47*(1), 1–3.
         These problems include the question of the necessity of using real
         users, as opposed to subject experts, in making relevance judgments,
         the possibility of evaluating individual components of the retrieval
         process, rather than the process as a whole, the kinds of aggregation
         that are appropriate for the measures used. . . . the difficulties in eval-
         uating interactive systems, and the kinds of generalization which are
         possible from information retrieval tests.

1996   The Dilemma of Measurement in Information Retrieval Research.
         David Ellis *47*(1), 23–36.
         The problem of measurement in information retrieval research is traced
         to its source in the first retrieval tests. . . . Finally, it is concluded that
         the original vision of information retrieval research as a discipline
         founded on quantification proved restricting for its theoretical and
         methodological development and that increasing recognition of this
         is reflected in growing interest in qualitative methods in information
         retrieval research in relation to the cognitive, behavioral, and affective
         aspects of the information retrieval interaction.

1996   Evaluating Interactive Systems in TREC. Micheline Beaulieu, Stephen
         Robertson, & Edie Rasmussen *47*(1), 85–94.
         The TREC (Text Retrieval Conference) experiments were designed
         to allow large-scale laboratory testing of information retrieval tech-
         niques. . . . groups within TREC have become increasingly interested
         in finding ways to allow user interaction without invalidating the
         experimental design.

1997   Writing with Collaborative Hypertext: Analysis and Modeling. Chaomei
         Chen *48*(11), 1049–1066.
         Markov chain models are derived from the empirical data on the use
         of the system.

Relevance

1964   The Notion of Relevance (I). Donald J. Hillman *15*(1), 26–34.
         Analysis of the problems of defining the mutual relevancies of queries
         and document-collections indicates that they essentially involve the
         problem of conceptual relatedness.

1964   The Consistency of Human Judgments of Relevance. A. Resnick, &
T.R. Savage *15*(2), 93–95.
The results showed that . . . humans are able to make such judgments
consistently.

1975   RELEVANCE: A Review of and a Framework for the Thinking on the
Notion in Information Science. Tefko Saracevic *26*(6), 321–343.
In the most fundamental sense, relevance has to do with effectiveness
of communication.

1987   Pictures of Relevance: A Geometric Analysis of Similarity Measures.
William P. Jones, & George W. Furnas *38*(6), 420–442.
A geometric analysis is advanced and its utility demonstrated through
its application to six conventional information retrieval similarity
measures and a seventh *spreading activation* measure.

1992   Psychological Relevance and Information Science. Stephen P. Harter
43(9), 602–615.
The objective sense of relevance as 'on the topic' is curious, in that it
is far removed from the way in which we use the words 'relevant' and
'relevance' in their everyday senses.

1994   User-Defined Relevance Criteria: An Exploratory Study. Carol L. Barry
45(3), 149–159.
The results indicate that the criteria employed by users included
tangible characteristics of documents . . . subjective qualities (e.g.,
agreement with the information provided by the document) and
situational factors. . . .

1995   Topical Relevance Relationships. I. Why Topic Matching Fails. Rebecca
Green *46*(9), 646–653.

Design and evaluation of operational information retrieval systems

1950   Multiple Coding and the Rapid Selector. Carl S. Wise, & James W. Perry
1(2), 76–83.
Punched cards have permitted us to take the first steps toward developing systems of information analysis capable of permitting searches
to be directed to new, unforeseen combinations of entities, concepts
and operations. . . . The coding method to be considered, as developed
by one of us (CSW) for 'Keysort' cards, permits coding to be based on
a vocabulary of 456,976 concepts, of which any sixteen (or less) may
be coded on a single card.

1959   Information Storage and Retrieval Using a Large Scale Random Access
Memory. J.J. Nolan *10*(1), 27–35.
The unit shown . . . is the first really large scale random access memory
device. It is the IBM RA-MAC. . . . The arm may be moved vertically
to any of the 50 disks and laterally to any of 100 concentric recording

tracks. Each track has a capacity of 1000 characters, 500 on the top surface and 500 on the bottom of each disk. Each disk, therefore, has a capacity of 100,000 characters, or a total of five million for the 50 disks.

1960   The MARLIS: A Multi-Aspect Relevance Linkage Information System—Present Position and Future Needs. B.C. Vickery *11*(2), 97–101.

1960   Exploitation of Recorded Information. I. Development of an Operational Machine Searching Service for the Literature of Metallurgy and Allied Subjects. Allen Kent *11*(2), 173–188.

1961   The Historical Development and Present State-of-the-Art of Mechanized Information Retrieval Systems. Charles P. Bourne *12*(2), 108–110. The last ten years of equipment development and the application of mechanization techniques are reviewed for each of several functionally separate approaches, such as punched-card systems, computer systems, and magnetic media systems.

1961   Selective Dissemination of New Scientific Information with the Aid of Electronic Processing Equipment. H.P. Luhn *12*(2), 131–138. Improvement of scientific communication is sought through machine assisted dissemination of new information. A service system is described in which a new document is characterized by a vocabulary or pattern of keywords. This pattern is then compared with the vocabularies or profiles characterizing each of the participants of the service. If a given degree of similarity exists between the two, the affected participants are notified by a card carrying an abstract.... Profiles are kept current by discarding patterns after they have reached a certain age.

1964   Boeing SLIP: Computer Produced and Maintained Printed Book Catalogs. Edward A. Weinstein, & Joan Spry *15*(3), 185–190.

1966   An On-Line Technical Library Reference Retrieval System. D.L. Drew, R.K. Summit, R.I. Tanaka, & R.B. Whiteley *17*(1), 3–7. In October 1964, Lockheed Missiles & Space Company (LMSC) started to experiment with an on-line reference retrieval system which uses a coordinate search strategy.... The current system and the second-generation design using a 'dialogue' are briefly described.

1969   MEDLARS: Report on the Evaluation of Its Operating Efficiency. F.W. Lancaster *20*(2), 119–142. The Medical Literature Analysis and Retrieval System (MEDLARS)... is a multipurpose system, a prime purpose being the production of *Index Medicus* and other recurring bibliographies. However, the present study concentrated on the evaluation of the *demand search* function (i.e., the conduct of retrospective literature searches in response to specific demands).

1975   Squibb Science Information System: Computerized Selective Dissemination, Current Awareness, and Retrospective Searching of Pharmaceutical Literature. S.J. Frycki, P.A. Roskos, C.F. Gerity, & I.S. Levett *26*(3), 174–183.

A combination of internally written PL/I and COBOL programs and the INQUIRE® data management system provide the software required to perform the operations on an IBM 370/155 computer with one megabyte of core storage. . . .

1979   Towards Everyday Language Information Retrieval Systems via Minicomputers. Colin Bell, & Kevin P. Jones 30(6), 334–339.

1983   The Use of Computer-Monitored Data in Information Science and Communication Research. Ronald E. Rice, & Christine L. Borgman 34(4), 247–256.

1984   Testing of a Natural Language Retrieval System for a Full Text Knowledge Base. Lionel M. Bernstein, & Robert E. Williamson 35(4), 235–247. . . . a retrieval system which combines use of probabilistic, linguistic, and empirical means to rank individual paragraphs of full text for their similarity to natural language queries proposed by users.

1986   An Intelligent System for Document Retrieval in Distributed Office Environments. Uttam Mukho-padhyay, Larry M. Stephens, Michael N. Huhns, & Ronald D. Bonnell. 37(3), 123–135.
MINDS (Multiple Intelligent Node Document Servers) is a distributed system of knowledge-based query engines for efficiently retrieving multimedia documents in an office environment of distributed workstations.

1988   Design Considerations for CD-ROM Retrieval Software. Edward M. Cichocki, & Susan M. Ziemer 39(1), 43–46.

1991   Nonmaterialized Relations and the Support of Information Retrieval Applications by Relational Database Systems. Clifford A. Lynch 42(6), 389–396.
A new query processing strategy called *nonmaterialized relations* is proposed which should significantly improve the ability of relational systems to support IR applications without requiring changes to the SQL query language interface.

1991   Extended Subject Access to Hypertext Online Documentation. Parts I and II: The Search-Support and Maintenance Problems. T.R. Girill, Thomas Griffin, & Robert B. Jones 42(6), 414–426.

1991   Imaging: Fine Arts. Howard Besser 42(8), 589–596.
We look at general image quality issues including image capture, resolution, and display, and then turn our attention to the uses of imaging technology for conservation and preservation purposes.

1995   Multimedia and Comprehension: The Relationship among Text, Animation, and Captions. Andrew Large, Jamshid Beheshti, Alain Breuleux, & Andre Renaud 46(5), 340–347.

1996   Evaluation of Interactive Knowledge-Based Systems: Overview and Design for Empirical Testing. F.W. Lancaster, Jacob W. Ulvila, Susanne M. Humphrey, Linda C. Smith, Bryce Allen, & Saul Herner 47(1), 57–69.

1996   Networked Information Retrieval and Organization: Issues and Questions. Joseph W. Janes, & Louis B. Rosenfeld 47(9), 711–715.

The creation of guides to resources on the Internet specific to particular subjects have [sic] raised a number of interesting questions and issues.

1996 The Effectiveness of the Electronic City Metaphor for Organizing the Menus of Free-Nets. Elaine G. Toms, & Mark T. Kinnucan 47(12), 919–931.

1998 Intelligent Information Agents: Review and Challenges for Distributed Information Sources. Donna S. Haverkamp, & Susan Gauch 49(4), 304–311.
There are many approaches, both theoretical and implemented, to using intelligent software agents for information retrieval purposes. These approaches range from desktop agents specialized for a single user to networks of agents used to collect data from distributed information sources, including Web sites.

1998 A Smart Itsy Bitsy Spider for the Web. Hsinchun Chen, Yi-Ming Chung, Marshall Ramsey, & Christopher C. Yang 49(7), 604–618.

1998 Architecture, Design, and Development of an HTML/JavaScript Web-Based Group Support System. Nicholas C. Romano, Jr., Jay F. Nunamaker, Jr., Robert O. Briggs, & Douglas R. Vogel 49(7), 649–667.

1998 Design Considerations in Instrumenting and Monitoring Web-Based Information Retrieval Systems. Michael D. Cooper 49(10), 903–919.

1998 Web Search Engines. Candy Schwartz 49(11), 973–982.

Indexing systems and techniques

1952 Unit Terms in Coordinate Indexing. Mortimer Taube, C.D. Gull, & Irma S. Wachtel 3(4), 213–218.
During the past six months, Documentation Incorporated has established an experimental coordinate index under a research program made possible by the Armed Services Technical Information Agency (ASTIA). . . . The procedure was not to re-index the reports but to use the subject headings on the TID and DSC cards as the basis for developing appropriate terms for two coordinate indexes, one for each group of cards. . . . The method of coordinate indexing was first described in two papers prepared about two years ago by Dr. Taube.

1958 The Thesaurus Approach to Information Retrieval. T. Joyce, & R.M. Needham 9(3), 192–197.
. . . the employment of a large number of terms, when indexing a collection of documents, must somehow take account of the existence of synonyms. On the other hand, the employment of a comparatively small number, particularly if the notions represented by the terms are not supposed to overlap, makes the indexing process considerably more difficult. These disadvantages can be avoided if a thesaurus is employed. . . .

1960   Keyword-in-Context Index for Technical Literature (KWIC Index).
        H.P. Luhn *11*(4), 288–295.
        A distinction is made between bibliographical indexes for new and
        past literature based on the willingness of the user to trade perfection
        for currency. Indexes giving keywords in their context are proposed
        as suitable for disseminating new information.

1965   An Experiment in Automatic Indexing. Fred J. Damerau *16*(4), 283–289.
        This report describes a method of indexing documents which is based
        on the assumptions, (1) that a subset of the words in a document can
        be an effective index to that document and, (2) that this subject can
        be approximated by selecting those words from the document whose
        frequencies are statistically unexpectedly high.

1976   Development of an Integrated Energy Vocabulary and the Possibilities
        for On-Line Subject Switching. R.T. Niehoff *27*(1), 3–17.
        Eleven vocabularies were analyzed and integrated.

1976   Classification from PRECIS: Some Possibilities. Phyllis A. Richmond
        27(4), 240–247.
        The PREserved Context Index System (PRECIS) developed for subject
        indexing for the British National Bibliography is discussed as a basis
        for various studies relating to classification which could be made from
        its initial phrases, strings, entries and back-up structure.

1987   Knowledge-Based Indexing of the Medical Literature: The Indexing
        Aid Project. Susanne M. Humphrey, & Nancy E. Miller *38*(3), 184–196.
        The system uses an experimental frame-based knowledge represen-
        tation language, FrameKit, implemented in Franz Lisp. The initial
        prototype is designed to interact with trained MEDLINE indexers
        who will be prompted to enter subject terms as slot values in filling
        in document-specific frame data structures that are derived from the
        knowledge-base frames.

1989   Cataloging and Expert Systems: AACR2 as a Knowledge Base. Roland
        Hjerppe, & Birgitta Olander *40*(1), 27–44.

1990   Automatic Derivation of Name Access Points in Cataloging. Elaine
        Svenonius, & Mavis Molto *41*(4), 254–263.
        Results show that approximately 88% of the access points selected by
        the Library of Congress or the National Library of Medicine could be
        automatically derived from title page data.

1993   A Method for Automatically Abstracting Visual Documents. Mark E.
        Rorvig *44*(1), 40–56.
        Visual documents—motion sequences on film, video-tape, and digital
        recordings—constitute a major source of information for the Space
        Agency.... This article describes a method for automatically selecting
        key frames from visual documents. These frames may in turn be used
        to represent the total image sequence. . . . The performance of the
        abstracting algorithm reduces 51 minutes of video sequences to 134
        frames: a reduction of information in the range of 700:1.

1994   GIPSY: Automated Geographic Indexing of Text Documents. Allison
        Gyle Woodruff, & Christian Plaunt 45(9), 645–655.
        Under this algorithm, words and phrases containing geographic place
        names or characteristics are extracted from a text document and
        used as input to database functions which use spatial reasoning to
        approximate statistically the geoposition being referenced in the text.

1995   Interactive Thesaurus Navigation: Intelligence Rules OK? Susan Jones,
        Mike Gatford, Steve Robertson, Micheline Hancock-Beaulieu, Judith
        Secker, & Steve Walker 46(1), 52–59.
        We discuss whether it is feasible to build intelligent rule- or weight-based
        algorithms into general-purpose software for interactive thesaurus
        navigation. . . . The results cause us to question many of the assumptions
        made by previous researchers in this area.

1998   A Texture Thesaurus for Browsing Large Aerial Photographs. We-Ying
        Ma, & B.S. Manjunath 49(7), 633–648.
        The salient components of this system include texture feature extraction,
        image segmentation and grouping, learning similarity measure, and a
        texture thesaurus model for fast search and indexing.

Indexing theory and evaluation

1956   Evaluation of Library Techniques for the Control of Research Materials.
        I.A. Warheit 7(4), 267–275.
        With the obvious deficiencies of traditional subject headings, jour-
        nal-type indexes, and classification schemes so ever-present, we are
        being offered three basic solutions: 1. punched cards, both edge-punched
        and field punched, and either hand or machine sorted, 2. the Batten or
        Peek-a-boo system, and 3. Coordinated Indexing using Uniterms. There
        are, in addition, some embryonic systems which store information
        magnetically, or on film, or by means of cathode ray tubes and which
        require expensive machines for their operation.

1957   Logical Structures in Language. Noam Chomsky 8(4), 284–291.
        Many of the problems that arise when one tries to construct a device
        that generates word sequences directly, vanish when one describes
        sentences in terms of phrase structure. . . .

1958   Delegation of Classification. R.A. Fairthorne 9(3), 159–164.
        Library classifications display lists of labels for texts, but rarely tell how
        to bestow the right label on the right text. . . . In other words, things
        are assumed created with name tags round their necks, and all we have
        to do is to catch them and read the tags. . . . This does not quite agree
        with practical experience. . . .

1962   Machinelike Indexing by People. Christine Montgomery, & Don R.
        Swanson 13(4), 359–366.

A study of several thousand entries in a classified bibliography . . . revealed that a large proportion of the title entries contained words identical to or synonymous with words of the corresponding subject heading. It is inferred that a major part of the bibliography studied could have been compiled by a machine procedure operating on titles alone, provided the machine were supplied with a suitable synonym dictionary.

1964　Use of Meta-Language in Information Retrieval Systems. W. Goffman, J. Verhoeff, & Jack Belzer *15*(1), 14–22.
The . . . meta-language terms here apply to the written records and their relations, rather than to the subject treated by these records. Examples of such terms are diagram, rehash, report, reprint, preprint, flow chart, analysis, description, discussion, table, definition, abstract. . . .

1969　Indexing Consistency and Quality. Pranas Zunde, & Margaret E. Dexter *20*(3), 259–267.
It is well known that any two indexers, indexing one and the same document individually, will select sets of indexing terms which are most unlikely to be identical.

1970　Terse Literatures: I. Terse Conclusions. Charles L. Bernier *21*(5), 316–319.
Prompt literatures of organized terse conclusions may increase ability to keep up in a subject, reduce need for translation, and make information available promptly.

1971　What Makes An Automatic Keyword Classification Effective? K. Sparck Jones, & E.O. Barber *22*(3), 166–175.

1973　An Experimental Framework For Observing the Indexing Process. Dan C. Clarke, & John L. Bennett *24*(1), 9–24.
An experimental framework for observing an indexing task carried out in a laboratory environment is described. The computer-based interactive Negotiated Search Facility served as the instrument which the experimenter used to monitor the indexing task and the tool which the indexer used to carry out the task.

1976　Machine Indexing: Linguistic and Semiotic Implications. Susan Artandi *27*(4), 235–239.

1977　On Indexing, Retrieval and the Meaning of About. M.E. Maron *28*(1), 38–43.

1979　A Sociological Approach to the Design of Information Systems. Donald F. Swift, Viola A. Winn, & Dawn A. Bramer *30*(4), 215–223.
Conventional information systems, founded on objectivist assumptions, are inappropriate for social scientists.

1982　Postcoordinate Retrieval: A Comparison of Two Indexing Languages. Ann H. Schabas *33*(1), 32–37.
This article reports on a comparison of the postcoordinate retrieval effectiveness of two indexing languages: LCSH and PRECIS. The effect of augmenting each with title words was also studied.

1985   The Five-Axiom Theory of Indexing and Information Supply. Robert Fugmann *36*(2), 116–129.

1986   Natural Language Processing in Information Retrieval. Tamas E. Doszkocs *37*(4), 191–196.
State-of-the-art information-retrieval systems are found to combine the functional capabilities of the conventional inverted file—Boolean logic—term adjacency approach, with statistical-combinatoric techniques pioneered in experimental information-retrieval research and formal natural-language processing methods and tools borrowed from artificial intelligence.

1986   Subject Access in Online Catalogs: A Design Model. Marcia J. Bates *37*(6), 357–376.
Design features presented are an access phase, including *entry* and *orientation*, a *hunting* phase, and a *selection* phase. An end-user thesaurus and a front-end system mind are presented as examples of online catalog system components to improve searcher success during entry and orientation.

1990   All the Right Words: Finding What You Want as a Function of Richness of Indexing Vocabulary. Louis M. Gomez, Carol C. Lochbaum, & Thomas K. Landauer *41*(8), 547–559.
The implications of index-word selection strategies for user success in interactive searching were investigated in two experiments. . . . The results demonstrate that searcher success is markedly improved by greatly increasing the number of names per object.

1992   A Gray Code Based Ordering for Documents on Shelves: Classification for Browsing and Retrieval. Robert M. Losee, Jr. *43*(4), 312–322.
Based on these requirements, information-theoretic considerations, and the Gray code, a classification system is proposed that can classify documents without human intervention. . . . The proposed system can incorporate both classification by subject and by other forms of bibliographic information, allowing for the generalization of browsing to include all features of an information carrying unit.

1994   Needs for Research in Indexing. Jessica L. Milstead *45*(8), 577–582.

1994   Some Issues in the Indexing of Images. Sara Shatford Layne *45*(8), 583–588.

1994   Indexing and Retrieval Performance: The Logical Evidence. Dagobert Soergel *45*(8), 589–599.
This article . . . defines the characteristics of indexing that affect retrieval—namely, indexing devices, view-point-based and importance-based indexing exhaustivity, indexing specificity, indexing correctness, and indexing consistency—and examines in detail their effects on retrieval.

1995   Automatic Thesaurus Generation for an Electronic Community System. Hsinchun Chen, Tak Yim, David Fye, & Bruce Schatz *46*(3), 175–193.

The experiment showed that the thesaurus was an excellent 'memory-jogging' device and that it supported learning and serendipitous browsing.

1997   Experiments with Automatic Indexing and a Relational Thesaurus in a Chinese Information Retrieval System. Tian-Long Wan, Martha Evens, Yeun-Wen Wan, & Yuen-Yuan Pao *48*(12), 1086–1096.
Two important issues have been explored: whether thesauri enhance the retrieval effectiveness of Chinese documents, and whether automatic indexing can compete with manual indexing in a Chinese information retrieval system.

1998   Indexing and Access for Digital Libraries and the Internet: Human, Database, and Domain Factors. Marcia J. Bates *49*(13), 1185–1205.

Search strategy and evaluation

1967   Analysis of Questions Addressed to a Medical Reference Retrieval System: Comparison of Question and System Terminologies. Barbara Flood *18*(4), 216–227.

1968   A Literature Search and File Organization Model. Ferdinand F. Leimkuhler *19*(2), 131–136.
A principle of sequential optimization in search theory distributes the search effort at each stage so as to maximize the probability of target detection with the effort expended thus far.

1974   Reference Question Analysis and Search Strategy Development by Man and Machine. Gerald Jahoda *25*(3), 139–144.
Question analysis and search strategy development . . . were characterized as nine decision making steps.

1978   Cost-Effectiveness Comparison of Manual and Online Retrospective Bibliographic Searching. Dennis R. Elchesen *29*(2), 56–66.

1979   Information Search Tactics. Marcia J. Bates *30*(4), 205–214.
Twenty-nine tactics are named, defined, and discussed. . . .

1980   Searching Biases in Large Interactive Document Retrieval Systems. David C. Blair *31*(4), 271–277.
The way that individuals construct and modify search queries on a large interactive document retrieval system is subject to systematic biases similar to those that have been demonstrated in experiments on judgments under uncertainty. These biases are shared by both naive and sophisticated subjects. . . .

1981   Online Searching: Measures that Discriminate among Users with Different Types of Experiences. Carol Hansen Fenichel *32*(1), 23–32.

1984   Online Searching Styles: A Case-Study-Based Model of Searching Behavior. Raya Fidel *35*(4), 211–221.

The model of operationalist and conceptualist searching styles describes searching behavior of experienced online searchers.

1985   An Investigation of Online Searcher Traits and Their Relationship to Search Outcome. Trudi Bellardo 36(4), 241–250.
The notion that searching performance can be predicted by or is dependent upon certain cognitive or personality traits has thus become highly suspect.

1986   Why are Online Catalogs Hard to Use? Lessons Learned from Information-Retrieval Studies. Christine L. Borgman 37(6), 387–400.

1988   A Study of Information Seeking and Retrieving. III. Searchers, Searches, and Overlap. Tefko Saracevic, & Paul Kantor 39(3), 197–216.
A concluding summary of all results is presented in Part III.

1989   Information-Seeking Strategies of Novices Using a Full-Text Electronic Encyclopedia. Gary Marchionini 40(1), 54–66.
Analysis of search patterns showed that novices used a heuristic, highly interactive search strategy.

1993   Effects of Search Experience and Subject Knowledge on the Search Tactics of Novice and Experienced Searchers. Ingrid Hsieh-Yee 44(3), 161–174.
The results showed that search experience affected searchers' use of many search tactics, and suggested that subject knowledge became a factor only after searchers have had a certain amount of search experience.

1993   Children's Information Retrieval Behavior: A Case Analysis of an OPAC. Paul Solomon 44(5), 245–264.

1994   What Is Used during Cognitive Processing in Information Retrieval and Library Searching? Eleven Sources of Search Information. Dee Andy Michel 45(7), 498–514.

1995   Searcher Response in a Hypertext-Based Bibliographic Information Retrieval System. Alexandra Dimitroff, & Dietmar Wolfram 46(1), 22–29.
... indicating that a hypertext-based approach to bibliographic retrieval could be appropriate for a variety of searcher experience levels.

1995   Children's Searching Behavior on Browsing and Keyword Online Catalogs: The Science Library Catalog Project. Christine L. Borgman, Sandra G. Hirsh, Virginia A. Walter, & Andrea L. Gallagher 46(9), 663–684.
The SLC approach overcomes problems with several searching features that are difficult for children in typical keyword OPAC systems: typing skills, spelling, vocabulary, and Boolean logic.

1996   Affective and Cognitive Searching Behavior of Novice End-Users of a Full-Text Database. Diane Nahl, & Carol Tenopir 47(4), 276–286.
Affective questions outnumbered cognitive and sensorimotor questions by two to one. This preponderance of affective micro-information needs during searching might be addressed by new system functions.

1996   Multiple Search Sessions Model of End-User Behavior: An Exploratory
        Study. Amanda Spink *47*(8), 603–609.

1997   Qualitative Exploration of Learners' Information-Seeking Processes
        Using Perseus Hypermedia System. Shu Ching Yang *48*(7), 667–669.

Interface design

1981   A Translating Computer Interface for End-User Operation of Hetero-
        geneous Retrieval Systems. I. Design. Richard S. Marcus, & J. Francis
        Reintjes *32*(4), 287–303.
        The interface allows users to make requests in a common language.
        These requests are translated by the interface into the appropriate
        commands for whatever system is being interrogated.

1982   A Computer Intermediary for Interactive Database Searching. I.
        Design. Charles T. Meadow, Thomas T. Hewett, & Elizabeth S. Aversa
        33(5), 325–332.

1986   Designing Menu Selection Systems. Ben Shneiderman *37*(2), 57–70.

1986   Transparent Information Systems Through Gateways, Front Ends,
        Intermediaries, and Interfaces. Martha E. Williams *37*(4), 204–214.

1991   System Design and Cataloging Meet the User: User Interfaces to
        Online Public Access Catalogs. Martha M. Yee *42*(2), 78–98.
        The following features of online public access catalogs are discussed:
        the demonstration of relationships between records, the provision of
        entry vocabularies, the arrangement of multiple entries on the screen,
        the provision of access points, the display of single records, and the
        division of the catalog into separate files or indexes.

1992   User-Friendly Systems Instead of User-Friendly Front-Ends. Donna
        Harman *43*(2), 164–174.

1993   An Expert System for Automatic Query Reformation. Susan Gauch,
        & John B. Smith *44*(3), 124–136.

1997   An Informal Information-Seeking Environment. David G. Hendry, &
        David J. Harper *48*(11), 1036–1048.
        When an *opportunistic* searcher encounters an *over-determined* informa-
        tion system, less than ideal search strategies often ensue. The mismatch
        can be addressed by reducing the determinacy of the system, thereby
        making it more amenable to informal problem-solving practices.

Information seeking and needs

1951   User Needs in a Microfacsimile Reader. G. Miles Conrad *2*(4), 201–204.

1960   Information Gathering Patterns and Creativity. Robert E. Maizell *11*(1),
        9–17.

The creative chemists differ from the 'noncreative' in that the former read more technical literature on the job . . . are less influenced in their independence of thought . . . are more inquisitive and have broader cultural interests.

1960   Mooers' Law: Or Why Some Retrieval Systems are Used and Others Are Not. Calvin N. Mooers *11*(3), ii.
Mooers' Law: An information retrieval system will tend *not* to be used whenever it is more painful and troublesome for a customer to have information than for him not to have it.

1962   The Process of Asking Questions. Robert S. Taylor *13*(4), 391–396.
Four levels of question formation may be isolated and analyzed. . . .

1966   The User's Place in an Information System. Edwin B. Parker *17*(1), 26–27.
We argued that information systems should be designed to maximize the amount of control by the receiver, and that, in general, the system should adapt to the receiver or user, rather than the user to the system.

1968   Some Questions Concerning 'Information Need.' John O'Connor *19*(2), 200–203.

1968   Psychology and Information. George A. Miller *19*(3), 286–289.
An aspect of the human use of information that has generally been overlooked in the automation of information services is the human tendency to locate information spatially.

1977   The Informative Act and Its Aftermath: Toward a Predictive Science of Information. Marilyn M. Levine *28*(2), 101–106.
A unit called the whomp is introduced which describes the net effect of receiving a message with both hard information and human stress points. Such messages can be said to 'produce' records.

1980   Toward Usable User Studies. Colin K. Mick, Georg N. Lindsey, & Daniel Callahan *31*(5), 347–356.
The model focuses on variables which can be manipulated by managers—primarily environmental and situational variables—rather than on variables describing individual attributes.

1981   Information-Processing Models of Cognition. Herbert A. Simon *32*(5), 364–377.
This article reviews recent progress in modeling human cognitive processes.

1985   The Dimensions of Perceived Accessibility to Information: Implications for the Delivery of Information Systems and Services. Mary J. Culnan *36*(5), 302–308.

1986   Managerial Support and the Use of Information Services. Francis W. Wolek *37*(3), 153–157.
The data support the belief that managers of high users of information give more encouragement for information use.

1991   Inside the Search Process: Information Seeking from the User's Perspective. Carol C. Kuhlthau *42*(5), 361–371.
The cognitive and affective aspects of the process of information seeking suggest a gap between the users' natural process of information use and the information system and intermediaries' traditional patterns of information provision.

1993   Environmental Scanning by CEOs in Two Canadian Industries. Ethel Auster and Chun Wei Choo *44*(4), 194–203.

1994   The Role of Attorney Mental Models of Law in Case Relevance Determinations: An Exploratory Analysis. Stuart A. Sutton *45*(3), 186–200.
This article examines the information seeking and evaluative behavior of attorneys as they search the corpus of law for primary authority. . . .

1996   The Impoverished Life-World of Outsiders. Elfreda A. Chatman *47*(3), 193–206.
Drawing upon a series of studies that examines the information world of poor people, the author discovers four critical concepts that serve as the basis for defining an impoverished life-world. These concepts are risk-taking, secrecy, deception, and situational relevance.

1998   User Satisfaction with Information Seeking on the Internet. Harry Bruce *49*(6), 541–556.

1998   Social Informatics in Information Science: An Introduction. Rob Kling, Howard Rosenbaum, & Carol Hert *49*(12), 1047–1052.
Social informatics (SI) refers to a multidisciplinary research field that examines the design, uses, and implications of information and communication technologies (ICTs) in ways that account for their interactions with institutional and cultural contexts.

1998   Work, Friendship, and Media Use for Information Exchange in a Networked Organization. Caroline Haythornthwaite, & Barry Wellman *49*(12), 1101–1114.

1999   Structures and Strategies of Interdisciplinary Science. Carole L. Palmer *50*(3), 242–253.
Interview data reveal that scientists undertake individual and cooperative boundary-crossing research. Four research modes are identified. . . .

Use of information systems

1966   Characteristics and Use of Personal Indexes Maintained by Scientists and Engineers in One University. G. Jahoda, Ronald D. Hutchins, & Robert R. Galford *17*(2), 71–75.

1974   Bibliographic Data Base Usage in a Large Technical Community. Donald T. Hawkins *25*(2), 105–108.

A survey of the bibliographic data bases used in satisfying some of the information requests received by the Bell Laboratories literature searching service was made.

1983   Usage Patterns of an Online Search System. Michael D. Cooper 34(5), 343–349.
This article examines the usage patterns of the ELHILL retrieval program of the National Library of Medicine's MEDLARS system.

1986   Observations of End-User Online Searching Behavior Over Eleven Years. Winifred Sewell, & Sandra Teitelbaum 37(4), 234–245.
Volume of searching is directly related to the convenient placement of the terminal in the work place. Slightly fewer than half of all potential searchers actually search for themselves. . . . End-users perform very simple searches, mostly using only the AND operator.

1992   An Empirically Grounded Model of the Adoption of Intellectual Technologies. Barbara M. Wildemuth 43(3), 210–224.
Detailed analysis of 43 user-developed computing applications resulted in a model consisting of five stages: Resource Acquisition, Application Development, Adoption/Renewal, Routinization/Enhancement, and External Adoption.

1995   High School Students' Use of Databases: Results of a National Delphi Study. Delia Neuman 46(4), 284–298.
The results confirm that the major issues related to schools' use of online and CD-ROM databases involve their role in students' development of the higher-order thinking skills necessary to plan, design, and conduct competent and credible research in the electronic information age.

1995   Context as a Factor in Personal Information Management Systems. Deborah K. Barreau 46(5), 327–339.
Seven managers were interviewed to observe how their electronic documents were organized, stored, and retrieved.

1996   The Modern Language Association: Electronic and Paper Surveys of Computer-Based Tool Use. Debora Shaw, & Charles H. Davis 47(12), 932–940.
Major changes in research habits included greater reliance on word processing and more work outside of libraries. Problems reported focused on access to computer-based resources, learning to use them, the need for instruction, and inconsistent interfaces.

1997   Factors That Influence the Use of Electronic Networks by Science and Engineering Faculty at Small Institutions. Part II. Preliminary Use Indicators. Peter Liebscher, Eileen G. Abels, & Daniel W. Denman 48(6), 495–507.
The article reports on five types of network use, E-mail, electronic discussion groups, accessing remote databases, accessing remote computer facilities, and file transfer.

1998   Topic Development in USENET Newsgroups. Larry N. Osborne 49(11), 1010–1016.

USENET topics are created, and they evolve, mutate, and become extinct in ways fundamentally different from spoken dialogue.

Scientific and scholarly communication

1963   **Is Information Retrieval Approaching a Crisis?** Yehoshua Bar-Hillel *14*(2), 95–98.
Scientists, so it is claimed, simply will not any longer be able to handle the flood of information.... It is a matter of simple fact, which I know from personal experience and observation and which the reader will be able to check similarly, that scientists did not spend on the average in 1962 more time on reading than they did 12 years ago, though printed scientific output has indeed almost doubled during this period.... There must therefore have been a way out between the horns of the dilemma. What is it? Everybody knows it: *Specialization.*

1969   **Stability in the Growth of Knowledge.** Manfred Kochen *20*(3), 186–197.
Merely more documents, more specialties, and more people who use or add to growing knowledge do not constitute an explosive situation.... The main idea behind the proposed logical structuring of this field is that the 'growing literature' organizes knowledge analogously to the way a learner, by creating models of his relevant environment, is able to take increasingly effective actions.

1972   **Interrelationships of Scientific Journals.** Francis Narin, Mark Carpenter, & Nancy C. Berlt *23*(5), 323–331.
A large amount of consistency was found between the citing characteristics of the journals in the different scientific fields, with quite clear boundaries between fields and a few well known cross disciplinary journals as cross field information links. The separate disciplines appear to relate to each other in an orderly manner, with a natural sequence: mathematics → physics → chemistry → biochemistry → biology.

1982   **Collaboration in Computational Musicology.** Miranda Lee Pao *33*(1), 38–43.
We suggest that such evidence supports the traditional belief that the humanist has a general tendency to work alone.

1990   **An Oasis Where Many Trails Cross: The Improbable Cocitation Networks of a Multidiscipline.** William Paisley *41*(6), 459–468.
The wide range of influences on the field of communication in its formative years is still reflected by its eclectic 'core literature'....

1992   **International Comparison of Departments' Research Performance in the Humanities.** A.J. Nederhof, & E.C.M. Noyons *43*(3), 249–256.

1993   **Communication Efficiency in Research and Development.** Patrick Wilson *44*(7), 376–382.
If communication in research and development is efficient, then the current cognitive situation in any specialty should fully reflect

all available relevant information. Available evidence suggests that communication in R & D is not in that sense efficient. . . .

1997 Modeling the Human Factors of Scholarly Communities Supported through the Internet and World Wide Web. Brian R. Gaines, Lee Li-Jen Chen, & Mildred L.G. Shaw *48*(11), 987–1003.

1998 Hotel Cosmopolitan: A Bibliometric Study of Collaboration at Some European Universities. Göran Melin, & Olle Persson *49*(1), 43–48.
There are some country variations, but within each country, the differences among the universities are small, if any.

Bibliometric laws and analysis

1955 Citation Indexes for Scientific Literature? W.C. Adair *6*(1), 31–32.

1960 The 'Half-life' of Some Scientific and Technical Literatures. R.E. Burton, & R.W. Kebler *11*(1), 18–22.
It is the purpose of this paper to examine the applicability of a *half-life* analogy to scientific literature, and to provide at least a small amount of data on which to base certain conclusions.

1962 Compilation of an Experimental Citation Index from Scientific Literature. Ben-Ami Lipetz *13*(3), 251–266.
Several forms of citation index were compiled with the aid of punched card techniques from 11,000 citations. . . .

1963 Bibliographic Coupling Between Scientific Papers. M.M. Kessler *14*(1), 10–25.
A single item of reference used by two papers was defined as a unit of coupling between them.

1963 New Factors in the Evaluation of Scientific Literature Through Citation Indexing. E. Garfield, & I.H. Sher *14*(3), 195–201.
More than one million citations from the scientific literature have been processed by the Citation Index Project at the Institute for Scientific Information.

1967 Bradford's Law and the Keenan-Atherton Data. Ole V. Groos *18*(1), 46.
Bradford's methods are applied to the Keenan-Atherton data. The results do not fit Bradford's Law.

1973 Co-citation in the Scientific Literature: A New Measure of the Relationship Between Two Documents. Henry Small *24*(4), 265–269.
A new form of document coupling called co-citation is defined as the frequency with which two documents are cited together.

1976 A General Theory of Bibliometric and Other Cumulative Advantage Processes. Derek de Solla Price *27*(5), 292–306.
A Cumulative Advantage Distribution is proposed which models statistically the situation in which success breeds success. It differs from the Negative Binomial Distribution in that lack of success, being a non-event, is not punished by increased chance of failure.

1978   Frequency-Rank Distributions. Bertram C. Brookes, & Jose M. Griffiths *29*(1), 5-13.

1978   **Highly Cited Old Papers and the Reasons Why They Continue to be Cited.** Charles Oppenheim, & Susan P. Renn *29*(5), 225-231.
... it was found that about 40% of the citations were for historical reasons, but that in the remaining 60% of the cases, the old paper is still begin [sic] actively used.

1978   **An Empirical Examination of Bradford's Law and the Scattering of Scientific Literature.** M. Carl Drott, & Belver C. Griffith *29*(5), 238-246.
The findings show that Bradford's Law is the reflection of some underlying process not related to the characteristics of the search mechanism or the nature of the literature. The authors conclude that there is instead a basic probabilistic mechanism underlying the law.

1981   **The Brillouin Measure of an Author's Contribution to a Literature in Psychology.** Bert R. Boyce, & David Martin *32*(1), 73-76.
An author whose presence holds a class together can be considered as an important element in this network of direct communication, and thus might well be a candidate for special funding consideration.

1981   **Author Cocitation: A Literature Measure of Intellectual Structure.** Howard D. White, & Belver C. Griffith *32*(3), 163-171.
The analysis assumes that the more two authors are cited together, the closer the relationship between them.

1982   **The Theoretical Foundation of Zipf's Law and Its Application to the Bibliographic Database Environment.** Jane Fedorowicz *33*(5), 285-293.
All of these—word frequency, citation frequency, and publication frequency—obey an ubiquitous distribution called Zipf's law.... The law has been shown to encompass many natural phenomena, and is equivalent to the distributions of Yule, Lotka, Pareto, Bradford, and Price.

1983   **Bibliometric Indicators versus Expert Opinion in Assessing Research Performance.** Michael E.D. Koenig *34*(2), 136-145.

1984   **Longitudinal Author Cocitation Mapping: The Changing Structure of Macroeconomics.** Katherine W. McCain *35*(6), 351-359.
Two types of scholarly 'migration' are observed, based on patterns of significant increases and decreases in correlations among authors—'active migration' (the individual refocussing of a scholar's efforts) and 'passive migration' (the result of reevaluation of an author's previous contributions in the context of the rise of new problem areas).

1985   **Private Acts and Public Objects: An Investigation of Citer Motivations.** Terrence A. Brooks *36*(4), 223-229.

1985   **The Dillon Hypothesis of Titular Colonicity: An Empirical Test from the Ecological Sciences.** J.A. Perry *36*(4), 251-258.
In a series of recent studies, J.T. Dillion [sic] has offered a hypothesis of titular colonicity as a guide for the analysis of scientific and scholarly quality. Formally stated. ... The presence of a colon in the title of a paper is the primary correlate of scholarship. ... The Hypothesis

appears to be supported. . . . Percent titular colonicity is higher in more prestigious and more widely respected journals . . . theoretical research is an order of magnitude more scholarly.

1986   A Relationship Between Lotka's Law, Bradford's Law, and Zipf's Law. Ye-Sho Chen, & Ferdinand F. Leimkuhler *37*(5), 307–314.

1987   Two Medical Literatures that are Logically but not Bibliographically Connected. Don R. Swanson *38*(4), 228–233.
This study demonstrates that certain unintended logical connections within the scientific literature, connections potentially revealing of new knowledge, are unmarked by reference citations or other bibliographic clues.

1987   Another Test of the Normative Theory of Citing. M.H. MacRoberts, & B.R. MacRoberts *38*(4), 305–306.
To put our findings in the terminology of citation analysis, we found that 'obliteration' . . . does not uniformly operate in science. We noted three patterns: (1) some work is used but is either never cited or is cited rarely, (2) some work is cited mainly or only through secondary sources, and (3) some work is credited every time it is used.

1990   Applications of the Theory of Bradford's Law to the Calculation of Leimkuhler's Law and to the Completion of Bibliographies. L. Egghe *41*(7), 469–492.
. . . this extension also has an application in determining the size and other properties of the complete unknown bibliography, based on the incomplete one.

1998   Visualizing a Discipline: An Author Co-Citation Analysis of Information Science, 1972–1995. Howard D. White, & Katherine W. McCain *49*(4), 327–355.

1998   Invoked on the Web. Blaise Cronin, Herbert W. Snyder, Howard Rosenbaum, Anna Martinson, & Ewa Callahan *49*(14), 1319–1328.
Where, how, and why are scholars invoked on the World Wide Web?

Literature domains

1980   Rationalization of Secondary Services: Measurement of Coverage of Primary Journals and Overlap between Services. J. Michael Brittain, & Stephen A. Roberts *31*(3), 131–142.

1987   A Study of Discourse Anaphora in Scientific Abstracts. Elizabeth Liddy, Susan Bonzi, Jeffrey Katzer, & Elizabeth Oddy *38*(4), 255–261.
Results show a mean use of 3.67 functioning anaphors per abstract in a random sample of 600 abstracts from two databases. Testing of rules indicates high feasibility of future algorithmic recognition of anaphoric uses of terms.

1990   Syntactic Patterns in Scientific Sublanguages: A Study of Four Disciplines. Susan Bonzi *41*(2), 121–131.
... it was found that the two disciplines representing the social sciences rarely differ significantly from each other, and the same is true of the two hard science disciplines. However, when hard and social sciences are compared, the differences are often significant.

1995   Domain Analysis, Literary Warrant, and Consensus: The Case of Fiction Studies. Clare Beghtol *46*(1), 30–44.

1995   Toward a New Horizon in Information Science: Domain-Analysis. Birger Hjørland, & Hanne Albrechtsen *46*(6), 400–425.
This new view of knowledge stresses the social, ecological, and content-oriented nature of knowledge. . . . The final section outlines important problems to be investigated, such as how different knowledge-domains affect the informational value of different subject access points in data bases.

1996   Using Corpus Statistics to Remove Redundant Words in Text Categorization. Yiming Yang, & John Wilbur *47*(5), 357–369.

1997   Disciplinary Variation in Automatic Sublanguage Term Identification. Stephanie W. Haas *48*(1), 67–79.
The research presented here describes a method for automatically identifying sublanguage (SL) domain terms and revealing the patterns in which they occur in text.

New documentary forms

1951   15 Years Experience with Auxiliary Publication. Watson Davis *2*(2), 87–89.
Auxiliary Publication, accomplished by deposit of a manuscript with the American Documentation Institute and the providing upon demand of photographic copies of that manuscript either in microfilm or photoprint form, is now an accepted and useful adjunct to scientific and scholarly publishing . . . was inaugurated in 1936 and has been in operation ever since. . . . Since its inauguration there have been 2,148 documents deposited through April 30, 1951.

1976   The Biblio-Profile—A Two-in-One Package of Information: Its Preparation, Production, Marketing, Uses. Lois F. Lunin *27*(2), 113–117.
... is a brief state-of-the-art report on a specific topic followed by a comprehensive bibliography.

1982   The Electronic Journal: A Progress Report. Murray Turoff, & Starr Roxanne Hiltz *33*(4), 195–202.
Four forms of electronic journals on the Electronic Information Exchange System (EIES) are described as divergent examples of potential journal forms that could become prominent in the future.

1985   The Personal Computer: Missing Link to the Electronic Journal?
        Donald Case *36*(5), 309–313.
        Noting developments in computer networking and microcomputer
        applications, the author contends that the widespread use of personal
        computers will accelerate the development of electronic publishing.

1995   The Electronic Medical Record: Promises and Problems. William R.
        Hersh *46*(10), 772–776.

1996   What is a Document? Rethinking the Concept in Uneasy Times. Linda
        Schamber *47*(9), 669–671.

1998   The Traditional Scholarly Journal Publishers Legitimize the Web.
        Robin Peek, Jeffrey Pomerantz, & Stephen Paling *49*(11), 983–989.
        The study identified that during 1997, traditional academic publishers
        made significant commitments to putting tables of content, abstracts,
        and the full-text of their print journals on the Web. At the same time,
        new services and organizations emerged that could ultimately compete
        with, or eliminate, the need for certain segments of the industry.

1999   An Analysis of Web Page and Web Site Constancy and Permanence.
        Wallace Koehler *50*(2), 162–180.

Libraries

1961   'Information Storage and Retrieval' and the Problems of Libraries.
        Verner W. Clapp *12*(3), 224–226.
        Meanwhile, because the current interest in information storage and
        retrieval is necessarily directed to the problems of specialist groups,
        it has scarcely touched those problems of libraries which arise from
        their efforts to serve many groups simultaneously. The new indexing
        methods are restricted in application and cannot yet replace the card
        catalog. . . .

1967   Libraries and Machines—A Review. Burton W. Adkinson, & Charles
        M. Stearns *18*(3), 121–124.
        The application of computers to library operations is discussed. . . . Three
        phases in automation of libraries are identified: the mechanization of
        conventional operations such as bibliographical control processes and
        administrative monitoring systems; the automation of search processes
        based on subject matter, and the move toward new and different kinds
        of services that computer technology may make possible. We are in
        the second phase, and snagged by the difficulty experienced by com-
        puters dealing with natural language and subjective ambiguities. To
        move forward through phase two will require a better dialog capacity
        between man and machine than presently exists. Before progressing
        into the third phase a better identification of the purposes that our
        files of information are to serve will be needed.

1971 Human Factors in the Design of an Interactive Library System. Caryl McAllister, & John M. Bell 22(2), 96–104.
ELMS (Experimental Library Management System) is an experimental system for total library management, operating on-line with an IBM 360 through IBM 2260 and 2741 terminals. . . . This paper discusses ELMS features that facilitate user interaction, and may prove useful in similar systems: techniques for tutoring the user . . . ; adaptability for the experienced user. . . .

1976 A Library Network Model. William B. Rouse 27(2), 88–99.
The performance on an interlibrary loan network is characterized. . . .

1979 Monograph Evaluation for Acquisitions in a Large Research Library. Charles B. Wenger, Christine B. Sweet, & Helen J. Stiles 30(2), 88–92.
A computerized method of assisting monograph collection development by correlating circulation with inventory statistics is presented.

1980 Directions in Library Networking. Henriette D. Avram, & Sally H. McCallum 31(6), 438–444.
Bibliographic control before and after MARC is reviewed. The capability of keying into online systems has brought an interdependence among libraries, the service centers that mediate between them, and the large utilities that process and distribute data.

1985 Cosmology and the Changing Role of Libraries: An Analogy and Reflections. Richard E. Lucier, & James F. Dooley 36(1), 44–47.
Immense cultural changes are taking place in contemporary society. . . . This article discusses these changes and the meaning and depth of their relationship to medical libraries. . . . Only by actively debating and creatively experimenting with these and other ideas will we be able to ensure our continued viability as information specialists and managers in the future.

1993 Users, User Interfaces, and Objects: Envision, a Digital Library. Edward A. Fox, Deborah Hix, Lucy T. Nowell, Dennis J. Brueni, William C. Wake, Lenwood S. Heath, & Durgesh Rao 44(8), 480–491.
All these efforts are leading not only to a usable prototype digital library but also to a set of nine principles for digital libraries. . . .

1994 Toward a User-Centered Information Service. Ruth C.T. Morris 45(1), 20–30.
This study suggests how an altered understanding of information can provide the basis for rethinking and potentially redesigning the library's mission. . . .

1996 Cost of Electronic Reference Resources and LCM: The Library Costing Model. Robert M. Hayes 47(3), 228–234.

1996 Organizational Dimensions of Effective Digital Library Use: Closed Rational and Open Natural Systems Models. Lisa Covi, & Rob Kling 47(9), 672–689.
We examine what constitutes effective DL use, how faculty members are using DLs, and how useful they find them.

1998   Speech Recognition for a Digital Video Library. Michael J. Witbrock, & Alexander G. Hauptmann *49*(7), 619–632.

Information policy and standards

1974   Cutting the NSF-OSIS Budget: Potential Disaster for Information Science and Technology. Joshua I. Smith *25*(2), 77–85.

1979   Assessing the Effect of Computer Augmentation on Staff Productivity. Harold E. Bamford, Jr. *30*(3), 136–142.

1980   A Progress Report on Information Privacy and Data Security. Gerard Salton *31*(2), 75–83.

1980   The Cultural Appraisal of Efforts to Alleviate Information Inequity. Glynn Cochrane, & Pauline Atherton *31*(4), 283–292.
This article suggests that action to alleviate information inequity should be guided by the principles of contextualism, incrementalism, motivation of information users, and more knowledge of the absorptive process that is unique to each cultural group.

1982   Information Resources as 'Goods' in the Life Cycle of Information Production. Karen B. Levitan *33*(1), 44–54.
The life cycle phases consist of generation, institutionalization, maintenance, enhancement, and distribution. An information resource stands at the midpoint of the life cycle, integrating and coordinating the various actors and activities of these phases.

1984   Fair Use versus Fair Return: Copyright Legislation and Its Consequences. Irving Louis Horowitz, & Mary E. Curtis *35*(2), 67–74.

1985   National Commission on Libraries and Information Science: A Brief Overview. Elinor M. Hashim *36*(6), 360–363.

1985   An Overview of Social Measures of Information. Michel J. Menou *36*(3), 169–177.
The rise of the information society calls for quantitative and qualitative measures of information activities at the subnational, national, and international levels. . . .

1987   Ethics and Information Science. Manfred Kochen *38*(3), 206–210.

1991   Information Technologies and Social Equity: Confronting the Revolution. Ronald D. Doctor *42*(3), 216–228.
As a society we are giving inadequate attention to ensuring that as new computer and telecommunications technologies become more pervasive, their benefits are distributed in ways that don't exacerbate existing disparities between the rich and poor.

1994   Standards for Indexing: Revising the American National Standard Guidelines Z39.4. James D. Anderson *45*(8), 628–636.

1994   The PEN Project in Santa Monica: Interactive Communication, Equality, and Political Action. Everett M. Rogers, Lori Collins-Jarvis, & Joseph Schmitz 45(6), 401–410.
An electronic communication system, especially if it is designed to accommodate open access via public terminals can allow information-disadvantaged groups (such as the homeless, and women) to exchange relevant information and engage in political action.

1996   Reading the Bones: Information Content, Value, and Ownership Issues Raised by the Native American Graves Protection and Repatriation Act. Teresa Olwick Grose 47(8), 624–631.
. . . questions have arisen concerning the nature of information as embodied in objects and the moral right to access that information.

1997   SGML and Related Standards: New Directions as the Second Decade Begins. James David Mason 48(7), 593–596.

1999   Catching a Ride on the NII: The Federal Policy Vehicles Paving the Information Highway. Patricia D. Fletcher, & Lisa K. Westerback 50(4), 299–304.

International information

1979   Deficiencies of Scientific Information Access and Output in Less Developed Countries. Michael D. Gordon 30(6), 340–342.

1984   Internationality of the Social Sciences: Implications for Information Transfer. J. Michael Brittain 35(1), 11–18.

1985   Information Transfer as Technical Assistance for Development. Marta L. Dosa 36(3), 146–152.

1985   Access to Primary and Secondary Literature from Peripheral or Less Developed Countries. Beth Krevitt Eres, & K.T. Bivins Noerr 36(3), 184–191.

1989   Sub-Saharan Africa and the Paperless Society: A Comment and a Counterpoint. Mutawakilu A. Tiamiyu 40(5), 325–328.
The issues . . . include the assessment of the relevance of the notion of a paperless library to Africa, the expected role of library and information services in narrowing the gap between the information rich and poor, and the possible inappropriateness of information technology and/or advice from developed countries.

1990   Databases on Optical Discs and Their Potential in Developing Countries. S. Nazim Ali 41(4), 238–244.
This in-house system will avoid cost associated with telecommunication tariffs and provide unlimited access.

1998   The Transition from 'National' to 'Transnational' Model and Related Measures of Countries' Performance. Michel Zitt, François Perrot, and Rémi Barré 49(1), 30–42.

The transition from a national science model in which the national language is used for publications and other communications, to a transnational model in which a single international language (English) is used and the market is dominated by Anglo-Saxon publishers, has continued in recent decades.

Economics of information

1982   Value-Added Processes in the Information Life Cycle. Robert S. Taylor *33*(5), 341–346.
Energy, time, and money must be invested to change useless data to productive knowledge, a value-added process.

1985   Experiments and Analysis of Information Use and Value In a Decision-Making Context. M.C. Yovits, & C.R. Foulk *36*(2), 63–81.

1989   Economics of Intellectual Property Rights in the International Arena. Yale M. Braunstein *40*(1), 12–16.

1989   The Value of Information: Approaches in Economics, Accounting, and Management Science. Aatto J. Repo *40*(2), 68–85.
It is concluded that economic approaches based on 'information theory' have not achieved significant practical results in a general sense, but 'classical' economic approaches can and should be used. . . .

1992   On the Market Value of Information Commodities. I. The Nature of Information and Information Commodities. Abbe Mowshowitz *43*(3), 225–232.
Information is used by marketplace actors to make decisions or to control processes. Thus, we define information as the ability of a goal-seeking system to decide or control.

1996   Simulation Model for Journal Subscription by Libraries. Richard E. Quandt *47*(8), 610–617.
Journal costs are thought to consist of 'first-copy costs' and 'additional copy costs'; when costs rise, subscription prices are increased by publishers and some libraries, being faced with fixed budgets, cancel some subscriptions. Publishers then find that the price charged is not correct and increase subscription prices again.

Curriculum

1971   Curricula in Information Science; Analysis and Development. Jack Belzer, Akkanad Isaac, Eugene Finkelstein, & James Williams *22*(3), 193–223.
The 7 [core master's] courses are: Introduction to Information Science, Systems Theory and Applications, Mathematical Methods in Information Science, Computer Organization and Programming Systems,

Abstracting/Indexing/Cataloging, Information and Communication Theory, and Research Methods.

1980   Education and Training for Computer-Based Reference Services: Review of Training Efforts to Date. Charles P. Bourne, & Jo Robinson *31*(1), 25–35.

1996   Impact of Distance Independent Education. Howard Besser, & Maria Bonn *47*(11), 880–883.

History

1989   Innovation, Pragmaticism, and Technological Continuity: Vannevar Bush's Memex. James M. Nyce, & Paul Kahn *40*(3), 214–220.
This material offers new perspective on how and why Bush published his 1945 essay 'As We May Think' in the *Atlantic Monthly*.

1992   The Other Memex: The Tangled Career of Vannevar Bush's Information Machine, The Rapid Selector. Colin Burke *43*(10), 648–657.

1997   The Origins of Information Science and the International Institute of Bibliography/International Federation for Information and Documentation (FID). W. Boyd Rayward *48*(4), 289–300.

1997   Bibliography of the History of Information Science in North America, 1900–1995. Robert V. Williams, Laird Whitmire, & Colleen Bradley *48*(4), 373–379.

Perspectives and special topic issues

Beginning in 1980, the *Journal* began publishing "Perspectives" issues, first under the editorship of **Susan Crawford** (four issues) and then, in the 33 subsequent issues to date, **Lois F. Lunin**. In each case the senior editor would work with a co-editor with a specialty in the designated topic of the Perspectives issue. Below, to reduce redundancy, only the co-editors are listed.

In 1994, the *Journal* began publishing a different kind of series—special topic issues. These would be under the Guest Editorship of a subject expert in the topic of the issue, and papers would be refereed in the same manner as regular articles in the journal.

Lists are provided below of both of these special series, arranged by year and issue number (in parentheses), up through the fifth issue of 1999, when this issue went to press. As the special section occupies most of the issue in which it is published, page numbers are not provided.

Perspectives issues

1980   (3) Online Systems in Science and Technology. Alan M. Rees

1980   (6) Library Networks and Resource Sharing. Barbara Markuson

1981   (2) Computer and Communications Technology. Manfred Kochen

1981   (5) Cognition: Human Information Processing. Belver C. Griffith

1982   (3) Secondary Information Services: Development and Future. Marianne Cooper

1982   (6) Systems Methodology and Information Research. Una Mansfield

1983   (3) Office Automation: Impact on Organization, User, Planner. Elliot Cole

1983   (6) Videodisc and Optical Disk: Technology, Research, and Applications. Judith Paris

1984   (3) Information Science: Retrospect and Prospect. Robert M. Hayes

1984   (5) Artificial Intelligence: Concepts, Techniques, Applications, Promise. Linda C. Smith

1985   (3) International Information Issues. Beth Krevitt Eres

1985   (6) The National Commission on Libraries and Information Science. Toni Carbo Bearman

1986   (4) Online Searching. Donald T. Hawkins

1986   (6) Telecommunications: Principles, Developments, Prospects. Larry L. Learn

1987   (1) The Federal Government and Health Information: Patterns, Impact, Expectations. Joseph F. Caponio

1988   (1) CD-ROM for Information Storage and Retrieval. Peter B. Schipma

1988   (2) Integrated Academic Information Management Systems (IAIMS). Marion J. Ball

1988   (5) Education of the Information Professional: New Dimensions, New Directions. Marianne Cooper

1989   (3) Hypertext. Roy Rada

1989   (5) Information Science and Health Informatics Education. Marion J. Ball

1990   (3) Evaluation of Scientific Information: Peer Review and the Impact of New Information Technology. Susan Crawford, & Charles T. Meadow

1990   (6) Author Co-Citation Analysis. Howard D. White

1991   (2) Integrated Information Centers within Academic Environments. George D'Elia

1991   (8) Imaging: Advanced Applications. Clifford A. Lynch

1992   (2) Human–Computer Interface. Donna Harman

1992   (8) Information Technology Standards. Michael B. Spring

1993   (4) **Knowledge Utilization.** Susan Salasin, & William Paisley

1993   (8) **Digital Libraries.** Edward A. Fox

1994   (5) **Redesign/Reengineering of an Information Services Division in a Major Health Sciences Institution.** Marion J. Ball, & Judith V. Douglas

1994   (8) **Indexing.** Raya Fidel

1994 (10) **Electronic Publishing.** Robin P. Peek

1995   (8) **The Chemist's Workstation.** Loren D. Mendelsohn

1995 (10) **Medical Informatics: Information Technology in Health Care.** William R. Hersh

1996   (3) **Costs and Pricing of Library and Information Services in Transition.** Eileen G. Abels

1996 (11) **Distance Independent Education.** Howard Besser, & Stacey Donahue

1997   (5) **Implementation and Evaluation of an Integrated Information Center in an Academic Environment.** George D'Elia

1998 (11) **Internet Issues.** Carol Tenopir

Special topic issues

1994   (3) **Relevance Research.** Thomas Froehlich

1994   (6) **Information Resources and Democracy.** Leah A. Lievrouw

1994   (9) **Spatial Information.** Myke Gluck

1996   (1) **Evaluation of Information Retrieval Systems.** Jean M. Tague-Sutcliffe

1996   (4) **Full-Text Retrieval.** MaryEllen C. Sievert

1996   (7) **Current Research in Online Public Access Systems.** Micheline Beaulieu, & Christine L. Borgman

1996   (9) **Electronic Publishing.** Robin Peek

1997 (4, 9) **History of Documentation and Information Science: Parts I, II.** Michael Buckland, & Trudi Bellardo Hahn

1997   (7) **Structured Information/Standards for Document Architectures.** Elisabeth Logan, & Marvin Pollard

1997 (11) **Current Research in Human–Computer Interaction.** Andrew Dillon

1998   (1) **Science and Technology Indicators.** Anthony F.J. Van Raan

1998   (3) **Management of Imprecision and Uncertainty.** Gloria Bordogna, & Gabriella Pasi

1998   (5) **Knowledge Discovery and Data Mining.** Vijay V. Raghavan, Jitender S. Deogun, & Hayri Sever

1998   (7) **Artificial Intelligence Techniques for Emerging Information Systems Applications.** Hsinchun Chen

1998   (9) **User-Centered Cooperative Systems.** Michael D. McNeese

1998 (12) **Social Informatics in Information Science.** Rob Kling, Howard Rosenbaum, & Carol Hert

1999 (1) **Youth Issues in Information Science.** Mary K. Chelton, & Nancy P. Thomas

1999 (4) **The National Information Infrastructure.** Patricia Diamond Fletcher, & John Carlo Bertot

4

Introduction to the *Encyclopedia of Library and Information Sciences, 3ʳᵈ* Edition

How to use this encyclopedia

Entries are arranged alphabetically in this encyclopedia (see end papers for alphabetical list). There is also a Topical Table of Contents. By scanning this list, the reader gets a sense at a glance of how the subjects are grouped and what related entry titles are to be found clustered together in the categories.

When seeking a topic that is more specific or detailed than appears in the titles of entries, or when looking for a proper noun, such as the name of a person, company, organization, standard, etc., look up the name or phrase in the index at the back of each volume. When searching the online form of the encyclopedia, search by title, author, or subject term(s).

In sum, relevant entries can be found by

1. Entry title (alphabetical arrangement of entries in the encyclopedia or listing in the end papers),

2. Subject category (Topical Table of Contents), or

3. Specific name or topic (index at the end of each volume).

If the first name or topic searched is not found, try several more variations—either different words or a different order of words. Most topics are described in several ways in the literature of a discipline, and the first term or phrase that comes to mind may not be the one used here.

First published as Bates, M. J. & Maack, M.N. (2010). Introduction to the *Encyclopedia of Library and Information Sciences, Third Edition*. In M. J. Bates & M. N. Maack (Eds.), *Encyclopedia of Library and Information Sciences* (3ʳᵈ ed., Vol. 1, pp. xiii–xx). Boca Raton, FL: CRC Press.

Scope of the encyclopedia

The title of the third edition of the *Encyclopedia of Library and Information Sciences* ends with the letter "s" because the encyclopedia is not limited to librarianship or information science. In fact, this work is designed to cover a spectrum of related information disciplines, including:

Archival Science

Bibliography

Document and Genre Theory

Informatics

Information Systems

Knowledge Management

Library and Information Science

Museum Studies

Records Management

Social Studies of Information

What unifies all of these disciplines is their interest in recorded information and culturally meaningful artifacts and specimens. If the universe of study in physics is the physical world, and in biology the study of living things, then the universe of study of all the information sciences and the disciplines of the cultural record is, in fact, that entire body of materials representing what cultures know and value. The information disciplines collect, organize, store, preserve, retrieve, transfer, display, and make available the cultural record in all its manifestations. These activities are essential for maintenance of and access to all kinds of cultural records, whether they are produced as a result of business, government, education, creative endeavors, or daily life.

To properly address the demand for effective management of cultural resources, information professionals must master the relevant resources and information technologies, and understand how human beings relate to and use the information. They need a broader perspective as well, understanding the political, social, and cultural ramifications of the choices made in collecting and managing information. The effective management of information, while often invisible to the broader society, in fact determines much of what can be known and is available to societies—both in terms of their cultural heritage and their contemporary needs for access to knowledge.

Identifying the subject matter of the information disciplines is a complex matter because most of them have both a professional and a theoretical component. Topics in research and theory are important, as are all the manifestations of professional practice—institutions, professional associations, and areas of professional practice. Three institutions of professional practice—libraries, archives, and museums—are central in the coverage of *ELIS-3*.

But theory and practice do not cover the entirety of these disciplines either. All of them exist in a social, cultural, political, and geographical environment. They have arisen out of an historical context that still influences their character today. To the extent possible, it is desirable to reflect these various influences in the coverage of the encyclopedia as well. Thus, there are entries on the cultural institutions of various nations of the world, entries on the history of the information disciplines, and entries on the political, social, and legal context of disciplinary interests.

One of our original objectives was to provide a series of country profiles that would include a discussion of libraries, archives, and museums and would also describe the education of professionals working in these three types of institutions. We had initially hoped to include all member countries of UNESCO, but this goal proved to be far too ambitious to carry out within a reasonable time frame. The lack of any budget for translation was also an inhibiting factor; as a result we were limited to those countries where we could identify authors who were able to produce a comprehensive and sophisticated overview entry in English. In many cases we could not find appropriate authors and in other cases we did have a contract with an author or a team of authors who had to withdraw from the project at a time when it was much too late to find a replacement (e.g. Argentina, Egypt, Italy, Lebanon, Portugal, New Zealand, Iran, Iraq, South Africa and Russia). Despite these disappointments, we have been able to produce cultural profiles of over 30 countries. Although Latin America and the Near East are underrepresented, we are pleased to have good coverage of Europe and a strong representation from all regions in Asia. We also have profiles from both Anglophone and Francophone Africa, and from a number of countries that were in the former Soviet bloc. In some cases, especially with countries in Africa, Eastern Europe and the former Soviet Republics, these essays may be the only place where a unified discussion of the country's cultural infrastructure may be found.

The many academic and professional dimensions addressed by the entries in the encyclopedia are evident in the list below of the main and sub-topic headings in the Topical Table of Contents. (See Topical Table of Contents for full list of entries arranged by subject.)

Information Disciplines and Professions
General Disciplines
Disciplinary Specialties
Cognate Disciplines
Career Options and Education

Concepts, Theories, Ideas
Key Concepts
Theories, Models, and Ideas

Research Areas
Cross-Disciplinary Specialties
Research Specialties

Institutions
Generic Institutions
Institution Types
Named Institutions
Ancillary Cultural Institutions
Collections

Systems and Networks
Information Systems
Network and Technology Elements

Literatures, Genres, and Documents
Literatures
Generic Resources
Named Resources
    Information organization tools
    Specific standards
    Projects
    Laws

Professional Services and Activities
Appraisal and Acquisition of Resources
Institutional Management and Finance
Organization and Description of Resources
Resource Management
User Services

People Using Cultural Resources
  General
  Population Groups
  Subject Areas

Organizations

National Cultural Institutions and Resources

History

A final bonus to be found in the encyclopedia are 37 entries that are being called **"ELIS Classics."** These are entries of historical or theoretical importance, usually by major individuals in library and information science, which appeared in earlier editions of the encyclopedia. Each "ELIS Classic" is so marked, and an Editor's Note explains why the entry was selected for inclusion.

The entries in this edition are almost entirely new. There are 565 entries, of which 37 are **ELIS Classics**, another 30 or so have been carried forward as is from an earlier edition, and about 80 are entries from earlier editions that have been updated to the present by their authors. All the remaining entries, more than 400, or about three-quarters of all entries, are totally new.

Encyclopedia authors

Many of the authors writing for the encyclopedia are major researchers or practitioners in the disciplines in which they are writing. They are highly esteemed in their fields, and many have been recipients of the highest honors. Noted scholars are well represented, and theories and models are often described by their originators. In addition, many *ELIS-3* authors are former presidents of associations, including the International Federation of Library Associations and Institutions (IFLA), the International Council on Archives (ICA), the American Society for Information Science and Technology (ASIS&T) and the American Association of Library and Information Science Education (ALISE); among our contributors there are also a number of former presidents of the American Library Association and its divisions. In addition, there are many contributors who are current or former directors of major institutions. Even though we were not able to secure entries from all those whom we asked, many thoughtful and insightful experts in the several disciplines willingly agreed to contribute. We are very proud of the array of authors within the encyclopedia.

What is not included in the encyclopedia

In general, the scope of the encyclopedia's entries is rather broad—whole research areas, sub-disciplines, and classes of resources or institutions are described. Thus, many smaller topics are addressed only within the context of entries on larger topics. Biographies of individuals have also been excluded, though key persons are mentioned throughout entries on other topics. The index at the back of the book can be very helpful in locating these specific references within entries on broader topics.

Because this broad, general approach was taken, specific institutions are profiled only where they are highly influential or constitute major examples. Many important institutions, such as national libraries and national archives and major museums, are given brief profiles in the country entries. Genres are very important to the information disciplines, and are covered in several general entries, as with "Internet Genres." A few highly important specific genres are addressed, such as "Archival Finding Aids," but most individual genres are not profiled.

Finally, the last topic that is not covered in *ELIS-3* is the subject matter of any entry that an author was unable to complete for the encyclopedia. Authors have grappled with the personal joys of newborn children, and the personal sadness of deaths of family members. Two authors themselves died during the time of preparation of this edition, and others were unable to complete their manuscript because of illness. Whenever possible, we have replaced entries when authors defaulted, but often there was not enough time remaining to allow that. So readers will no doubt be surprised at a random scattering of topics that do not appear in the encyclopedia, and should, such as Digital Libraries, Economics of Information, Scientific and Scholarly Communication, Information Security, Website Design and Management, Library Public Services, Digital Reference, the Internet, History of Archival Institutions, Data Archives, Social Computing, Geoinformatics, Legal Informatics, Knowledge Organization Systems, Science Museums, History Museums, and Curating the Arts. This problem of missing topics was also acknowledged by our predecessor Allen Kent, editor of the First Edition of *ELIS*. Kent stated in 1973: "I have prepared this presentation to make sure the lessons of Diderot-d'Alembert are recalled in terms of encyclopedia-making as an exercise in the art of the possible" (Kent, 1973, pp. 602–604).

Background and development of the encyclopedia

The first edition of *ELIS*, under the Editorship principally of Allen Kent and Harold Lancour, was published between 1968–1982. The 33 volumes of Edition 1 were published in alphabetical sequence during those years. After the "Z" volume appeared in 1982, a number of supplements were published at roughly the rate of two per year, up to and including volume 73, which appeared in 2003 (Kent & Lancour, 2003). Miriam Drake was appointed Editor for the second edition, which appeared in 2003, both online and in paper. The second edition came out at one time in four large-format volumes, with a supplement in 2005 (Drake, 2005). Kent and Lancour covered a wide range of librarianship, information science, and some computer science topics. Drake, an academic library director, emphasized academic libraries, and the *ELIS-2* volumes contained many profiles of major academic libraries and professional library associations.

Our objective with the Third Edition of *ELIS* is to reflect the growing convergence among the several disciplines that concern themselves with information and the cultural record. Historically, library and information science, archival science, museum studies, and the several other information disciplines listed earlier have tended to develop fairly independently, and to have relatively little interaction. In recent years, as all kinds of recorded information has migrated into digital form, the problems and challenges facing the several professions have converged. At the theoretical level, the growing societal attention to information and cultural heritage in recent years has led to a growing recognition that the several disciplines can both learn from and contribute to each other in theory development and research as well. A more detailed discussion of this convergence is available in Bates (2007).

We saw the audience for the encyclopedia as principally (though not entirely) consisting of, 1) the educated lay person interested in one or more of its topics, 2) students learning about a topic, and 3) professionals and researchers in the several fields who want to learn about something new, or to be refreshed on a familiar topic.

It was quite a challenge to develop an encyclopedia addressing some ten disciplines; even one discipline raises many questions of what should and should not be in an encyclopedia about that discipline. We first assembled a large group of superb individuals to constitute our Editorial Advisory Board. (See listing in the front matter.) We sought out leaders and experts from all the areas we are addressing in the encyclopedia, and their suggestions for topics were immensely valuable.

Editor Bates also reviewed all of the entries in the many prior volumes from the First and Second Editions. Prior topics addressed were listed and grouped, to get a sense of what issues might arise in selecting topics. This review also revealed a number of "gems," entries we later carried forward as is, either as ELIS Classics or otherwise.

Early on, an important question arose: Would we identify each discipline and select a set of analogous topics for each? Both the Editors and the Editorial Advisory Board agreed that we did not want parallel "silo" groupings of entries on the several disciplines. The purpose of this edition is to address these related disciplines in a way that both demonstrates the unities across the fields, as well as recognizes their uniquely distinguishing characteristics. Thus, the choice of topics reflects this persistent duality; some authors address a topic across the disciplines, other authors specialize in what they know best. In many cases, but not all, what has been learned in one field can be applied in others. Fund-raising techniques used in one non-profit area can usually be utilized in another non-profit area. On the other hand, only librarians are likely to need information about serials vendors, and only museums professionals have to address trafficking in art objects. Throughout the topic selection process, countless judgments were made in an effort to produce a useful encyclopedia for all these information disciplines, as well as an encyclopedia that, through the convergence of topics, educates readers about neighboring fields. Prospective authors also helped in defining topic coverage in their areas of expertise.

Topic identification, selection, and grouping was a gradual and iterative process that, for the Editor, felt like a snake shedding skins. Dozens of spreadsheets of topic lists were created, each one moving a bit closer to a final set of entries and a final grouping of entries for maximum clarity and usefulness. The reader will, of course, be the final judge of usefulness and ease of use of the encyclopedia.

A word on categorizing and classifying the topics

We are well aware that in library and information science, unlike other disciplines, the organizing and formatting of information is of great importance; it is at the heart of much that we do in the field. So this introduction cannot be complete unless we address these questions with respect to *ELIS-3*.

A rigorous classification of topics would require the most consistent language possible, and an array of entries both mutually exclusive and jointly exhaustive. However, the realities of philosophical and practical issues in the several disciplines meant that we could not always achieve such purity. The coverage of entries, and their corresponding titles, represents the sum total of a negotiation process between editors and authors, and between editors and the current debates in the fields around a given topic.

In particular, the most contentious area we encountered was archives and records management. U.S. traditions separate these two fields, and their practitioners often have different backgrounds and philosophies of work. U.S. records managers tend to see a disciplinary merger as infeasible and undesirable. However, leaders of a dynamic movement in Australia have argued for the combination of these two fields into one. Under the banner of the "Records Continuum Model," these authors view the separation of the fields as artificial and out of date.

How, then, to reconcile these conflicting views? In the end, authors were chosen from around the world to expound on various aspects of these two disciplines, so that the several positions would be present in the encyclopedia. The authors were encouraged both to recognize the other positions and to present their own models of the professions. Readers, therefore, will sometimes experience a somewhat internally conflicting body of exposition on the several topics.

These differences are expressed even in the titles of the entries. Karen Anderson, the author asked to write on archival education, chose to title her entry, "Careers and Education in Archives and Records Management." The philosophy she subscribes to would combine these two areas into one. On the other hand, the author of the entry on records management education, Carol Choksy, chose to title her entry "Careers and Education in Records and Information Management." The American field separates archives from "records and information management," and uses the latter name for the disciplinary area. Forcing the content of these two titles to be more "mutually exclusive"—that is, for the words "records management" to appear in only one of the two titles—in the time-honored tradition of classification system design, would, in fact, warp the actual coverage of the entries, as intended by the authors, and as reflecting the realities of the field today.

For those familiar with the organizational principles developed by the great library classificationist, S.R. Ranganathan, the final subject topic listing above can be seen to reflect his fundamental set of facets: Personality, Matter, Energy, Space, and Time. The Disciplines, Concepts, and Research Areas represent the Personalities of the disciplines; the Institutions, Systems, Networks, and Literatures represent the Matter of the disciplines; and the Professional Services and Activities, the People Using Cultural Resources, and the Organizations represent the Energy of the disciplines. The National profiles represent geography, or Space, and, finally, History represents Time. Overall, the Topical Contents List presents the broad topic areas being listed from first to last, in the order of Ranganathan's P-M-E-S-T.

Of course, in some regards, every entry addresses most or all of Ranganathan's facets. He expected people to combine several facets to describe any individual book. However, the Topical Table of Contents groups the entries by the aspects that they each emphasize. So an entry on Information Retrieval Systems, though it expresses some of the general multi-faceted approach to the subject matter taken by information retrieval experts in their research, nonetheless emphasizes the *systems* themselves, as is reflected in the entry title. This entry is therefore grouped with other entries on various types of information systems. Seeing all these systems entries together in the Topical Contents List enables the reader to see at a glance the wide variety of kinds of information system that have been created. An information system is a kind of matter in the world of the information disciplines, and is therefore grouped with the other "matter"-based topical areas, such as institutions, networks, and literatures.

Quality controls

Authors were selected and invited to write who knew the topic well. (In fact, researching authors and inviting them to write—and replacing those who declined—took many months of full-time work for the Editors, especially for Bates.) Authors were urged to double check the accuracy of their entries to meet the societal expectation of accuracy for an encyclopedia. Though they were welcome to expound on the theories and approaches that they preferred, authors were also required to represent common points of view on their topic in their entries, not only the perspective that the author favored.

All new entries were reviewed by an outside expert reviewer, as were updated entries that were extensively revised. Referees were specifically asked to note any errors or mistakes that they were aware of in the entries they reviewed. Entries were checked for plagiarism with searches on sentences in Google. Spot checks for accuracy were done on facts within entries and references at the ends of entries by our research assistants, who were partially funded by small grants from the UCLA Council on Research. Most entries were read several times—by the editors, fact checkers, and the reviewers. Nonetheless, checking the accuracy of every fact was infeasible.

The future

It can be said that the information disciplines are in an analogous position to that of the social sciences, but displaced a century later. The social sciences, after being seen as marginal and insubstantial in the 19th century, finally came into their own in the 20th century as mature, influential disciplines, holding their rightful place in the full spectrum of academic and professional expertise. We believe that the information disciplines, which, during the 20th century, were often at the margins of the university and of society, are coming into their own at last in the 21st century. In societies feeling overwhelmed with information, and at sea among dozens of information technologies, the information disciplines should at last take their rightful place as serious, substantial and important fields among the many forms of endeavor in society. The Editors hope that this encyclopedia may contribute to that revolution.

ACKNOWLEDGEMENTS

The breadth and depth of the final complete encyclopedia would not have been possible without Associate Editor Mary Niles Maack's contributions. Her international interests, knowledge of the library field, experience as a historian, her skilled editing, and her many suggestions complementary to mine have made this a much richer and better encyclopedia than would have been otherwise possible. I also wish to thank the Editorial Advisory Board for their advice, suggestions of topics and authors, and their support. Of special help, above and beyond the call of duty, were Howard White, David Bearman, Martha Morris, Barbara Nye, Virginia Walter, and Michele Cloonan.

I also wish to thank the Taylor & Francis Editors, Claire Miller and Susan Lee, who supported and humored the Editors and authors throughout years of work on the encyclopedia. I also wish to express my deep appreciation to Stephen C. Maack, whose skilled efforts were immeasurably valuable during a crucial period of work on the encyclopedia. Also vital were the efforts of our assistants Barbara Birenbaum and Faye Baker as plagiarism and fact-checkers, and Carol Perruso-Brown as text editor. Finally, I thank the hundreds of authors who wrote entries, and the hundreds more reviewers who refereed the entries. Without them, there would be no encyclopedia.

—*Marcia J. Bates, Editor-in-Chief*

The breadth, scope, and intellectual framework of the encyclopedia is due to the vision, dedication and perseverance of Marcia Bates, who has devoted over three years of her life to this enormous project. In addition to establishing the table of contents, she has spent countless hours identifying authors, corresponding with them, and editing their work. My editorial responsibilities have been mostly confined to the historical entries, the entries on libraries and librarianship, the entries on associations, and the country profiles. The latter set of entries were in many ways the most demanding and rewarding. I am especially grateful to two of our board members, T.D. Wilson and Elena Macevičiūtė, who were very helpful in identifying authors from Central and Eastern Europe. My sincere thanks also go to Peter Lor, the former Secretary General of the International Federation of Library Associations and Institutions (IFLA), who generously helped to locate potential authors and reviewers.

I am also very grateful for the excellent work of my UCLA graduate assistants Stasa Milocevic and Kim Anderson. Stasa not only checked facts but also provided very insightful review comments on many manuscripts, and Kim aided in the preliminary editing of several entries. Finally my heartfelt appreciation goes to my husband Stephen Maack. At a time when my teaching overload at UCLA would have made it impossible for me to continue as Associate Editor, he put aside his work for his firm, Reap Change Consultants, and devoted many hours to securing association entries, following up with overdue manuscripts, updating spreadsheets, and doing extensive editing on a number of key entries. I simply could not have continued in this endeavor without his love, support, and hard work.

—*Mary Niles Maack, Associate Editor*

REFERENCES

Bates, M.J. (2007). Defining the information disciplines in encyclopedia development. *Information Research, 12*(4), paper colis29. Retrieved from http://InformationR. net/ir/12-4/colis/colis29.html.

Drake, M.A. (Ed.). (2005). *Encyclopedia of Library and Information Science,* (2nd ed., 4 Vols., 1 Suppl). New York: Marcel Dekker.

Kent, A. (1973). Is ELIS worthy of the name? Yes! The editor replies. *Wilson Library Bulletin 47,* 602–604.

Kent, A., & Lancour, H., et al. (Eds.). (2003). *Encyclopedia of library and information science,* (1st ed., 73 Vols.). New York: Marcel Dekker.

5

Topical table of contents, *Encyclopedia of Library and Information Sciences,* 3ʳᵈ Edition

WHAT IT DOES FOR YOU

While the main arrangement of entries in the whole encyclopedia is alphabetical by entry title, this "Topical Table of Contents" groups together similar or related encyclopedia entries in categories, in order to make it easier to find clusters of entries of interest on a subject. The labeled sections contain the full topical contents list, providing titles and authors of each entry.

After the information for each entry, *"See also"* references point to related topics that might also be of interest—some to whole categories (*"See also* Topical Category 8: People Using Cultural Resources") and some to individual entries (*"See also* Special Libraries . . ."). *"See"* references lead from a likely, but unused, title for a topic to the title that is actually used for the topic ("Filtering Software *See* Internet Filtering Software . . .").

First published as Bates, M. J. (2010). Topical table of contents. In M. J. Bates & M. N. Maack (Eds.), *Encyclopedia of library and information sciences* (3ʳᵈ ed., Vol. 1, pp. xxi–xlv). Boca Raton, FL: CRC Press.

Topics

1: Information Disciplines and Professions

See also Topical Category 7: *Professional Services and Activities*

1: General Disciplines

See also Topical Category 1.4: *Career Options and Education*

Archival Science / *Elizabeth Shepherd*

Information Science / *Tefko Saracevic*

   See also Bibliometric Overview of Information Science / *Howard D. White*

Information Systems / *E. Burton Swanson*

Knowledge Management / *Kimiz Dalkir*

Library and Information Science / *Leigh S. Estabrook*

Museum Studies / *Marjorie Schwarzer*

Publishing Studies / *Robert E. Baensch*

2: Disciplinary Specialties

1: Archival Science

Diplomatics / *Luciana Duranti*

   See also Provenance of Archival Materials / *Shelley Sweeney*

Film and Broadcast Archives / *Karen F. Gracy and Karen E. King*

   See also Film Archiving: History / *Steve Ricci*

Sound and Audio Archives / *Mark Roosa*

Visual and Performing Arts Archives / *Francesca Marini*

   See also Arts Literatures and Their Users / *Lisl Zach*

2: Informatics

Bioinformatics / *Carrie L. Iwema and Ansuman Chattopadhyay*

   See also Biological Information and Its Users / *Kalpana Shankar*

Biomedical Informatics / *Nancy K. Roderer, Catherine K. Craven, and Harold P. Lehmann*

   See also Health Sciences Librarianship / *Judy Consales and Elaine R. Martin*

Business Informatics / *Markus Helfert*

   See also Business Information and Its Users / *Eileen G. Abels*

Chemoinformatics / *Peter Willett*

   See also Chemistry Literature and Its Users / *Judith N. Currano*

Community Informatics / *Kate Williams and Joan C. Durrance*

   See also Everyday Life Information Seeking / *Reijo Savolainen*

Digital Humanities / *Julia Flanders and Elli Mylonas*

   See also Humanities Literatures and Their Users / *Stephen E. Wiberley Jr.*

   See also Text Encoding Initiative (TEI) / *Edward Vanhoutte and Ron Van den Branden*

1: Information Disciplines and Professions, (cont'd.)

3: Cognate Disciplines, (cont'd.)

Social Epistemology / *Steve Fuller*

Technical Writing / *Michael J. Salvo*

Telecommunications / *Richard A. Thompson*

4: Career Options and Education

Accreditation of Library and Information Science Education / *Karen L. O'Brien*

Careers and Education in Archives and Records Management / *Karen Anderson*

Careers and Education in Information Systems / *Paul Gray and Lorne Olfman*

Careers and Education in Library and Information Science / *Jana Varlejs*

Careers and Education in Museum Work / *Marjorie Schwarzer*

Careers and Education in Records and Information Management / *Carol E. B. Choksy*

iSchools / *Ronald L. Larsen*

2: Concepts, Theories, Ideas

1: Key Concepts

Access in a Digital Age / *Betty J. Turock and Gustav W. Friedrich*

Accessibility / *Lori Bell*

    See also Blind and Physically Handicapped: Library Services / *Jane Caulton and Stephen Prine*

    See also Deaf and Hearing Impaired: Communication in Service Contexts [*ELIS Classic*] / *Warren R. Goldmann*

Bibliographic Control / *Robert L. Maxwell*

    See also Cataloging / *Arlene G. Taylor and Daniel N. Joudrey*

    See also Descriptive Cataloging Principles / *Elena Escolano Rodriguez*

    See also Anglo-American Cataloging Rules (AACR) / *J. H. Bowman*

Credibility and Cognitive Authority of Information / *Soo Young Rieh*

    See also Cultural Memory / *Robert DeHart*

Custody and Chain of Custody / *Bernadette G. Callery*

    See also Provenance of Archival Materials / *Shelley Sweeney*

Data and Data Quality / *Thomas C. Redman, Christopher Fox, and Anany Levitin*

    See also Information Systems Failure / *Chris Sauer and Gordon B. Davis*

Decision Making *See* Information Use for Decision Making / *Edward T. Cokely, Lael J. Schooler, and Gerd Gigerenzer*

Half-life of Literatures *See* Information Obsolescence / *David Nichols, Ian Rowlands, Paul Huntington, and Hamid R. Jamali*

Heritage *See* Intangible Heritage / *John Feather*

    See also Cultural Memory / *Robert DeHart*

Information / *Marcia J. Bates*

2: Theories, Models, and Ideas

2: Concepts, Theories, Ideas, (cont'd.)

2: Theories, Models, and Ideas, (cont'd.)

See also Citation Analysis / *Howard D. White*

Information Overload / *Tonyia J. Tidline*

Information Practice / *Crystal Fulton and Jean Henefer*

   See also Information Behavior / *Marcia J. Bates*

   See also Information Needs / *Charles M. Naumer and Karen E. Fisher*

Information Search Process (ISP) Model / *Carol Collier Kuhlthau*

   See also Information Searching and Search Models / *Iris Xie*

Information Society / *Frank Webster*

Information Systems Failure / *Chris Sauer and Gordon B. Davis*

   See also Data and Data Quality / *Thomas C. Redman, Christopher Fox, and Anany Levitin*

   See also Software Reliability Management / *Yashwant K. Malaiya*

Information Technology Literacy / *James W. Marcum and Denise I. O'Shea*

   See also Digital Divide / *Mark Warschauer*

Information Theory / *Paul B. Kantor*

Informetric Laws / *Ronald Rousseau*

   See also Information Scattering / *Suresh K. Bhavnani and Concepción S. Wilson*

Knowledge Management Models / *Kimiz Dalkir*

Library Anxiety / *Diane Mizrachi*

   See also Information Literacy Instruction / *Esther S. Grassian and Joan R. Kaplowitz*

Organization Theories / *Evelyn Daniel*

Qualitative Research Methods in Library and Information Science [*ELIS Classic*] / *Brett Sutton*

Records Continuum Model / *Sue McKemmish, Franklyn Herbert Upward, and Barbara Reed*

Sense-Making / *Brenda Dervin and Charles M. Naumer*

Subject Cataloging Principles and Systems / *Theodora L. Hodges and Lois Mai Chan*

   See also Topical Category: 6.3: Literatures, Genres, and Documents: Named Resources

   See also Classification Theory / *Clare Beghtol*

3: Research Areas

1: Cross-Disciplinary Specialties

Information Arts / *Christiane Paul and Jack Toolin*

Linguistics and the Information Sciences / *John C. Paolillo*

   See also Multilingual Information Access / *Douglas W. Oard*

   See also Natural Language Processing for Information Retrieval / *Elizabeth D. Liddy*

Philosophy and the Information Sciences / *Jonathan Furner*

See also Epistemology / *Paul K. Moser*

See also Knowledge / *Paul K. Moser and Arnold vander Nat*

Sociology of the Information Disciplines / *Michael F. Winter*

2: Research Specialties

1: Bibliometrics, scientometrics

Citation Analysis / *Howard D. White*

See also Information Obsolescence / *David Nicholas, Ian Rowlands, Paul Huntington, and Hamid R. Jamali*

Citer Motivations [*ELIS Classic*] / *Terrence Brooks*

Information Scattering / *Suresh K. Bhavnani and Concepción S. Wilson*

See also Informetric Laws / *Ronald Rousseau*

Webometrics / *Mike Thelwall*

See also Informetrics / *Judit Bar-Ilan*

2: Information behavior and searching

Information Searching and Search Models / *Iris Xie*

See also Information Search Process (ISP) Model / *Carol Collier Kuhlthau*

See also Wayfinding and Signage / *Dennis O'Brien*

Information Use for Decision Making / *Edward T. Cokely, Lael J. Schooler, and Gerd Gigerenzer*

See also Decision Sciences / *Sven Axsäter and Johan Marklund*

Knowledge Sharing Mechanisms / *Kah Hin Chai*

Learning and Information Seeking / *Louise Limberg and Mikael Alexandersson*

See also Independent (Free-Choice) Learning / *Lynn D. Dierking*

Online Catalog Subject Searching / *Danny C.C. Poo and Christopher S. G. Khoo*

See also Online Public Access Catalogs (OPACs) / *Kevin Butterfield*

Personal Information Management (PIM) / *William Jones*

Reading and Reading Acquisition / *Brian Byrne*

Reading Disorders / *H. L. Swanson and Olga Jerman*

Wayfinding and Signage / *Dennis O'Brien*

3: Information Organization and Description

See also Topical Category 7.3: Professional Services and Activities: Organization and Description of Resources

Indexing: History and Theory / *Bella Haas Weinberg*

Latent Semantic Indexing / *Dian I. Martin and Michael W. Berry*

Metadata and Digital Information / *Jane Greenberg*

Moving Image Indexing / *James M. Turner*

Ontologies and Their Definition / *Jos de Bruijn and Dieter Fensel*

See also Taxonomy / *Andrew Grove*

Semantic Interoperability / *Marcia Lei Zeng and Lois Mai Chan*

3: Research Areas, *(cont'd.)*

4: Institutions

4: Institutions, (cont'd.)

2: Institution Types, (cont'd.)

1: Archives, (cont'd.)

Corporate Archives / *Philip Mooney*
Institutional Records and Archives / *David Farneth*
National Archives / *Helen Forde*
State Archives / *David B. Gracy II and Adam D. Knowles*
University Archives / *William J. Maher*

2: Libraries

See also History of Libraries / *John Mark Tucker and Edward A. Goedeken*
See also Renaissance Libraries [*ELIS Classic*] / *Lawrence S. Thompson*
See also Subscription Libraries [*ELIS Classic*] / *David E. Gerard*
Academic Libraries / *Susan C. Curzon and Jennie Quiñónez-Skinner*
See also Academic Librarianship / *Barbara B. Moran and Elisabeth Leonard*
College Libraries / *Thomas G. Kirk, Jr.*
See also Academic Librarianship / *Barbara B. Moran and Elisabeth Leonard*
Corporate Information Centers / *Barbara M. Spiegelman and Nancy Flury Carlson*
Hospital Libraries / *Rosalind K. Lett*
See also Health Sciences Librarianship / *Judy Consales and Elaine R. Martin*
National Libraries / *Ian McGowan*
One-Person Libraries / *Judith A. Siess*
Presidential Libraries / *Sharon Fawcett*
Public Libraries / *Barbara H. Clubb*
See also History of Public Libraries [*ELIS Classic*] / *Frank B. Sessa*
See also Public Librarianship / *Kathleen de la Peña McCook and Katharine J. Phenix*
School Libraries / *Blanche Woolls*
See also School Librarianship / *Blanche Woolls*
Shared Libraries / *Ruth E. Kifer and Jane E. Light*
Special Libraries / *David Shumaker*
See also One-Person Libraries / *Judith A. Siess*
See also Special Librarianship / *Susan S. DiMattia*
State Libraries and State Library Agencies / *Barratt Wilkins*
Zoological Park and Aquarium Libraries and Archives / *Vernon N. Kisling, Jr.*

3: Museums

See also Cabinets of Curiosities / *Margaret A. Lindauer*
See also History of Museums / *John E. Simmons*
Art Museums / *David Gordon*
See also Corporate Art Collections / *Laura Matzer*

5: Systems and Networks

1: Information Systems

Clinical Decision-Support Systems / *Kai Zheng*

Collaborative Systems and Groupware / *David Jank*

Collection Management Systems / *Perian Sully*

Decision Support Systems / *Marek J. Druzdzel and Roger R. Flynn*

    See also Clinical Decision-Support Systems / *Kai Zheng*

Document Information Systems / *K. van der Meer*

    See also Corporate Records Management / *Barbara M. Cross and Barbara E. Nye*

    See also Modeling Documents in Their Context / *Airi Salminen*

Filtering Software *See* Internet Filtering Software and Its Effects / *Lynn Sutton*

Geographic Information Systems (GIS) / *Timothy F. Leslie and Nigel M. Waters*

    See also Geographical Literature: History [*ELIS Classic*] / *Nora T. Corley*

Information Retrieval Support Systems / *Yiyu Yao, Ning Zhong, and Yi Zeng*

Information Retrieval Systems / *Ray R. Larson*

Integrated Library Systems (ILS) / *Emily Gallup Fayen*

    See also Library Automation: History / *Robert M. Hayes*

Interactive Multimedia in Museums / *Nik Honeysett*

    See also **Museum Websites and Digital Collections** / *David Bearman and Jennifer Trant*

Internet Filtering Software and Its Effects / *Lynn Sutton*

Knowledge Management Systems / *Dick Stenmark*

Online Public Access Catalogs (OPACs) / *Kevin Butterfield*

    See also Catalogs and Cataloging: History [*ELIS Classic*] / *Eugene R. Hanson and Jay E. Daily*

    See also Online Catalog Subject Searching / *Danny C.C. Poo and Christopher S. G. Khoo*

Personal Bibliographic Systems (PBS) / *Dirk Schoonbaert and Victor Rosenberg*

Recommender Systems and Expert Locators / *Derek L. Hansen, Tapan Khopkar, and Jun Zhang*

    See also Web Social Mining / *Hady J. Lauw and Ee-Peng Lim*

Search Engines / *Randolph Hock*

    See also Internet Search Tools: History to 2000 / *Dale J. Vidmar and Connie J. Anderson-Cahoon*

    See also Search Engine Optimization / *Nicholas Carroll*

Thesaurus Management Software / *Leonard Will*

2: Network and Technology Elements

Authentication and Authorization / *David Millman*

Data Transmission Protocols / *Chuan Heng Foh*

Information Retrieval Protocols: Z39.50 and Search & Retrieve via URL / *William Moen*

Information Storage Technologies / *Scott L. Klingler*

    See also CD-ROM in Libraries / *Norman Desmarais*

    See also Reprography in Libraries / *Julia M. Bauder*

Intranets / *Kathleen Swantek*

Library Portals and Gateways / *Frank Cervone*

    See also Electronic Resource Management in Libraries / *Timothy D. Jewell*

    See also Institutional Repositories / *Joseph Branin*

Network Management / *Robert J. Sandusky*

Optical Character Recognition (OCR) / *Matthew C. Mariner*

Semantic Web / *Kieron O'Hara and Wendy Hall*

    See also Extensible Markup Language (XML) / *Kevin S. Clarke*

    See also Resource Description Framework (RDF) / *Nicholas Gibbins and Nigel Shadbolt*

    See also Unicode Standard / *Joan M. Aliprand*

Wireless Services in Libraries / *Christinger Tomer*

World Wide Web (WWW) / *Christinger Tomer*

    See also World Wide Web Consortium (W3C) / *Terrence Brooks*

6: Literatures, Genres, and Documents

1: Literatures

Business Literature: History [*ELIS Classic*] / *Edwin T. Coman, Jr.*

Children's Literature / *Judith V. Lechner*

Economics Literature: History [*ELIS Classic*] / *Arthur H. Cole and Laurence J. Kipp*

Geographical Literature: History [*ELIS Classic*] / *Nora T. Corley*

Grey Literature / *Joachim Schöpfel and Dominic J. Farace*

Mathematics Literature: History [*ELIS Classic*] / *Barbara Schaefer*

Medical Literature: History [*ELIS Classic*] / *William K. Beatty*

2: Generic Resources

1: Genres and Document Types

Citation indexes and the Web of Science / *Marie E. McVeigh*

Digital Images / *Melissa Terras*

Internet Genres / *Kevin Crowston*

Popular Literature Genres / *Barry Trott*

Primary Records: Future Prospects [*ELIS Classic*] / *G. Thomas Tanselle*

Reference and Informational Genres / *Thomas Mann*

2: Professional Documentary and Organizational Tools

Approval Plans / *Robert F. Nardini*

6: Literatures, Genres, and Documents, *(cont'd.)*

2: Generic Resources, (cont'd.)

2: Professional Documentary and Organizational Tools, *(cont'd.)*

Archival Finding Aids / *Su Kim Chung*

> *See also* Encoded Archival Description (EAD) / *Daniel V. Pitti*

Controlled Vocabularies for Art, Architecture, and Material Culture / *Murtha Baca*

> *See also* Cataloging Cultural Objects (CCO) / *Elizabeth O'Keefe and Maria Oldal*

Folksonomies / *Jonathan Furner*

Markup Languages / *Airi Salminen*

> *See also* Extensible Markup Language (XML) / *Kevin S. Clarke*

Records Retention Schedules / *Barbara E. Nye*

3: Standards

Information Technology Standards for Libraries / *Christinger Tomer*

> *See also* Information Retrieval Protocols: Z39.50 and Search & Retrieve via URL / *William Moen*

International Standards for Archival Description / *Wendy M. Duff and Sharon Thibodeau*

> *See also* International Records Management Standards ISO 15489 and 23081 / *Barbara Reed*

> *See also* Encoded Archival Description (EAD) / *Daniel V. Pitti*

Knowledge Organization System Standards / *Stella G. Dextre Clarke*

Library Standards in Higher Education / *William Neal Nelson*

Thesaurus Standards *See* Knowledge Organization System Standards / *Stella G. Dextre Clarke*

3: Named Resources

1: Information Organization Tools

Anglo-American Cataloging Rules (AACR) / *J. H. Bowman*

Bliss Bibliographic Classification Second Edition / *Vanda Broughton*

> *See also* Bliss Bibliographic Classification First Edition *[ELIS Classic]* / *Jack Mills*

Cataloging Cultural Objects (CCO) / *Elizabeth O'Keefe and Maria Oldal*

> *See also* Controlled Vocabularies for Art, Architecture, and Material Culture / *Murtha Baca*

Colon Classification (CC) / *M. P. Satija and Jagtar Singh*

> *See also* Facet Analysis *[ELIS Classic]* / *Douglas J. Foskett*

Dewey Decimal Classification (DDC) / *Joan S. Mitchell and Diane Vizine-Goetz*

Functional Requirements for Bibliographic Records (FRBR) / *Sara Layne*

Library of Congress Classification (LCC) / *Lois Mai Chan and Theodora L. Hodges*

7: Professional Services and Activities

See also Topical Category 1: *Information Disciplines and Professions*

1: Appraisal and Acquisition of Resources

Archival Appraisal and Acquisition / *Barbara Reed*

Archivists and Collecting / *Richard J. Cox*

Automated Acquisitions / *Patricia A. Smith and Margo Sasse*

Collection Development in Public Libraries / *Cynthia Orr*
> *See also* Popular Literature Genres / *Barry Trott*

Collection Development in the ARL Library / *David F. Kohl*

Digital Content Licensing / *Paul D. Callister and Kathleen Hall*

Museum Collecting and Collections / *Robert B. Pickering*

2: Institutional Management and Finance

Archival Management and Administration / *Michael J. Kurtz*

Business Process Management / *Suvojit Choton Basu, Prashant P. Palvia, and Leida Chen*

Competitive Intelligence / *Stephen H. Miller*

Disaster Planning and Recovery for Cultural Institutions / *Sheryl Davis, A. Patricia Smith-Hunt, and Kristen Kern*

Fund-Raising on the Internet for Non-Profit Organizations / *Pieter Boeder and Bettina Hohn*

Impact Assessment of Cultural Institutions / *Sara Selwood*
> *See also* Productivity Impacts of Libraries and Information Services / *Michael Koenig and Laura Manzari*

Library Architecture and Design / *Charlene S. Hurt and Thomas L. Findley*
> *See also* Library Architecture: History / *Nan Christian Ploug Dahlkild*

Library Fund-Raising and Development / *Susan K. Martin*

Managing an Information Business / *Gloria Dinerman*

Museum Accreditation Program / *Leah Arroyo and Julie Hart*
> *See also* American Association of Museums (AAM) / *Elizabeth E. Merritt*

Museum Architecture and Gallery Design / *Volker M. Welter*
> *See also* Exhibition Design / *Lee H. Skolnick, Dan Marwit, and Jo Ann Secor*

Museum Management / *Gary Edson*

Strategic Planning in Academic Libraries / *Sheila Corrall*

Unions in Public and Academic Libraries / *Kathleen de la Peña McCook*

3: Organization and Description of Resources

Archival Arrangement and Description / *Joanne Evans, Sue McKemmish, and Barbara Reed*

Archival Documentation / *Gavan McCarthy and Joanne Evans*

Back-of-the-Book Indexing / *Nancy C. Mulvany*

Cataloging / *Arlene G. Taylor and Daniel N. Joudrey*

4: Resource Management

5: User Services

7: Professional Services and Activities, (cont'd.)

5: User Services, (cont'd.)

See also Online Library Instruction / *Beth Evans*

Marketing Library and Information Services / *Dinesh K. Gupta and Réjean Savard*

Museum Education *See* Independent (Free Choice) Learning

Online Library Instruction / *Beth Evans*

See also Distance Learning and Virtual Libraries / *Edward D. Garten*

Photocopying *See* Reprography in Libraries / *Julia M. Bauder*

Reference Services / *Linda C. Smith*

See also Reference and Informational Genres / *Thomas Mann*

Storytelling / *Brian Sturm*

See also Children's Literature / *Judith V. Lechner*

Usability Testing of User Interfaces in Libraries / *Sharon L. Walbridge*

Volunteer Services in Cultural Institutions / *Barbara Cohen-Stratyner*

8: People Using Cultural Resources

1: General

Independent (Free-Choice) Learning / *Lynn D. Dierking*

See also Learning and Information Seeking / *Louise Limberg and Mikael Alexandersson*

Internet and Public Library Use / *Corinne Jörgensen*

Reading Interests / *Catherine Sheldrick Ross*

See also Sociology of Reading / *Martine Poulain*

Visitor Studies / *Susan Foutz and Jill Stein*

See also Museums and Their Visitors: Historic Relationship / *Samuel J. Redman*

2: Population Groups

See also African-Americans and U.S. Libraries: History / *Cheryl Knott Malone*

See also Latinos and U.S. Libraries: History / *Romelia Salinas*

Everyday Life Information Seeking / *Reijo Savolainen*

See also Leisure and Hobby Information and Its Users / *Jenna Hartel*

Information Needs and Behaviors of Diasporic Populations / *Ajit K. Pyati*

Information Needs and Behaviors of Populations in Less Developed Regions / *Innocent I. Ekoja*

See also Information Technology Project Implementation in Developing Countries [*ELIS Classic*] / *Athanase B. Kanamugire*

Lesbian, Gay, Bisexual, and Transgender Information Needs / *Patrick Keilty*

Older Adults' Information Needs and Behavior / *Kirsty Williamson and Terryl Asla*

Students' Information Needs and Behavior / *Heidi Julien*

Youth Information Needs and Behavior / *Melissa Gross*

See also Homework Centers / *Cindy Mediavilla*

9: Organizations

Association for Library Service to Children (ALSC) / *Virginia A. Walter*

Association of College and Research Libraries (ACRL) / *Mary Ellen K. Davis and Mary Jane Petrowski*

Association of Library Trustees, Advocates, Friends and Foundations (ALTAFF) / *Sally Gardner Reed*

Association of Research Libraries (ARL) / *Lee Anne George and Julia Blixrud*

Association of Specialized and Cooperative Library Agencies (ASCLA) / *Barbara A. Macikas*

Australian Library and Information Association (ALIA) / *Sue Hutley, Jane Hardy, and Judy Brooker*

Bibliographical Society [London] / *Julian Roberts*

Bibliographical Society of America (BSA) / *Hope Mayo*

Canadian Heritage Information Network (CHIN) / *Shannon Ross and Paul M. Lima*

Canadian Library Association (CLA) / *Donald I. Butcher*

CENDI / *Bonnie C. Carroll, Gail M. Hodge, and Kathryn R. Johnson*

Center for Research Libraries (CRL) / *Mary I. Wilke*

Chartered Institute of Library and Information Professionals (CILIP) / *Tim Owen*

Coalition for Networked Information (CNI) / *Joan K. Lippincott*

Copyright Clearance Center (CCC) / *Edward Colleran*

Council on Library and Information Resources (CLIR) / *Chuck Henry and Deanna Marcum*

EDUCAUSE / *Peter B. DeBlois and Jarret S. Cummings*

Institute of Museum and Library Services (IMLS) / *Jeannine Mjoseth*

International Association of Sound and Audiovisual Archives (IASA) / *Ilse Assmann*

International Association of Technological University Libraries (IATUL) / *Nancy Fjällbrant and Alice Trussell*

International Communication Association (ICA) / *Ronald E. Rice*

International Council of Museums (ICOM) / *Patrick J. Boylan*

International Council on Archives (ICA) / *Perrine Canavaggio and Marcel Caya*

International Council on Knowledge Management (ICKM) / *Franz Barachini and Albert Boehm*

International Federation for Information and Documentation (FID) / *Forest W. Horton, Jr.*

International Federation of Library Associations and Institutions (IFLA) / *Ross Shimmon, Peter J. Lor, Sofia Kapnisi, Sjoerd Koopman, and Stuart Hamilton*

International Federation of Television Archives (FIAT/IFTA) / *Steve Bryant*

International Organization for Standardization (ISO) / *Alan Bryden and Catherine Dhérent*

International Records Management Trust (IRMT) / *Anne Thurston*

International Society for Knowledge Organization (ISKO) / *Ingetraut Dahlberg*

JSTOR / *Kevin M. Guthrie*

Library and Information Technology Association (LITA) / *Andrew K. Pace*

Library Leadership and Management Association (LLAMA) / *Robert Allen Daugherty*

10: National Cultural Institutions and Resources

10: National Cultural Institutions and Resources, (cont'd.)

See also Louvre / *Bette W. Oliver*

Germany: Libraries, Archives and Museums / *Claudia Lux*

Greece: Part I Libraries / *Anestis Sitas and Mersini Moreleli-Cacouris*

Greece: Part II Archives / *Nestor Bamidis*

Greece: Part III Museums / *Maria Economou*

Hungary: Libraries, Archives and Museums / *Katalin Radics*

India: Libraries, Archives and Museums / *Krishan Kumar, V. Jeyaraj, and Ramesh C. Gaur*

Israel: Libraries, Archives and Museums / *Snunith Shoham and Silvia Schen-kolewski-Kroll*

Japan: Libraries, Archives and Museums / *Masaya Takayama, Yoriko Miyabe, Toru Koizumi, and Hiroyuki Hatano*

Kazakhstan: Libraries, Archives and Museums / *Leslie Champeny, Joseph Luke, Anna Bergaliyeva, and Olga Zaitseva*

Kenya: Libraries, Museums and Archives / *Irene Muthoni Kibandi, Pancras Kimaru, Caroline Kayoro, Philomena Kagwiria Mwirigi, Sophie Ndegwa, Nelson Otieno Karilus, Charles Nzivo, Linda Mboya, and Lilian Gisesa*

Lithuania: Libraries and Librarianship / *Elena Macevičiūtė*

Mexico: Libraries, Archives and Museums / *Jesús Lau*

Moldova: Archives, Museums and Libraries / *Hermina G. B. Anghelescu, Silviu-Andrieş Tabac, and Elena Ploşniţâ*

Netherlands: Archives, Libraries and Museums / *Eric Ketelaar, Frank Huysmans, and Peter van Mensch*

Peru: Libraries and Library Science / *Sergio Chaparro-Univazo*

Poland: Libraries and Archives / *Jadwiga Woźniak-Kasperek*

Saudi Arabia: Libraries, Archives and Museums / *Ayman Shabana*

Senegal: Libraries, Archives and Museums / *Bernard Dione and Dieyi Diouf*

Serbia: Libraries, Archives and Museums / *Staša Milojević*

Slovakia: Libraries, Archives and Museums / *Jela Steinerová, Juraj Roháč, and Gabriela Podušelová*

South Korea: Archives and Libraries / *Hyun-Yang Cho, Eun Bong Park, Soyeon Park, Jae-Hwang Choi, Seong Hee Kim, and Jong-Yup Han*

Spain: Libraries, Archives and Museums / *Lawrence J. Olszewski*

Switzerland: Libraries, Archives and Museums / *Jean Frédéric Jauslin and Andreas Kellerhals*

Tanzania: Libraries, Archives, Museums and Information Systems / *Janet Kaaya*

Thailand: Libraries, Archives and Museums / *Robert D. Stueart*

Tunisia: Libraries, Archives and Museums / *Ali Houissa*

Ukraine: Libraries / *Lyudmila Shpilevaya*

United Kingdom: Part I Libraries and Librarianship / *David J. Muddiman*

See also British Library / *Andy Stephens*

United Kingdom: Part II Museums and Museology / *Sarah Elliott and Peter Davis*

United Kingdom: Part III Archives and Archival Science / *Helen Forde*

United States: Part I Archives and Archival Science / *Cheryl L. Stadel-Bevans and Danna Bell-Russel*

11. History

11. History, (cont'd.)

History of Paper / *Sidney Berger*

History of Public Libraries [*ELIS Classic*] / *Frank B. Sessa*

History of Records and Information Management / *William Benedon*

History of the Book / *Jonathan Rose*

History of Three Basic Printing Processes / *Sidney E. Berger*

> *See also* Incunabula [*ELIS Classic*] / *John P. Immroth and Romano Stephen Almagno*

Hypertext and Hypercard: Early Development [*ELIS Classic*] / *Susan K. Kinnell and Carl Franklin*

Illumination [*ELIS Classic*] / *Abdul Moid*

Incunabula [*ELIS Classic*] / *John P. Immroth and Romano Stephen Almagno*

Intellectual Freedom and the American Library Association (ALA): Historical Overview [*ELIS Classic*] / *Judith F. Krug*

> *See also* Censorship and Content Regulation of the Internet / *Peng Hwa Ang*

Internet Search Tools: History to 2000 / *Dale J. Vidmar and Connie J. Anderson-Cahoon*

> *See also* Search Engines / *Randolph Hock*

Knowledge Management: Early Development / *Michael Koenig and Ken Neveroski*

Latinos and U.S. Libraries: History / *Romelia Salinas*

Library Architecture: History / *Nan Christian Ploug Dahlkild*

> *See also* Library Architecture and Design / *Charlene S. Hurt and Thomas L. Findley*

Library Automation: History / *Robert M. Hayes*

> *See also* Integrated Library Systems (ILS) / *Emily Gallup Fayen*

Library College: Prototype for a Universal Higher Education [*ELIS Classic*] / *Louis Shores*

Library of Congress: History / *John Y. Cole*

> *See also* Library of Congress: 21st Century / *John Y. Cole*

Library Science in the United States: Early History / *John V. Richardson, Jr.*

Machine Readable Cataloging (MARC): 1961–1974 [*ELIS Classic*] / *Henriette D. Avram*

> *See also* Machine Readable Cataloging (MARC): 1975–2007 / *Sally H. McCallum*

Museums and Their Visitors: Historic Relationship / *Samuel J. Redman*

> *See also* Visitor Studies / *Susan Foutz and Jill Stein*

Renaissance Libraries [*ELIS Classic*] / *Lawrence S. Thompsotn*

SMART System: 1961–1976 [*ELIS Classic*] / *G. Salton*

Subscription Libraries [*ELIS Classic*] / *David E. Gerard*

User-Centered Revolution: 1970–1995 [*ELIS Classic*] / *Diane Nahl*

> *See also* User-Centered Revolution: 1995–2008 / *Diane Nahl*

Word Processing: Early History [*ELIS Classic*] / *Daniel Eisenberg*

6

Many paths to theory: The creative process in the information sciences

The creative process, by its very nature, is unpredictable and surprising. Nonetheless, one can develop skills that will promote and enhance creativity, and will increase the likelihood of producing fruitful ideas. My present purpose is to say something about this process, based on my own experience. Also, I will describe several research ideas that I think are promising and that I wanted to develop but did not have the time to pursue before my retirement. I hope that readers will draw on both the skills and the ideas presented here to produce further progress in the information sciences.[1]

Part 1: Being open to, and using, ideas

First skill: learn to be open to ideas

First, be open to ideas and research possibilities. It is here, at the beginning, that many people get stopped before they really get started. It helps to recall Sigmund Freud's psychological concepts of the id, the ego, and the

---

1 At this time of fluidity in disciplinary boundaries, I use several terms to describe the fields to which these ideas apply, depending on the orientation and emphasis of the work described. "Library and information science (LIS)," "information science," and "the information sciences" are all used here. I discuss these field distinctions in several other publications (see, e.g., Bates 1999, 2007a, 2015).

First published as Bates, M. J. (2016). Many paths to theory: The creative process in the information sciences. In D. H. Sonnenwald (Ed.), *Theory development in the information sciences*. Austin: University of Texas Press.

superego in the human mind. The id is the child, the autochthonous root of behavior, unpredictable and seemingly uncontrollable. The ego is the adult manager, seeing you through life safely—judicious and mature. The ego does experiment, but cautiously and thoughtfully. The superego is the controller, the guilt-tripper, the part of your mind telling you to follow social rules and religious edicts.

All of these parts of the mind develop in their own way and time, and work together, more or less well, to produce the acting adult that one becomes. Ideally, when one is doing research, the id produces ideas, the ego manages the ideas productively for the benefit of the individual and society, and the superego insures that the work is done and reported on ethically.

Unfortunately, what happens more often than not is that the super-ego clamps down on the idea production by the id, along with all the other things in the id that the superego suppresses, as people grow from childhood to adulthood. We have all seen the "uptight" individual, afraid to try anything, super-polite and tightly regulated. In this age of looseness and ease about so many things, we tend to pity these people and wish they would tell that superego of theirs to lighten up. However, other people who are not uptight generally and are quite normal in behavior can still be uptight when it comes to idea generation. It is not so easy sorting out just how you do and do not want that id to be allowed to influence your life. Unfortunately, as we "put away childish things," we often put away as well the wonderful fecundity of the id.

I have observed a number of researchers in my life who, I believe, would never ever allow a stray idea to wander up from their id into their ego as they work. In these people, *everything is controlled*; perhaps a better term would be *locked down*—so tightly that a fleeting creative thought would never be allowed into the thinking mind. The result is a predictability, a dryness, a purged-of-color-and-personality quality to their work. If there is an obvious interpretation, they will find it. They are not just analytical; they are analytical/boring. In most cases analytical is good; analytical/boring is not.

Now, in order to let in ideas from the id, one needs to be confident one can handle the ideas and not make a fool of oneself. This is usually one of the biggest reasons for suppressing that id. But here is where we can apply a really valuable idea, and that is this: keep in mind that when you are coming up with wild, silly, ridiculous ideas, *no one else ever has to know.* You have complete control over what you do and do not say or write. You can try out ideas, play with them, even write about them, but, still, no one else need ever know, if you conclude that these ideas are non-starters.

In other words, do not slam the door on the ideas from the id, but let them in, sit down with them, play with them, and then decide whether you want to do anything further with them, such as taking them out for a visit with the rest of the world.

So how do you become aware of the ideas from the id (or from anywhere in your creative psyche)? At first they may come to you as just fleeting—the sort of thing you ordinarily knock out of your mind without thinking. Well, do not knock them out. Stop and think. Why does this occur to me now? What connection can there be between this goofy thought and the research I was just thinking about? The idea that floats into your mind often constitutes an analogy. People have asked me where I got the idea of "berrypicking" for my paper of that name (Bates, 1989). But moving around through different information sources, getting a bit here, a bit there, reminded me of picking huckleberries in the forest with a boyfriend of mine when I lived in Washington state. I allowed that thought in, instead of dismissing it.

Another thing to keep in mind is that creativity requires fecundity, abundance. In her lifetime, a woman produces several hundred eggs, and their associated menstrual cycles, and men produce billions of sperm, all in order to produce just the one or two or three children you ultimately have. Think of how much you and your spouse produced, in order to have just a few children. Ideas are like that, too. You will have *many* ideas in order to produce the few that you actually concentrate on. There may be a thousand ways to think about a particular issue, but only three ways to solve it. To have any hope of solving it, you need to think of a lot of ideas, not just one or two.

Once you allow ideas in, you should have many; more and more should start coming, once you are open to them. It should not be necessary for you to think that every good researchable topic you come up with must be the result of years of striving to find and shape just one or two ideas. If so, you are doing something wrong—such as not letting those ideas in when they first appear. Generally, research does not do well on a paucity model. After all, research and the development of theoretical concepts require multiple big and little moments of creativity to reach fruition. What may look like a single idea at the end is actually often the result of many successive original thoughts and insights.

And if an idea does not work out, do not beat yourself up about it. You are not *wrong* because an idea does not work out. The idea just did not work out. Most of them do not. Move on. That is another advantage of having lots of ideas. You have plenty to spare.

Second skill: Draw on a variety of research traditions

I often encounter doctoral students who, with a laser-like focus, want to know exactly and only the courses they need to take to get through the doctoral program in the shortest time possible. Given the expense, and the time away from other work and family, this is an entirely understandable sentiment, but, in fact, I do not think this approach is productive for the rest of one's career. Doctoral work should be a time to explore intriguing areas, to take courses that may not be obviously related to your work. This should be a time to trust, and to follow up on, that tickle of interest you feel in anything from whole other disciplines, specialties within disciplines, or simply a research question that has been studied by someone in one or more other fields—or our own. To pursue those interests often involves taking or auditing courses in other departments, or taking not-obviously-relevant courses in your own department, or doing an independent study course. Again, your unconscious mind may be directing you toward research and theory that you may not only have a talent for, but which you may also be able to combine in new and creative ways with other knowledge that you already have.

I emphasize this idea of wide exposure to research areas and topics because I believe that this is one of the most productive ways to gain new insights and identify intriguing new topics for research. Each specialty and discipline necessarily brings a whole set of assumptions and established knowledge with it; these are the "paradigms" (Kuhn, 2012) that we hear so much about, which are particularly important in the social sciences. When you are exposed to several different paradigms in your studying, you may well see important inconsistencies or conflicts between those ideas. Thinking about these conflicts can point to new questions that neither paradigm is addressing. There can also be a valuable synergy between the different approaches. It is not uncommon for two disciplines to address essentially the same topic, but from their separate perspectives—and often without being aware that the other discipline has a long line of research on that very question.

Here is a case from my own life history. I wrote about "information search tactics" (Bates, 1979b) and "idea tactics" (Bates, 1979a) after bringing together ideas from several fields. I originally became interested in techniques of information searching when I found myself to be rather incompetent and slow-witted in reference practice exercises in my library education program. A doctoral student serving as a teaching assistant for our reference lab class took fiendish delight in crafting test reference questions whose answers could not be found in any of the "obvious" places. Other students were much quicker in finding answers to these difficult questions than I was. One day, my friend found answers to twenty-two questions in an hour; I found answers to just two! I was an excellent student in general; why was I so poor on this core librarian task? I soon found that there was little instruction available on this topic in the library literature. Ability to do good reference work seemed to be just assumed by everyone. Note the research opening here: a topic that had not been much addressed in the literature, and yet that was crucial to professional performance as a librarian. Note also that I did not slink away in shame at not being as good as my classmates at this skill; instead I took it as an interesting challenge to find helpful ideas on the matter.

I gathered together everything I could find in the library literature, and then I branched out to psychology, because thinking of other places to look for hard-to-find information was a kind of creative process, and I knew that psychology addressed creativity. I had long been interested in military history and had read books on the subject and liked to watch war movies. So I was aware of the concepts of strategy and tactics. It made sense to me that one should have a strategy for the overall search, and then, as one made various moves to complete the search, one would apply various helpful tactics within the search to increase the chance of success in finding the desired information. So, improbably, I brought military theory together with psychology to address a topic in library and information science. Another word for helpful tactics is "heuristics," a term that cropped up in various fields such as psychology, computer science, and operations research. (I had taken courses in all those areas, by the way, as a doctoral student.) So looking at that concept was productive as well.

The insight about tactics and strategies to promote effective searching provided the framework for my thinking on steps to improve searching. Once I had that idea, I explored all the ways I could think of that would help promote good searching, ending up with several broad categories of tactics: monitoring, file structure, search formulation, term, and so on (over thirty tactics in total). Note that I also drew on my knowledge of librarianship, specifically, the organization of information, in suggesting tactics, for example, that involved experimenting with different terms for a subject—going broader, narrower, and so forth. Thus, several literatures were brought to bear on the topic. These ideas also proved productive for several subsequent articles I wrote (one of which received the Association for Information Science and Technology Best *JASIST* Paper of the Year Award) and led to others in the field doing research on tactics (e.g., Hsieh-Yee, 1993; Xie, 2000).

Incidentally, around the time I was working on these papers, I had an experience that further supports my argument about the value of exposing yourself to varied fields when you are researching a question. In the late 1970s, I was teaching at the University of Washington, and used to sit in on some of the lectures of visiting speakers in the Philosophy Department. One time, Paul Grice, a very eminent philosopher of language and reasoning, came to speak. What he described in his lecture was his developing work on logical reasoning. He distinguished between formal reasoning and the informal tricks we use to get to an answer to what we are reasoning about. He said he was currently working on the informal reasoning aspect. During the question period, I asked him if he would say that he was working on a conception of the *heuristics of reasoning.* "Well," he said, "'heuristics' is not exactly a term that comes tripping off my tongue." Clearly, he did not know the term. This was a classic example of different fields addressing the same topic under different names. This is not to question in any way Grice's very important and original contributions to this area in philosophy. But might his thinking have changed in any way had he been aware of the several other fields also studying heuristics, each from its own distinctive perspective?

So far, I have presented the idea of drawing on a variety of research traditions as a way to enrich your own work and generate interesting new ideas. But in some cases, the work in related fields is so highly relevant that to *not* address that related work constitutes intellectual failure. A prime example of relevant material typically not referenced in our field is, ironically, the work on relevance by a psychologist and an anthropologist. Dan Sperber and Deirdre Wilson's book *Relevance: Communication and Cognition* (Sperber & Wilson, 1995), originally published in 1986, has generated an enormous amount of interest and debate in many social science fields, as well as in philosophy, and has been cited about 1,500 times, according to the Thomson Reuters Web of Science (wokinfo.com). Only a handful of those citations, however, are from the field of information science, despite the hundreds of articles that address the subject of relevance in information science. Two information scientists who have recognized the significance of the Sperber and Wilson work, and incorporated it into their thinking, are Stephen Harter (1992) and, more recently, Howard White (2007a, 2007b). Otherwise, however, despite occasional references, information science seems to have ignored this high-impact work. Given the recent popularity of "soft," more humanistic approaches to library and information science (LIS), it is puzzling that their subtle approach has not been generally taken up in the field. As long as we in information science like to feel that we "own" the idea of relevance, we had better deal with the many implications of the work on this topic elsewhere in academia. The longer we avoid it, the more behind-the-curve we will appear to be when we finally do take it seriously.

Third skill: Read deeply, not just widely

As I have argued, familiarizing yourself with research traditions in several fields can lead to productive new ideas and approaches in your thinking. However, this suggestion comes with a caveat: if you want to use the ideas drawn from other disciplines in your own work, take the time to really understand what you are reading. As noted earlier, one of the things that makes these encounters between disciplines productive is that the disciplines take different approaches to the same questions. But you cannot just borrow the vocabulary, discuss your own field's work using the new vocabulary, and be successful in making a contribution. The conflicts between the ideas as pursued in the different disciplines often lie at a deeper level than merely the vocabulary. In short, when you study a different approach from another field, you need to understand that approach all the way down, that is, all the way to the underlying philosophical perspective driving that field. Different worldviews and philosophical assumptions separate the various fields. If you want to contribute to the information sciences using those models from other fields, you have to do some serious thinking about how to adapt and integrate those ideas with those of your home field. You need to understand deeply the other field's worldview and how it plays out in the specific topic of interest before you can successfully adapt it to (one of the several) information science paradigms operating (Bates, 2005).

Jenna Hartel's (2010a) work on the use of ethnographic methods in the information sciences represents a recent example of reading deeply and working hard to integrate a methodology widely used in the social sciences with the needs of information-seeking research. Ethnographic techniques are well developed in anthropology and sociology, and have been used by some researchers in the information sciences. What Hartel has contributed to this important methodological philosophy is a deep understanding of information seeking, keeping, and using. The ethnographic methods are not native to the information sciences, and the information orientation is not well known to the other social sciences using ethnographic methods. Her interest in the extensive information gathering and collecting of passionately devoted gourmet cooks surprised some of her sociology professors. As sociologists, they had not ever thought much about *the ethnography of information.* Bringing those two together is a difficult task, requiring extensive reading in philosophy and theory, and years of analytical work. Hartel described her experiences in a series of articles (Hartel, 2010a, 2010b, 2011) that are combinations of research results and expositions of ethnographic methods as applied to the information worlds of the people being observed.

Fourth skill: Relate your work appropriately to that of others

One of the standard requirements of an academic article is that the author should show how the work fits in with other research and thinking in the area. This is not an idle requirement of fussy journal editors; it is done for a reason. Work that is published should not only be new, but also be *shown* to be new. One does this by describing prior work and explaining how the current work relates to the prior work and yet advances beyond it. The literature review in an article (1) *contextualizes the research* presented in the article, so the reader can see where the work fits with other work in the field; (2) demonstrates that the author is *current with other work in the area,* and so is not rediscovering things already known; (3) *affords recognition* to those authors for the earlier work; and (4) *claims its own new territory.*

So the literature review, which may be seen superficially as a boring prelude to the real action of the article where the research is described, is actually a socially very significant part of the article, and needs to be thought about carefully.

I bring this up here in a chapter on creativity in the information sciences because the creativity that produces new ideas or methods can come into conflict in a variety of ways with providing recognition to others for their work. How much credit is due to my thinking, and how much credit is due to others for their thinking?

Why is recognition important? The currency of business is the bottom line—the amount of money made. No matter how clever your product or service, if you do not make money on it, you are not a successful business-person. The currency of science, on the other hand, is *recognition* for work done and discoveries made. People go into research to discover new things and advance science or scholarship. The mark of success, therefore, is to have other talented people, who also know that research area well, recognize and esteem your work, and value your presence in the field. Esteem is shown, for academics, through successful publication in good venues, through promotions, through awards, through naming things after their discoverer, and so on. It is all about recognition for one's talents by one's respected colleagues.

I got a lesson in this importance early in my career. I was writing the article to be later titled "Rigorous Systematic Bibliography" (Bates, 1976). After the article was, for the most part, finished, I was reviewing it and realized suddenly that I had drawn throughout the article on the work of Patrick Wilson, one of my professors at UC Berkeley, where I got my doc-torate. In effect, the article represented a way to operationalize, or put into practice, the much more theoretical ideas to be found in Wilson's book *Two Kinds of Power* (Wilson, 1968). I had so thoroughly absorbed Wilson's ideas that I did not recognize that I was working through methods of applying his ideas to the practice of bibliography in the article. I had not submitted the article yet, and, somewhat embarrassed, I rewrote it to include full and repeated references to his work. After a bit, I realized that I had gone too far, and was not highlighting what was original in my own contribution in the article, so I rewrote it again, with what finally felt like a good and accurate balance.

The relationship between our own work and that of others can be a very touchy matter, and it is better to recognize that, and work it through, rather than deny what is happening. One makes a good-faith effort at reviewing the earlier literature, and one cites it when it is relevant. I suspect that the intense hostile animus that shows up in some journal article reviews arises from the reviewer feeling that you have somehow violated, wittingly or unwittingly, his or her intellectual turf. Perhaps I take this matter of giving credit where credit is due a little too seriously, but I have seen many occasions where people did not deal well with the challenge of getting the credit right.

One time a senior professor wrote to me enthusiastically about how much he liked one of my articles and how influential it had been for him. Later, he wrote a whole book on the subject, and I looked to see the role my work played in it. He cited the reference to my paper seriously incorrectly, and, to read the relevant section, you would think my work had virtually no effect on his, and was at best peripheral to his work. In another case, I was asked to review an article for a double-blind journal (i.e., neither author nor reviewer knows the identity of the other) in which the author had used almost the identical research design and research question as I had in a major research project, but had not related it to my study, or referenced it. I pointed to the articles that I had published out of my study, and asked that they be appropriately referenced and related to the results of the work in the article being reviewed. Later, after the article had been published, I saw that the senior professor who wrote it now cited my article but still did not reference or discuss any of the parts that were actually overlapping. The reader would have no idea of the similarities.

I have read many articles that used my work but did not reference it or that referenced my popular article "The Design of Browsing and Berrypicking Techniques" (Bates, 1989) instead of referencing the article of mine that actually dealt with the content of the referencing article. By referencing an irrelevant article of a person, recognition appears to have been given, when, in fact, it has not. The new work presented in the article is not connected at all to the actual relevant prior work that should have been discussed in the article. One reviews prior work in order, among other things, to compare the results and discuss any differences. By citing a different article than the relevant one, one avoids addressing altogether the actual relationship between the current and cited research. This happened to me so many times that I once sat down, with the help of the Thomson Reuters Web of Science, to review every reference to my work in the literature in order to tally the actual frequency of this problem. However, not wanting to be the skunk at the garden party and offend my professional colleagues, I gave up on this endeavor.

It can be disappointing to discover that someone else has been there before you—but they did get there, and you owe it to them to acknowledge it. Further, if the works differ in results, some discussion of this is in order; such discussion and debate is what science is supposed to be about. If your own work is any good, you will also be contributing work of value; it does not undermine your own contribution to recognize those who have gone before.

Let us turn now to those promising ideas to be put forward. I will have more to say about being creative in information science research as I go along.

Part 2: Promising ideas

In this section, I discuss a number of ideas that I think have promise, and that I am unlikely to be able to pursue in my lifetime. I encourage students and established researchers to take up these ideas and see where they lead. Keep in mind that these are *ideas*; they are not fully researched or substantiated. There will be gaps, unreferenced literature, and so on. Actual in-depth work with these ideas might lead to different conclusions than I anticipate. In short, much remains to be done, but most of these should have the potential to be good dissertation or other research project topics.

Area 1: Information seeking at the human and machine interface

Here I will describe just three of many topics that could be mentioned as promising ideas to follow up in the area of information seeking at the human and machine interface. However, to introduce those ideas, I need to pick a fight—just a little one—with my colleagues, in this book and elsewhere, from the field of human-computer interaction (HCI). I believe that the full contribution that information science can make to this area is generally not recognized outside the field, particularly by researchers in HCI. Many in HCI come from a psychology background, and when they address the experience of people at a computer interface, they bring a psychologist's desire to discover fundamental principles about human psychology in carrying out the work of using computers. The objective is to understand the whole HCI experience; it is assumed that the most fundamental discoveries will apply across most or all application areas. From that standpoint, *information seeking* on computers is just a single application area, and therefore not of much intrinsic interest to HCI researchers.

I would argue, in reply, that information seeking is a much larger area than generally recognized and has distinctive, important features that need to be understood and designed for in order to produce the best HCI in information-seeking contexts. General HCI principles are not enough for optimal results.

Gregory Bateson (1968) made an interesting distinction between what he called "value seeking" and "information seeking" (p. 178–179). In value seeking, a person has an idea in mind and goes out into the world to shape that world so as to produce a result that matches the idea in mind. If one wants bacon and eggs for breakfast, one does certain things in the world with pigs, chickens, and a stove, with the end result that one has a plate of bacon and eggs. In information seeking, on the other hand, one goes out into the world to discover things so as to create the idea in mind to match or reflect what is in the world. This is a simplified distinction, obviously, but a powerful one. The nature of our actions in the world will be very different when we have a plan in mind to impose on the world, versus when we are open to the world imposing some part of its character or shape on our own minds. If we move to a part of the world where bacon and eggs are not to be found, and we are hungry, we face a very different type of challenge in our effort to learn about things that might satisfy our hunger.

Information seeking is not just an application field, like engineering, or retail services; rather, it is the other side of action. It is about gaining knowledge, and there are many characteristics of that behavior that distinguish it from value seeking. First, information seeking makes one vulnerable. Information, by many definitions, is surprise. Not all surprises are fun or desirable. Therefore, the seeker opens him- or herself up to risk, to the potential need to reorganize or reorient his or her hard-earned knowledge. The consequences of this risk ripple throughout the behavior we call information seeking. The behavior ranges all the way from avoiding information to actively seeking it out in cases where our lives are on the line or our passions are engaged with a fascinating subject.

Second, by definition, *you are seeking something that you do not know.* How, then, does one specify the sought information? Whole courses in information science are devoted to this question. Not only the designs of classifications and indexing vocabularies are at issue, but also courses on the interaction between people and information. We know quite a bit about what people do to their own queries to make them understandable to information systems and to information professionals. These propensities have huge consequences for the design of the information-seeking interface in information systems.

Some years back in the Department of Information Studies at the University of California, Los Angeles (UCLA), one of my colleagues taught a course on HCI in information systems, and I taught a course on user-centered design of information systems. By then, information systems were overwhelmingly online, so one might expect quite a bit of overlap in the contents of the two courses. But there was, in fact, virtually no overlap, because I emphasized the design issues that were specific to information seeking/searching, and she taught about HCI in general. *There is a distinct body of knowledge in information science around information seeking on computer interfaces.* Approaches assuming that general HCI knowledge applies without an understanding of the information-seeking part are underperforming in the quality of design for this central human process. My objective in the following sections is to describe some HCI areas involving information searching and retrieving that have interested me, and that I believe would be of value to pursue further.

Topic 1: Design for real browsing What is generally called a "browsing capability" in online systems is nothing of the sort. Overwhelmingly, as currently manifested, this capability in online systems consists of being able to *scan* down the page, with the help of the scrolling function. That systematic, top-to-bottom or left-to-right scanning behavior is not browsing. Think about standing in a bookstore or at a magazine stand, or shopping in a bazaar, for that matter; it is not about systematic scanning. Your eyes dart all around. You glimpse here, then there, then way over there. Things catch your eye. You look at one, then at another. If something is really interesting, you pause and take a serious look at it, then select it, or move on. Rarely do you run your finger along the books, or items in the bazaar, in a systematic way, studying one, then the one right next to it, then the one right next to that, and so on. The eye darts around in browsing; it does not scan, as I have argued in detail (Bates, 2007b). Things catch your eye because you first do a gross glance that does not analyze the visual scene in detail; then you put your closer attention to the things that pass your crude filter (Wolfe, 1994).

I have argued that "browsing is seen to consist of a series of four steps, iterated indefinitely until the end of a browsing episode: (1) glimpsing a field of vision, (2) selecting or sampling a physical or informational object within the field of vision, (3) examining the object, (4) acquiring the object (conceptually and/or physically) or abandoning it. Not all of these elements need be present in every browsing episode, though multiple glimpses are seen to be the minimum to constitute the act" (Bates, 2007b). Thus, design for browsing in online systems would necessarily be very different from the current provision of the capability to scan. First, the screen needs to be large enough to allow the eye to take glimpses, attending to one part, then to another part. Second, there should be a variety of options available to the user—and not just on pull-down menus, which require choosing to pull down the menu! Instead, the many options should be available *at the same time* on the screen—just as a scene glimpsed by our forebears in the forest contained many points of interest. Icons scattered on the screen, representing different types of search capability, could simultaneously present to the user a rich array of options for searching. A single search box, à la Google, has been viewed as the sine qua non for searching—the easiest, most simplified approach. But what if we found a way to make the search itself interesting? Let people browse through different capabilities, and different classes of metadata or taxonomies. There are many ways to structure such a system; the problem is that *we have not ever taken seriously the desirability of designing for true browsing.*

Topic 2: Interface design specific to searching For a very long time, there has been a pervasive assumption among computer scientists that an information system search interface can simply be superimposed on any body of searchable data. Set up the information so that various elements can be searched, then superimpose some sort of search engine—most any kind will do—and you have your information system. But, as I argued in "The Cascade of Interactions in the Digital Library Interface" (Bates, 2002), there are, in fact, many layers of design needed beneath and in front of the interface that culminate in design imperatives for the resulting interface. The information itself—its content, structure, medium, types of indexing and metadata, and so on—influences how one can best find desired results within the body of information, and therefore how search should be designed in the interface for the user, as well as behind the scenes in the computer. Likewise, everything on the searcher side of the interface—subject area of interest, type of query, level of skill as a searcher, and so on—interacts with the interface design in a productive or unproductive way. Bad design at just one of these several layers can block the effective functioning of all the other layers (Bates, 2002).

Here is an example contrasting two situations. In situation A, you have a database of biological reports addressing the various concerns of biologists, from studying animals in their natural habitat to using parts of nature for research in other areas, such as agriculture, animal breeding, and environmental concerns. In situation B, you have a database of articles published in the humanities literature on the several fields encompassed by that term—national literatures, philosophy, religion, and so forth.

Extensive research and practical experience have demonstrated that faceted vocabularies provide the most appropriate means for indexing humanities articles (for the explanation, see Bates, Wilde, and Siegfried, 1993). The Getty Research Institute (associated with the Getty Museum) consequently committed itself to creating faceted vocabularies, using them to index its extensive database production. In contrast, scientific databases do best with the one- to three-word phrases known as descriptors, which have been prompted by classical indexing theory and embodied in the technical standards developed for them (National Information Standards Organization, 2005). With faceted indexing, the whole query is composed of terms drawn from each of several distinct facets, while with scientific literature, the search is composed of descriptors presented to the system in Boolean combinations, either implicitly or explicitly. If searchers are to take advantage of these differences, the interfaces must be designed differently for the two systems—a single simple search box, in particular, will not do.

Examples of this sort could be proliferated indefinitely. If the one database contains technical reports with a certain standard structure in introductory matter (author, affiliation, abstract, etc.), and the other database consists of humanities articles using typical bibliographic rules for humanities articles (e.g., University of Chicago Press rules rather than American Psychological Association rules), then there are different ways of coding and representing these bibliographic entities that can either promote successful search or get in the way of it. And I have not even mentioned other types of information, such as video, image, multi-lingual, and other variant forms of data.

When designing for search, one must design the interface all the way down to the actual content itself. A single, standard search interface, superimposed on the huge actual variety of types of information and information organizational schemes, can just about be guaranteed to sub-optimize. Furthermore, there is a huge variety of types of queries characteristically associated with different types of needed information. For example, art databases are searched not only by art history scholars but also by designers and artists looking for inspiration in the images they find, and by schoolchildren for their assignments. The kinds of search interface design features that each of these groups could most benefit from differ. Different capabilities are needed for each. To optimize information search, all these various design layers need to be recognized, understood, and designed for in an interface that nonetheless feels simple and natural to the end user. This is a high standard, and one that has been mostly ignored outside of information science. There are abundant opportunities here, with the right funding and sufficient imagination, to see how complex much of information searching actually is.

Topic 3: Question and answering—the 55 percent rule Up through the 1990s, there was considerable interest and research in library and information science on what was called the "55 percent rule." Researchers addressed the question of whether the answers provided at library reference desks were accurate. This was usually done by sending testers to library reference desks or by calling library reference departments to ask typical reference questions of the librarians, then checking the accuracy of the answers provided. To the researchers' horror, in study after study, the accuracy rate came out, surprisingly, to a fairly consistent figure of around 55 percent. How could this be? Responding to reference queries is the bread and butter of a substantial portion of professional librarians, the so-called public services staff, and to compile an accuracy rate so low was concerning, to say the least. Why did study after study get results around 55 percent?

I first got interested in this question when asked to review a book by Frances Benham and Ronald Powell (1987) that reported results of two separate such studies. The bottom-line results, the accuracy rates for answering questions in the two studies, were 52.73 and 58.73 percent— remarkably similar, considering variations in the sampling of the two studies (Benham & Powell, 1987, p. 136). (There is a large literature on this subject, which will not be reviewed here.)

Something about the consistency of these results across many studies troubled me, however. If training in the reference interview and teaching reference librarians about the typical reference resources available to provide answers to these questions really were the chief influencing factor in performance, then why was there not *more* variation? Surely, smaller libraries with poor reference collections and, probably, staff with less or no professional training should produce poorer results, and the larger, better staffed and stocked libraries should produce much better results. Yet the patterns were very similar.

Then I learned about the US Internal Revenue Service call sites, where people can call the agency to inquire about aspects of the tax laws. Studies had been done over several years, and there was some controversy about the nature of the sampling and test questions; these questions were resolved through a thorough review of methodology. By 1989, with good, verified methodology, the test call survey report stated: "IRS' overall... results for the 1989 tax filing season showed that IRS telephone assistors responded correctly 62.8 percent of the time to the survey's tax law test questions" (US General Accounting Office, 1990, p. 1).

That result rang a bell for me. *If this completely different context could produce a result so similar to that for libraries, then the problem was likely due to something other than simply inadequate library resources or professional training.* As I explored the literature around this question, I came to suspect that the problem concerned any information question-asking situation that was at all complex. In other words, I was forming the hypothesis that in *any* situation where there is a lot of detail or context necessary to fully understand the issue, there will be a 55 percent general rate of accuracy. Indeed, another of the reports on the IRS tests states: "For questions that required IRS assistors to probe callers for more information in order to sufficiently understand the question, the accuracy rate was 56 percent compared to 90 percent for questions where probing was not required" (US General Accounting Office, 1987, p. 2). *I hypothesized that this is a general human communication problem, not a library-specific problem.*

This rate is a source of embarrassment wherever it crops up, because it seems so low. I suspect, however, that the error rate is high because it is impractical for both questioner and assistor or librarian to carry out the necessary amount of probing needed to insure that the question is actually understood in all relevant respects, and can therefore be answered correctly. The questioner has in mind a vast amount of context surrounding the question, and, often, has no way of knowing which particular elements of that context the assistor needs to know in order to produce a correct answer. On the other hand, the assistor does not know that context, and may not realize that some key invalidating characteristic of the context might, if the assistor knew about it, change the assistor's response to accommodate it. There is probably no easy answer to this situation, because to fully insure that all factors have, in fact, been taken into account might require a very long interaction. Most of the time—say, about 55 percent of the time—that extra long interaction would be wasted or unnecessary, because the initial response would be correct, but the rest of the time, a much higher investment might be needed to provide higher accuracy. (This is not to suggest, however, that more complex questions may not, in themselves, and apart from context, be more subject to error, on the part of both assistor and requester, just due to their complexity.)

My research assistant and I did quite a bit of literature searching around this question, looking at informational interactions in law and medicine, as well as in general. There is a whole subliterature about inaccuracy and failure to communicate in physician-patient interactions alone! I planned to devote part of the summer of 1994 to explicating this whole issue of question-asking accuracy. Unfortunately, I was not able to follow up, because that spring the campus administration proposed shutting down our school, and two miserable years followed, wherein I served as department chair as we fought back to keep the program alive, though, ultimately, as a part of a new, combined entity: the Graduate School of Education and Information Studies.

On the basis of what I had found in the literature, and on what, to me, seemed the uniting themes of that research, I planned to further research the hypothesis about information-seeking interaction that summer, and would probably conclude by making the argument stated above. I felt that one particularly telling argument would be this: notice that in cases before a court, where absolute accuracy in all respects is the goal of the questioning of witnesses, a long string of questions is posed to the witness, probing, in numbing detail, every aspect of the situation being discussed. "Was the stop sign visible from where you stopped? Were there any leaves or tree branches blocking your line of sight to the sign? Was the overhead streetlight shining on it?" And on and on and on. The practice of law has demonstrated the need to go far beyond the basic initial interaction in order to get all the relevant facts, to get the details needed to push the answer beyond the point of being just 55 percent accurate. Indeed, even when this probing is done, there may be still much more to the story than comes out in court (Finnegan, 1994).

The relatively poor accuracy results were embarrassing for librarians and for the IRS. I think it is important to realize that in dense, complex question-asking situations, this accuracy problem is probably built into the nature of human knowledge and interaction. We have no hope of improving the librarian success rate without attributing the problem to its likely true sources in general human communication, rather than to library training and resources alone. This topic needs to be further explored, argued, tested. . . .

Area 2: Information organization: Event indexing

At its very heart, all indexing theory is built around the core idea that we index *nouns, or conceptual objects*. As the technical standard for thesaurus development states, an "indexing term" is "the representation of a concept in an indexing language, generally in the form of a noun or noun phrase" (National Information Standards Organization, 2005, p. 6).

I wish to challenge that core assumption to produce a result that can still fit within the larger framework of indexing but that might enrich the possibilities available now. Indexing is generally described as an effort to identify and flag the "subjects" of a document. Those subjects are, implicitly, the topics "covered" or "discussed" in the document. As I write, workers for the San Francisco Bay Area Rapid Transit, or BART, are striking to get more favorable contract terms. Conventionally, a newspaper index would index articles on this event as "strikes and lockouts," "labor unions," or "collective bargaining." These are all noun phrases and certainly represent the topics addressed in the articles, as per our usual understanding of indexing.

But a strike is also an *event* that takes place in time and spools out through time. A little bit of reflection will suggest that while an event can be considered a topic of discussion, if we really want to understand the event, and be able to search for it, should we not develop a way of indexing that is more true to the nature of that event?

Here is one way to think about it. Compare *narrative* and *expository* writing. A *narrative* is a story that takes place through time. It may be fiction or nonfiction. *Exposition* of a topic elaborates and explains that topic. *Standard indexing theory is designed for exposition, not narrative.* Indeed, the question of how to index fiction has been raised many times, and there are no easy answers regarding either the ease of indexing fiction or the value of doing so (Beghtol, 1994; Pejtersen, 1978; Saarti, 1999).

However, here I want to stick to questions of indexing nonfiction, as in newspaper and magazine indexing. Compare topic indexing and event indexing (see table 2.1). In the *topic-indexing* condition, we are addressing a document or document portion devoted to expounding on a topic. The bulk of the text involves description and explanation. If the text is reasonably coherent, it can be considered to be addressing one or more particular topics. Classical indexing attempts to identify those topics and provide good, consistent descriptions of them through the development, control, and application of appropriate index terms.

Now, what do we have in the *event-indexing* condition? First, we have narrative, which is the telling of a story of some sort, spooling out in time. Furthermore, the narrative describes, above all, a *situation*. A situation is a complex of conditions, circumstances, events, and actors existing at a point in time, and often, through time. Within the context of a situation, one or more events may occur. An event is an occurrence, a complex of actors and circumstances that change one condition to another within a noticeably short time frame.

In the example of the Bay Area transit strike, the *situation* is that of contract negotiations between the labor union and BART management. The strike is an *event* that occurs during the contract negotiations.

Now, suppose we are a newspaper needing to index the ongoing events associated with the contract negotiations. We create a template specifically for situations and events. We use the classical aspects for event description that have been identified in journalism: who, what, where, when, why, how. *When* should include duration, that is, beginning and ending times.

Thus, within the larger *situation*, we code for various *events* that occur within it (table 2.2). In indexing, the newspaper creates a situation name and assigns a code number to that situation. At the beginning of contract negotiations, an indexer fills in the names of the parties—the union and the management—notes the dates, location, and so on. Then, as the negotiations continue through various events, the coding and names automatically populate over to the next event, except where the event itself changes things, and then those features are changed by indexers.

Since events are about *things that happen, that is, action,* we should explore what indexing might look like if *verbs* were actually indexed, too.

TABLE 2.1. *Contrasting topic and event indexing*

TOPIC INDEXING	EVENT INDEXING
Exposition	Narrative
Topic	Situation
Description/explanation	Event

TABLE 2.2. *Situation and event coding for indexing*

SITUATION: BART CONTRACT NEGOTIATIONS 2013			CODE: 54025
EVENT	INITIATE TALKS	GO ON STRIKE	RETURN TO TALKS
Who			
What			
Where			
When			
Why			
How			

So, verb indexing terms might be words like "initiate [talks]," "strike," "settle," "conclude [talks]," and so on. Alternatively, a separate set of index terms that are nominalized verbs could be carved out to be used specifically to index events. These would be terms like "initiation of talks," "strike," "settlement," "conclusion [of talks]," and so forth. For a newspaper, or for legal, law enforcement, or medical records, having a separate action category of verbs or of verb-based nouns might be useful for identifying contents and enabling discovery by searchers. For example, an ongoing lawsuit is between two parties, and the case name and parties involved will be scattered throughout the records of the suit. But if one is looking for a particular action—say, to discover when a request was made to enjoin someone from acting—then a search on "enjoin" or "injunction" might be helpful. Such event-indexing terms might be extracted from existing thesauri or created anew as specifically verb index terms to draw attention to the sequence of actions that characterizes events and situations that spool out through time. In a certain sense, events have been masked, or submerged, within the broader scope of classical subject indexing. The approach suggested here highlights the distinctive features of events and situations, that is, of *narratives*, as distinct from *exposition*.

Area 3: Information densities

The term "information densities" actually covers a number of possible areas to pursue. I will develop them here as well as I am able, given that I have not actually pursued this area in depth.

Topic 1: Information whomps In March 1977, Marilyn Levine published an article in the *Journal of the American Society for Information Science* called "The Informative Act and Its Aftermath: Toward a Predictive Science of Information." Levine argued that important events in a society have an impact that can be quantified in (among other ways) the number of books written on that subject or event. The event has an emotional, economic, political (etc.) impact, and people respond by adapting to the event, discussing it, and reorganizing their thinking and their lives in order to accommodate the event and move on. One of the important ways that human beings do all these things is to produce new information expressing their reactions, ideas, and solutions. They then share these ideas with the rest of society through publications, which promotes the absorption and integration of the collective experience into society and people's thinking. Today, unlike in 1977, we see some of this reaction process through the immediate production of "tweets" and other forms of brief communication. But the process of reacting occurs in a deeper and more thorough way as well through the publication of books, articles, and other written communications.

Levine compared the resignation of President Nixon to the resignations of a New York City mayor and of several state governors. She calculated the information "whomps," or the collective impact of the news, as a multiple of the number of people affected in the country, state, or city by the resignation times the level of stress associated with the resignation and the single hard bit of information, namely, that the resignation had occurred. Obviously the resignation of a US president affects more people than the resignation of a governor or mayor. She then looked at the number of books that appeared in subsequent years in the then-standard listing, the *Cumulative Book Index* (H. W. Wilson Company, 1898–1999; the index ceased publication in 1999), and she found that the number of books subsequently appearing about each of the events was roughly proportional with the calculated number of whomps. In other words, societies reflect the impact of important events through publication; the higher the impact, the more publications that result.

Now, I hasten to point out that, in my opinion, there are many methodological and theoretical problems with Levine's article. The initial literature review is superficial and does a poor job of linking together the various theories and bodies of research that she cites. Further, there are methodological issues with how she counts whomps. One might argue that a resignation of a state governor affects more than the citizens of that state, and challenge her basis for calculating the amount of hard information and the degree of stress associated with each type of event. In short, there are a lot of issues with this article, and I doubt that it would be published today in this form. But I am glad that the editor (Art Elias) published the article, because the core idea is a powerful one, and deserving of more attention.

The discussion and disputation of ideas and events is core to any literate society. It seems to be a reasonable hypothesis that the amount of writing and discussion around certain events or issues would be roughly proportional to their importance to our social discourse. Why is there no bibliometric science today focusing on the measurement and discussion of the number of information whomps associated with various events, or debates about why a particular event produced fewer or more publications than one might have expected? Surely, this kind of measurement would be of interest to historians, social scientists, and information scientists. (We may be returning to these questions through another route, for example, the growing attention recently to discovering the significance of various trends in social media. See, e.g., Mike Thelwall's work on Twitter [Thelwall, Buckley, & Paltoglou, 2011; Thelwall et al., 2010].) Thousands of articles have been produced on scientific reputations as measured by number of citations. Though our academic egos are no doubt involved in such questions, surely the study of publication rates as reflections of societal upheaval, stress, and progress is of at least as great importance?

The situation with Levine's idea illustrates another point about creativity in research. Yes, the ideas could have been better developed, and perhaps the colloquialism "whomp" put some people off. But this scrappy little diamond in the rough of an idea is still a diamond—waiting (for the past thirty-five years!) for someone to take it seriously and develop it well and thoughtfully. The lesson: learn to differentiate the several parts of an issue or topic, and select out and develop the good parts. Do not just reject the whole package because some parts of it make you uncomfortable. Derek J. de Solla Price (1975, 1986) studied some basic statistics about publication in the history of science and came up with a remarkable range of important new knowledge about the nature of science and scientific publication patterns. Levine had the great insight to introduce the concept of whomps; whomps are still waiting for their Price for full development.

One final point: I am sure that many people would reject this topic out of hand because of the age of the article. In some contexts, materials that are more than a few months or years old are derisively rejected as being hopelessly out of touch. That is a common way that people unnecessarily limit themselves. In fact, some older things are out of date, and other older things are highly relevant, and can still be the stimulus for further creative developments. One must develop discrimination and selectivity, not just wave something away because it is not currently trendy. *Look for the intrinsic value, not just the current fashion, then reshape the material for the present.* Given the current social media context, and the constant attention to "trending" topics, Levine's idea seems even more relevant—and possibly more easily measured—today than it was at the time of original publication. Even with the demise of the *Cumulative Book Index*, one can still measure the societal impact of various issues and events in book publication, as well as in many other media. Take the earlier idea and adapt it to the current context. Bingo! You have a new area of research.

Topic 2: Information investment As many in the information sciences know, Fritz Machlup (1972) argued that a very large portion of all human economic activity centers on the production and use of information. The book blurb on the publisher's website for his book *The Production and Distribution of Knowledge in the United States* summarizes it nicely: "Machlup's cool appraisal of the data showed that the knowledge industry accounted for nearly 29 percent of the U.S. gross national product, and that 43 percent of the civilian labor force consisted of knowledge transmitters or full-time knowledge receivers. Indeed, the proportion of the labor force involved in the knowledge economy increased from 11 to 32 percent between 1900 and 1959—a monumental shift" (Princeton University Press, 2013).

Robert Hayes, professor and former dean at the Graduate School of Library and Information Science at UCLA, produced some research in the early 1980s that I have long felt has much more potential than has actually been developed. A much more recent discussion of this area can be found in Koenig and Manzari's (2010) encyclopedia entry entitled "Productivity Impacts of Libraries and Information Services." Hayes and Erickson (1982), according to Koenig and Manzari, used

> the Cobb-Douglas production function to estimate the value added by information services. In the basic Cobb-Douglas formula, the value of goals [*sic:* goods?] and services sold is calculated to be the product of a constant times the values of different inputs, labor, capital, and so forth, each raised to a different power (exponent). The exponents are solved for by seeing which exponents best fit a number of separate cases. In the Hayes and Erickson formulation, the value added V in manufacturing industries is a function of labor L, capital K, purchase of information services I, and purchase of other intermediate goods and services X. (Koenig & Manzari, 2010, p. 4311)

Hayes and Erickson found that those industries that thrived best over the period he studied were those, such as pharmaceuticals, that had very much larger investments in information relative to the other Cobb-Douglas factors than other industries had. Koenig and Manzari note that Yale Braunstein continued the work, making some modifications in the formula, yet still came out with much the same result—that there is "substantial underinvestment in the purchase of information" (Koenig & Manzari, 2010, p. 4311).

I leave it to economists to explicate these ideas—but if the premise holds true that there is huge underinvestment in information in the economy, this should have huge implications for many disciplines, including ours. The argument should be researched and developed further, and the results published in the economics literature, as well as the information sciences literatures.

Topic 3: Distilled information There is yet another sense in which we seldom discuss information, yet that may have interesting implications for theory in the information sciences as well as, ultimately, for practice. This sense concerns the idea of information as a distilled product of thought and communication.

Books, journal articles, newspaper articles, blog entries—even micro-messages such as "tweets"—are all different forms of work products. Often, they have serious economic value or economic implications. Someone put some brainpower into the creation of them. Furthermore, having created the text, images, and so on, human beings have engaged in various forms of economic production to make them available, whether investing in a printing press and paper or in a computer server to hold this work product ready for access on the Internet.

Both the intellectual investment and the physical investment made so these products will be available represent the information economy in the Machlup and Hayes and Erickson senses. We invest a substantial portion of both our cognitive and physical energies as a society in creating, distributing, and maintaining (in libraries, on the web) these intellectual products. The quantity and quality of the work that goes into these information products varies tremendously, of course. But the same thing about quality and quantity of work could be said of retail clothing, drugs, or many other products of human activity.

But let us focus particularly on these intellectual products. An academic book, for example, may represent a decade, or a lifetime, of research, study, analysis, and writing. We may think of the resulting book as a distillation of all that massive amount of work by an individual. The encyclopedia that I recently edited with Mary Niles Maack (Bates & Maack, 2010) had about seven hundred authors, each of whom spent some substantial chunk of time in the preparation and writing of journal-article-length entries. I was retired and spent four years full time as editor-in-chief of the encyclopedia, while Maack spent many weeks and months as co-editor, taken from her otherwise hectic schedule as a professor, to produce the resulting seven-volume, 5,742-page encyclopedia. On the usual rule of thumb of two thousand hours of work for an individual for a year, the editing alone took five or more person-years. If each of the 565 articles took a month of person-time, which may be an underestimate, forty-seven or more years of person-time went into the creation of the encyclopedia entries. The resulting seven-volume set, taking up a little more than a cubic foot of volume, represents over fifty full-time person-years of human thought, writing, and editing effort.

In fact, each book or other information product can be thought of as such a distillation. *In the network of human relationships and social activity, each document is a node of distillation, a point of intensification of human labor and intellection.* We have developed means of condensing much thought into small packages, and we store those nodes of information all around us. Picture one of those economic atlases that shows the countries of the world, with clear plastic layers that may be laid over the map, each layer representing something of economic interest, such as agricultural production, manufacturing production, and so on. Lay over that map an additional layer that shows the location of information stores. If this is done, then server farms, libraries, bookstores, websites, and many more resources all become points of great intensification of information in the economic map of society.

In my view, the huge amount of energy invested in bibliometric studies of various kinds should include the study of the distribution of these information densities in society, followed by the study of the amount of use that is made of them, where they are underutilized according to formulas such as Cobb-Douglas, and so on.

Concluding thoughts

The role of information in the economy and in social relations is so integral to all we do that it is like the air we breathe. Sometimes we do not see it at all because we take it so much for granted. But it has great meaning and social impact in terms of the mental and physical processing and organizing of our world. There are so many more ways to think about and study the role of information than we have fully engaged in; we have the prospect of huge disciplinary development in the information sciences if we can only start to see the (informational) air we breathe.

REFERENCES

Bates, M.J. (1976). Rigorous systematic bibliography. *RQ, 16*(1), 7–26.
Bates, M.J. (1979a). Idea tactics. *Journal of the American Society for Information Science, 30*(5), 280–289.
Bates, M.J. (1979b). Information search tactics. *Journal of the American Society for Information Science, 30*(4), 205–214.
Bates, M.J. (1989). The design of browsing and berrypicking techniques for the online search interface. *Online Review, 13*(5), 407–424.
Bates, M.J. (1999). The invisible substrate of information science. *Journal of the American Society for Information Science, 50*(12), 1043–1050.
Bates, M.J. (2002). The cascade of interactions in the digital library interface. *Information Processing & Management, 38*(3), 381–400.

Bates, M.J. (2005). An introduction to metatheories, theories, and models. In K.E. Fisher, S. Erdelez, & L. McKechnie (Eds.), *Theories of Information Behavior* (pp. 1–24). Medford, NJ: American Society for Information Science and Technology.

Bates, M.J. (2007a). Defining the information disciplines in encyclopedia development. *Information Research, 12*(4), paper colis29. Retrieved from http://InformationR.net/ir/12-4/colis/colis29.html.

Bates, M.J. (2007b). What is browsing—really? A model drawing from behavioural science research. *Information Research, 12*(4), paper 330. Retrieved from http://InformationR.net/ir/12-4/paper330.html.

Bates, M.J. (2015). The information professions: Knowledge, memory, heritage. *Information Research, 20*(1), paper 655. Retrieved from http://InformationR.net/ir/20-1/paper655.html.

Bates, M.J., & Maack, M.N. (Eds.). (2010). *Encyclopedia of Library and Information Sciences* (3rd ed., 7 Vols). New York: CRC Press.

Bates, M.J., Wilde, D.N., & Siegfried, S. (1993). An analysis of search terminology used by humanities scholars: The Getty Online Searching Project report no. 1. *The Library Quarterly, 63*(1), 1–39.

Bateson, G. (1968). Information and codification: A philosophical approach. In J. Ruesch and G. Bateson (Eds.), *Communication: The social matrix of psychiatry* (pp. 168–211). New York: Norton.

Beghtol, C. (1994). *The classification of fiction: The development of a system based on theoretical principles.* Metuchen, NJ: Scarecrow Press.

Benham, F., & Powell, R.R. (1987). *Success in answering reference questions: Two studies.* Metuchen, NJ: Scarecrow Press.

Finnegan, W. (1994). Doubt. *New Yorker, 49*(48), 48–67.

Hartel, J. (2010a). Managing documents at home for serious leisure: A case study of the hobby of gourmet cooking. *Journal of Documentation, 66*(6), 847–874.

Hartel, J. (2010b). Time as a framework for information science: Insights from the hobby of gourmet cooking. *Information Research, 15*(4), paper colis715. Retrieved from http://InformationR.net/ir/15-4/colis715.html.

Hartel, J. (2011). Visual approaches and photography for the study of immediate information space. *Journal of the American Society for Information Science and Technology, 62*(11), 2214–2224.

Harter, S.P. (1992). Psychological relevance and information science. *Journal of the American Society for Information Science, 43*(9), 602–615.

Hayes, R.M., & Erickson, T. (1982). Added value as a function of purchases of information services. *Information Society, 1*(4), 307–338.

Hsieh-Yee, I. (1993). Effects of search experience and subject knowledge on the search tactics of novice and experienced searchers. *Journal of the American Society for Information Science, 44*(3), 161–174.

Koenig, M., & Manzari, L. (2010). Productivity impacts of libraries and information services. In M.J. Bates and M.N. Maack (Eds.), *Encyclopedia of library and information sciences* (3rd ed., pp. 4305–4314). New York: CRC Press.

Kuhn, T.S. (2012). *The structure of scientific revolutions* (4th ed.). Chicago, IL: University of Chicago Press.

Levine, M.M. (1977). The informative act and its aftermath: Toward a predictive science of information. *Journal of the American Society for Information Science, 28*(2), 101–106.

Machlup, F. (1972). *The production and distribution of knowledge in the United States.* Princeton, NJ: Princeton University Press.

National Information Standards Organization. (2005). *Guidelines for the construction, format, and management of monolingual controlled vocabularies* (ANSI/NISO Z39.19-2005). Bethesda, MD: NISO Press.

Pejtersen, A.M. (1978). Fiction and library classification. *Scandinavian Public Library Quarterly, 11*(1), 5–12.

Price, D.J. de Solla. (1975). *Science since Babylon.* New Haven, CT: Yale University Press.

Price, D.J. de Solla. (1986). *Little science, big science—and beyond.* New York: Columbia University Press.

Saarti, J. (1999). Fiction indexing and the development of fiction thesauri. *Journal of Librarianship and Information Science, 31*(2), 85–92.

Sperber, D., & Wilson, D. (1995). *Relevance: Communication and cognition* (2nd ed.). Malden, MA: Blackwell.

Thelwall, M., Buckley, K., & Paltoglou, G. (2011). Sentiment in Twitter events. *Journal of the American Society for Information Science and Technology, 62*(2), 406–418.

Thelwall, M., Buckley, K., Paltoglou, G., Cai, D., & Kappas, A. (2010). Sentiment strength detection in short informal text. *Journal of the American Society for Information Science and Technology, 61*(12), 2544–2558.

US General Accounting Office. (1987). *Tax administration: Accessibility, timeliness, and accuracy of IRS' telephone assistance program* (GAO/GGD-88-17). Retrieved from http://www.gao.gov/assets/210/209820.pdf.

US General Accounting Office. (1990). *Tax administration: Monitoring the accuracy and administration of IRS' 1989 test call survey* (GAO/GGD-90-37). Retrieved from http://www.gao.gov/assets/220/211963.pdf.

White, H. D. (2007a). Combining bibliometrics, information retrieval, and relevance theory, Part 1: First examples of a synthesis. *Journal of the American Society for Information Science and Technology, 58*(4), 536–559.

White, H. D. (2007b). Combining bibliometrics, information retrieval, and relevance theory, Part 2: Some implications for information science. *Journal of the American Society for Information Science and Technology, 58*(4), 583–605.

Wilson, P. (1968). *Two kinds of power: An essay on bibliographical control.* Berkeley, CA: University of California Press.

Wolfe, J.M. (1994). Guided search 2.0: A revised model of visual search. *Psychonomic Bulletin & Review, 1*(2), 202–238.

Xie, H. (2000). Shifts of interactive intentions and information-seeking strategies in interactive information retrieval. *Journal of the American Society for Information Science, 51*(9), 841–857.

7

An introduction to metatheories, theories, and models

Introduction

The objective of this chapter is to provide a general introduction to some key theoretical concepts of use in library and information science (LIS) research. First, the three terms in the title—metatheory, theory, and model—are defined and discussed. Next, an extended example is provided of a case in which a researcher might consider and test various models or theories in information-seeking research. Next, metatheories are considered at greater length, and the distinction is made between nomothetic and idiographic metatheories. Finally, 13 metatheoretical approaches in wide use in LIS are described. Explanatory texts are referenced, as well as example studies using each approach. The discussion is necessarily brief and simplifying.

Definitions

It is important, first, to distinguish the terms metatheory, theory, and model. These concepts are often confused and used interchangeably. They

First published as Bates, M. J. (2005). An introduction to metatheories, theories, and models. In K. E. Fisher, S. Erdelez, & L. McKechnie (Eds.), *Theories of information behavior* (pp. 1–24). Medford, NJ: American Society for Information Science and Technology.

should not be, as understanding the distinctions among them can help in thinking about theoretical aspects of LIS.

- Metatheory A theory concerned with the investigation, analysis, or description of theory itself. (*Webster's Unabridged Dictionary*)

- Theory (a) The body of generalizations and principles developed in association with practice in a field of activity (as medicine, music) and forming its content as an intellectual discipline. . . . *Webster's Unabridged Dictionary*) (b) A system of assumptions, accepted principles, and rules of procedure devised to analyze, predict, or otherwise explain the nature or behavior of a specified set of phenomena. (*American Heritage Dictionary*, 1969). (See also Reynolds, 1971.)

- Model A tentative ideational structure used as a testing device. . . . (*American Heritage Dictionary*, 1969). (See also Lave & March, 1975.)

Metatheory can be seen as the philosophy behind the theory, the fundamental set of ideas about how phenomena of interest in a particular field should be thought about and researched (see also Wagner & Berger, 1985; Vakkari, 1997). The term has not been used much in LIS, but it is rapidly becoming more important to our understanding. In earlier years, the underlying philosophy behind research in the field could be identified as coming from few directions—from a general humanities approach and a general scientific approach. In recent years, however, more and more metatheoretical approaches have been developed within the field and borrowed from other fields. The result has been that we now have a confusion of many approaches competing for attention.

The concept of a metatheory has a lot of overlap with the term "paradigm," which was given its modern understanding in science by Thomas Kuhn (1996). In the terms used here, Kuhn considered a paradigm to be the metatheory, the theory, the methodology, and the ethos, all combined, of a discipline or specialty. So paradigm would have a broader meaning than metatheory. At the same time, metatheory is absolutely core to any paradigm, and is defining of a paradigm in many senses.

Theory, as defined in definition (a), can be thought of as the entire body of generalizations and principles developed for a field, as in "the theory of LIS." Second, and more of interest for this paper, is the concept of a single theory. A theory is a system of assumptions, principles, and relationships posited to explain a specified set of phenomena. Theories often carry with them an implicit metatheory and methodology, as in the "rules of procedure" in definition (b). However, for most purposes, the core meaning of theory centers around the idea of a developed understanding, an explanation, for some phenomenon.

Models are of great value in the development of theory. They are a kind of proto-theory, a tentative proposed set of relationships, which can then be tested for validity. Developing a model can often help in working through one's thinking about a subject of interest. Indeed, there is not always a sharp dividing line between a model and a theory about the same phenomenon. Models sometimes stand as theoretical beacons for years, guiding and directing research in a field, before the research finally matures to the point of producing something closer to a true theory.

In science, a classic sequence of development has been characterized as "description, prediction, explanation." That is, the first task when studying a new phenomenon is to describe that phenomenon. It is difficult to think about something if you know very little about it. So description comes first. Second, once one knows something about a phenomenon, it should be possible to predict relationships, processes, or sequences associated with the phenomenon. Third, based on the testing of predictions, one should be able to develop an explanation of the phenomenon, that is, a theory. Theories can always be overturned by later theories; even when a theory has been well tested it is always possible that later research will provide a more thorough, deeper explanation for the phenomenon of interest.

Models are most useful at the description and prediction stages of understanding a phenomenon. Only when we develop an explanation for a phenomenon can we properly say we have a theory. Consequently, most of "theory" in LIS is really still at the modeling stage.

In the next section, an example proto-theory, or model, is analyzed, and means of testing the model are discussed. However, some metatheories explicitly eschew the value and possibility of generalizing the studied reality of a situation in order to create a theory. Ethnomethodology, for example, "never bought into the business of theorizing, it was iconoclastic, it would not theorize foundational matters" (Button, 1991, pp. 4–5). Rather, ethnomethodologists "generally decline to theorize about the social world, preferring instead to go out and study it" (Ritzer, 2000, p. 75). At a minimum in the following discussion, one must assume a metatheoretical position that allows for and legitimates models and theories. So the following discussion cannot be applicable to every possible metatheoretical position.

Example using these terms

Let us take, as an example, the Principle of Least Effort. This is probably the most solid result in all of information-seeking research. Specifically, *we have found that people invest little in seeking information, preferring easy-to-use, accessible sources to sources of known high quality that are less easy to use and/ or less accessible.* Poole (1985) did a meta-analysis of 51 information-seeking studies, in which he found this proposition strongly confirmed. (He also has a good discussion of theory in LIS.)

So ease of use and accessibility of information seem to be more important to people than quality of information. But what is the *explanation* for this phenomenon? Why are people unwilling to invest that little bit of extra energy in order to get information that they themselves would acknowledge is of better quality? We do not really have a theory. We have described the phenomenon; further, we have found this to be the case in many different environments with many different types of people, so it is a result that appears to be highly generalizable. Consequently, we can also confidently make predictions from these results. For example, we can predict that when we study a new group of people, they will probably also invest little energy in information seeking, and prefer easy-to-use, accessible resources.

So, through description and prediction we have modeled the Principle of Least Effort. Though we often represent models in diagrams that display relationships, we do not have to do so. In this case, our model can be described in a sentence (see the italicized statement above). (For some examples of models presented in diagram form, see Bates, 2002; Gaines, Chen, & Shaw, 1997; Metoyer-Duran, 1991; Wang & White, 1999; Wilson, 1999.) So the Principle of Least Effort is an observed behavior, one we have observed widely enough to confidently model as a principle. But we do not yet have an explanation—so we do not yet have a theory.

How can we move this research from being a model to being a theory? First, we can hypothesize various possible explanations based on work we find elsewhere in the field or in other fields. Here are some I have thought of:

1) People "satisfice" in all realms of life, including information seeking. The idea of satisficing comes from Simon (1976), who argued that in decision making, people make a good enough decision to meet their needs, and do not necessarily consider all possible, or knowable, options. Translated to the language of LIS, for example, using Dervin's concept of "Sense-Making" (Dervin, 1983, 1999), we could hypothesize that people make sense of their situations based on what they know and can learn

easily. Their Sense-Making need only be adequate to continue with life; it does not need to be so perfect or extensive as to enable them to make sense of everything.

2) People underestimate the value of what they do not know, and overestimate the value of what they do know. People have difficulty imagining what the new information would be that they do not know, while what they do know is vivid and real to them. Consequently, they underinvest in information seeking. See Gilovich, Griffin, and Kahneman (2002) and Kahneman and Tversky (2000) for work on distortions in decision making and choice.

3) Gaining new knowledge may be emotionally threatening in some cases. Gregory Bateson once described what he called "value-seeking" and "information-seeking" (Ruesch & Bateson, 1968, pp. 178–179). In value-seeking, a person has an idea in mind of something that he or she wants. Suppose one wants some eggs and toast to eat, for example. One then goes out into the world, does various things involving chickens, grain, cooking, and baking, with the end result that one has a breakfast of eggs and toast. Thus, one has done things to parts of the world in order to make the world match the plan one has in mind. In information seeking, on the other hand, according to Bateson, the directionality is reversed; one acquires information from the world in order to impress it on one's own mind.

However, new knowledge can always bring surprises, sometimes, uncomfortable ones. If "we are what we know," if our sense of self is based, in part, on our body of knowledge of the world, then to change that knowledge may be threatening to our sense of self.

4) Information is not tangible, and objects are. Intangible things seem less real to us, therefore less valuable. Consequently, we invest more in acquiring tangible than intangible things.

Each hypothesis above is not a complete explanation. For instance, *why* do people satisfice? However, if we were to test this satisficing hypothesis and we learned that people do satisfice in information seeking, we would have an explanation that tells us more than just the *observed fact* of least effort. We would then be able to place this result in the context of all the other research in other disciplines that has observed that people satisfice in a variety of circumstances, and could then draw on that research to develop tentative explanations (tentative theories) that go deeper than the satisficing explanation alone.

In fact, Simon's satisficing may be, in effect, another name for Zipf's Principle of Least Effort (1949). Poole (1985) believed his results fit well with Zipf's earlier work. Zipf had a more extensively conceptualized understanding of least effort, one that constitutes a preliminary explanation, i.e., theory, and which contributes to a better understanding of least effort than we usually articulate in LIS. To Zipf, according to Poole, least effort was technically the "least average rate of probable work" (Poole, 1985, p. 90). That is, people do not just minimize current work associated with some activity, because they could eventually do a total of much more work in the end. Rather, they make a considered estimate of all likely work associated with a given effort, now and in the future, and do the amount of work now that they estimate will best reduce their overall effort, now and later combined (Poole, 1985).

How could we test these four hypotheses listed? In each case one or more studies could be designed in order to attempt to discover which, if any, of these explanations is operating in people's information seeking. For example, in an experimental approach to Hypothesis 2, people could be placed in a realistic situation where they have certain information and do not have other information. They have to expend units in order to "purchase" additional clues or hints to solve the test problem. There are other ways they can expend those same units. The experimental subjects assign their units according to their best judgment. Afterward, they are given the information they did not have earlier. Do they now rate higher or lower the value of the information that they had not had in the test situation? On what basis do they assign value at each step of the experiment?

In an observational approach to Hypothesis 3, people could be studied in real information-seeking situations—suppose in three different types of situations: 1) finding information about a disease diagnosed in a family member, 2) researching a paper in a required course on a topic of little interest, 3) finding out more about a hobby or avocation (Hartel, 2003). Searching could be observed and the subjects interviewed about their feeling reactions to their situation and the acts of information seeking in which they engage. Do they avoid new information or seek it eagerly? Are there signs of anxiety and threat around discovering new information? Do people have different responses to the different types of situation, and why?

In the example above, we started with a descriptive finding—the widely observed tendency of people to prefer easy-to-use and accessible information sources over harder to get, higher quality sources of information. This "Principle of Least Effort" has been so widely observed that we were able to make confident predictions about where else it might appear as well. But we still had no explanation, no theory as to why this phenomenon occurs (except possibly in Zipf's original research, 1949). We hypothesized four possible explanations, and considered ways in which these theories could be tested. Testing might then lead to further tentative theories that would explain this phenomenon still more deeply.

Sources of metatheories

In the preceding section much was made of models and theories. What about metatheories? Where do they fit in? As Kuhn observed, in most natural sciences most of the time, there is a single predominant paradigm out of which researchers identify and test research questions. Metatheories about the nature of research and the desirable methods for each discipline are embedded in those paradigms. In the social sciences, however, it is more common to have a general paradigm for a field, which describes the domain of interest for that discipline—the operations of the mind for psychology, for example—but more than one metatheory, or philosophy of research, competing for the loyalties of researchers within that discipline. In the case of psychology, in the 1960s and 1970s there was a split between an older, behaviorist metatheory for the study of psychology (Skinner, 1992 reprint), and a newer, information processing approach (Chomsky, 1959; Anderson, 1995). The split went so deep that the latter approach came to be known by a different name, cognitive science. Over the last 10 to 15 years another metatheory, by the name of evolutionary psychology, has challenged the information processing approach (Barkow, Cosmides, & Tooby, 1992).

In the sciences, a new paradigm usually revolutionizes the field, that is, the new paradigm reconfigures all prior learning around a new core metatheory and body of research results. Examples have been plate tectonic theory in geology and molecular biology in biology. In the social sciences, however, several metatheories may continue side by side. Sometimes a metatheory will simply die out and other times it will grow and change, and still compete for the interest of researchers.

In the late 20th and early 21st centuries, there has been a proliferation of metatheories in the social sciences generally, and, certainly, in LIS as well. In our society, in general, old ways of thinking are breaking up and breaking down; supposed eternal verities are falling right and left, from the fall of the Berlin Wall and all that it meant about rigid social structures in East and West, to social boundaries that formerly split communities by race, gender, religion, and other long-standing, stable divisions. Even the eternal verities of forms of writing—the book, the journal, the newspaper article—are being shaken up in the new world of Internet information.

Under these circumstances, we should perhaps not be surprised that basic metatheoretical assumptions about what research is or should be are also breaking down and being challenged by newer approaches. I think it is also the case that different people have different cognitive styles, certain ways of thinking that are natural to them. We are all drawn to the sort of research and thinking that works best for us, that is most harmonious with the way our minds work. Wagner and Berger (1985) call these "orienting strategies."

In earlier, more rigid times, it tended to be the case that only certain orienting strategies were considered legitimate in a given field at one time. Heaven help the psychology doctoral student who wanted to take a qualitative approach back in the heyday of behaviorism, for example. Many talented people were forced out, simply because they had the wrong cognitive style for the intellectual spirit of the times. Now, there is generally more tolerance for different approaches, although there is still some tendency to argue that one's own preferred approach is the one true or best philosophy of research, and everything else is bunk.

I believe that the intensity of these struggles arises, in part, out of the different cognitive styles people have, which then draw them to corresponding different orienting strategies. It just feels so right to follow one's preferred approach that it just must be the case that the other guys are all wrong. However, I believe that every orienting strategy brings us something valuable, if we are only open to learn what it has to offer.

Thus, it is likely that there will continue to be several approaches in LIS to studying the phenomena of interest to our field. When one takes up a particular approach, however, it is important to understand the philosophy and some of the history behind the development of a particular research approach. That way, there will be a smooth and logically consistent passage from philosophy to theory and methodology.

The nomothetic-idiographic contrast

First, we need to make a distinction between what are known as nomothetic and idiographic approaches to research. These two are the most fundamental orienting strategies of all.

- *Nomothetic*—"Relating to or concerned with the study or discovery of the general laws underlying something" (*Oxford English Dictionary*).

- *Idiographic*—"Concerned with the individual, pertaining to or descriptive of single and unique facts and processes" (*Oxford English Dictionary*).

The first approach is the one that is fundamental to the sciences. Science research is always looking to establish the general law, principle, or theory. The fundamental assumption in the sciences is that behind all the blooming, buzzing confusion of the real world, there are patterns or processes of a more general sort, an understanding of which enables prediction and explanation of the particulars.

The idiographic approach, on the other hand, cherishes the particulars, and insists that true understanding can be reached only by assembling and assessing those particulars. The end result is a nuanced description and assessment of the unique facts of a situation or historical event, in which themes and tendencies may be discovered, but rarely any general laws. This approach is the one that is fundamental to the humanities. (See an excellent discussion of these science/humanities theoretical differences in Sandstrom & Sandstrom, 1995; see also discussion in Bates, 1994.)

For the last couple of centuries, the social sciences have been the crossroads where these two approaches intersect, the ground over which the nomothetic and idiographic orienting strategies have fought. One of the common narratives of the 20th century was of the academic social science department, say, political science or economics, being invaded by newcomers with a mathematical or scientific approach to their subject, in opposition to the prior discursive, idiographic approach. In the late 20th century, that narrative was often reversed, when postmodernist theorists came into departments and superseded the more nomothetically oriented researchers who had been there previously.

LIS has not been immune to these struggles, and it would not be hard to identify departments or journals where this conflict is being carried out. My position is that both of these orienting strategies are enormously productive for human understanding. Any LIS department that definitively rejects one or the other approach makes a foolish choice. It is more difficult to maintain openness to these two positions, rather than insisting on selecting one or the other, but it is also ultimately more productive and rewarding for the progress of the field.

Metatheories in LIS

The purpose of this section is to present brief descriptions of a number of the more popular metatheories that are being expressed in LIS these days. The arraying of these approaches in a common framework may be helpful for beginners in understanding the range of research approaches taken in LIS.

There are many metatheories operating in the field currently. There is disagreement between proponents of various metatheories, and there are also various interpretations and descriptions of any one metatheory.

Furthermore, researchers become interested in new approaches as they appear in the field, and may change metatheories and methodologies during their career. Examples given below should be seen as just that, examples; researchers should not be assumed to be always unequivocally associated with a single metatheoretical approach.

It should also be understood that what is presented below is a *personal, idiosyncratic, and simplifying* selection. See Cool (2001); Hjørland (1998, 2000); Pettigrew, Fidel, and Bruce (2001); and Talja, Tuominen, and Savolainen (2005) for other categorizations of metatheories.

For expositions and debates on metatheory and methodology in LIS, see Bar-Ilan and Peritz (2002); Bates, J. A. (2004); Bates, M. J. (1999); Case (2002); Crabtree et al. (2000); Dervin (1999, 2003); Dick (1995, 1999); Ellis (1992); Fidel (1993); Given and Leckie (2003); McClure and Hernon (1991); McKechnie (2000); Pettigrew and McKechnie (2001); Powell (1997, 1999); Sandstrom and Sandstrom (1995, 1998); Sonnenwald and Iivonen (1999); Talja (1999, 2001); Thomas and Nyce (1998); Trosow (2001); Wang (1999); and Westbrook (1994).

With the description of each metatheory below, example applications are provided where possible, and textual sources explaining or elaborating on the various metatheories are also suggested. The listing begins with idiographic approaches in numbers 1–5, mixed approaches in numbers 6 and 7, and primarily nomothetic approaches in numbers 8–13.

1) A *historical* approach, in which understanding of the present is seen to arise out of an understanding of the past social, political, and economic events and processes, which have led to current conditions. For historical methods and issues, see Barzun and Graff's classic work (1992), as well as Appleby, Hunt, and Jacob (1994), and Rayward (1996). For examples of historical research in LIS, see Hildenbrand (1996), Maack (2000), and Wiegand and Davis (1994).

2) A *constructivist* approach, arising out of education and sociology, in which individuals are seen as actively constructing an understanding of their worlds, heavily influenced by the social world(s) in which they are operating. According to Kuhlthau (1993), educational constructivist theory built on the work of Dewey (1933, 1944), Kelly (1963), and Vygotsky (1978), among others, while, according to Ritzer (2000), sociological constructivist theory arose from Schutz (English translation 1967, original 1932), Berger and Luckmann (1990 reprint), and the closely related ethnomethodological work of Garfinkel (1967). Major proponents of this approach in LIS have been Dervin (1983, 1999) and Kuhlthau (1993).

3) A *constructionist or discourse-analytic* approach, with both humanities and social sciences roots, in which it is assumed that the discourse of a

society predominately conditions the responses of individuals within that society, including the social understanding of information. According to Talja, Tuominen, and Savolainen (2005), constructionism sees "language as constitutive for the construction of selves and the formation of meanings." Further, "We produce and organize social reality together by using language." This metatheory arose from the work of Bakhtin (Holquist, 2002) and Foucault (1972), among others. Frohmann (1994) and Talja (1999) have expounded on the use of this approach in LIS. This approach has been applied in LIS by Budd and Raber (1996), Frohmann (2001), and Talja (2001), among others. A non-LIS, but highly relevant example can be seen in Hayles (1999).

4) A *philosophical-analytic* approach, in which the classical techniques of the discipline of philosophy, namely extremely rigorous analysis of ideas and propositions, are brought to bear on information-related matters. Certainly, the field of philosophy itself expresses and represents many different theoretical orientations and metatheories. However, despite the many differences among philosophers, there is a fairly universal and well-understood form of analysis and argumentation that is characteristic of the discipline as a whole. Philosophers who have come into LIS, or philosophers outside the field who have addressed LIS-related questions inevitably bring with them this mode of analysis and discourse. For a classic example of this, read Patrick Wilson's still-relevant discussion on the nature of the subject of a book (Wilson, 1968, pp. 69–92). See also Blair (2003), Cooper (1971), Dretske (1981), Fuller (2002), and Wilson (1977, 1983).

5) A *critical theory* approach, in which the hidden power relations and patterns of domination within a society are revealed and debunked (Ritzer, 2000, p. 140ff). Michael Harris (1986) was an early practitioner in LIS. More recently, others have joined the debate, critiquing the roles of librarians, the kinds of research done in LIS, and so on. See Carmichael (1998), Chu (1999), Day (2001), Roma Harris (1992), Pawley (1998), Radford (2003), and Wiegand (1999).

6) An *ethnographic* approach, originating in anthropology, but now used throughout the social sciences, involving the use of a variety of field techniques, such as observation, documentation, and interviewing. These techniques are intended to enable the researcher to become immersed in a culture, identify its many elements, and begin to shape an understanding of the experience and world views of the people studied (Fielding, 1993). In LIS, see, for example, Chatman (1992), Kwasnik (1992), Pettigrew (2000), and Wilson and Streatfield (1981). A related, popular approach is grounded theory development (Glaser & Strauss, 1967). See Ellis (1993), Ellis and Haugan (1997), Kwasnik (1991), and Mellon (1986). Sandstrom and Sandstrom

(1995) discuss the ways in which both nomothetically and idiographically oriented researchers have used ethnographic methods.

7) A *socio-cognitive* approach (Hjørland, 2002), in which both the individual's thinking and the social and documentary domain in which the individual operates are seen to influence the use of information. See also Jacob and Shaw (1998). Paisley presaged this viewpoint in his 1968 "Information needs and uses" review of scientists working within 10 social and information system contexts (Paisley, 1968). More recently, see Case (1991), Covi (1999), and Kwasnik (1991). The nature of context has been discussed in detail by Dervin (1997), and the nature of situation by Cool (2001). Because of the centrality in information studies of 1) information, 2) information technology, and 3) people's use of these, the interplay among these three elements is arguably at the heart of most social research in information studies.

Hjørland and Albrechtsen (1995) call the analysis of information and its social formation in a community of thought "domain analysis." Other roots of the domain analytic approach can be seen in the areas of historical and descriptive bibliography in librarianship (Bowers, 1994; Updike, 2001), as well as in recent developments around genre theory (Berkenkotter & Huckin, 1993; Vaughan & Dillon, 1998; Orlikowski & Yates, 1994).

The field of social informatics also focuses on the interactions among people, social environments, information technology, and documentary forms. See Bishop and Star's review (1996), as well as work by Kling and McKim (2000), and Palmer (2001). This metatheory shares some of both the nomothetic and idiographic orientations.

8) A *cognitive* approach, arising out of cognitive science, in which the thinking of the individual person operating in the world is the dominating focus of research on information seeking, retrieval, and use (Bates, 1979; Belkin, 1990; Belkin, Oddy, & Brooks, 1982; Ellis, 1989; Ingwersen, 1992, 1999). See Newell and Simon (1972) and Anderson (1995) for expositions of this approach.

9) A *bibliometric* approach, in which the analysis of the statistical properties of information is seen to provide understanding of value for both the design of information provision and the theoretical understanding of social processes around information, including historical processes. The earliest theory was provided by Bradford (1948) and Zipf (1949). More recent major work has been done by Brookes (1968), Price (1986), Small (1999), and White and McCain (1998), among others. Much of this work has been made possible through the existence of citation indexes (Garfield, 1983).

10) A *physical* approach to information transfer, dating principally from the 1950s and 1960s interest in signaling and physical communication

generated by the development of Claude Shannon's information theory (Cherry, 1966; Miller, 1951; Pierce, 1961; Shannon & Weaver, 1975; Wiener, 1961).

11) An *engineering* approach to information, in which it is assumed that human needs and uses of information can best be accommodated by successive development and testing of ingenious systems and devices to improve information retrieval and services. The fundamental test of validity for the engineering approach is an operational one, namely, "Does it work?" Thus a major method of developing new knowledge in engineering is through "proof of concept" work, in which an experimental system or device is developed and tested, improved, tested some more, and so on. For theory of engineering, see Dahlbom, Beckman, and Nilsson (2002) and Simon (1981). For applications in LIS, see Croft and Thompson (1987), Hendry and Harper (1997), Kraft and Petry (1997), Over (2001), and Salton and McGill (1983). Variations on this approach are found in artificial intelligence (Minsky, 1968; Russell & Norvig, 1995) and natural language processing (Allen, 1995; Chowdury, 2003; Liddy et al., 1993).

12) A *user-centered design* approach, in which the development and human testing of information organization and information system designs is seen as a path to both scientific understanding and improved information access. User-centered design takes the "Does it work?" engineering question one step farther, and asks, "Does it work so well that people can concentrate on what they are doing rather than on operating the system or device?" Classic work in this area is by Norman (1990) and Nielsen (1993). A great deal of design work relevant to LIS goes on in human-computer interaction research (Carroll, 2002; Rogers, 2004). A number of people in LIS focus on user-centered design, for example, Ackerman (2000), Bates (1990, 2002), Dillon (1994, 1995), Hildreth (1989), and Marchionini (1995). See also Marchionini and Komlodi (1998).

13) An *evolutionary* approach, in which the insights of biology and evolutionary psychology are brought to bear on information-related phenomena (Barkow, Cosmides, & Tooby, 1992; Wright, 1994). This approach is just beginning to appear in LIS. See Bates (2005, 2006), Madden (2004), and Sandstrom (1994, 1999).

Each of the metatheories above is some part philosophy and some part methodology. However, the historical, philosophical-analytic, ethnographic, bibliometric, engineering, and design approaches are primarily methodology with some philosophy attached, while the others, the constructivist, discourse-analytic, critical theory, socio-cognitive, cognitive, physical, and evolutionary approaches are driven more by philosophical and theoretical orientations, which have methodological implications.

Summary and conclusions

The objective of this chapter has been to introduce the concepts of metatheory, theory, and model, and distinguish them for the purposes of doing research in information seeking. An example result, the Principle of Least Effort, has been analyzed and discussed in relation to the three concepts. Methods of bringing this model closer to the status of a theory have been suggested.

The sources of metatheories in the social sciences have been discussed, and the nomothetic-idiographic distinction has been explained. Finally, 13 metatheories operating in LIS have been described. Sources for each metatheory and examples of its application have been presented.

ACKNOWLEDGMENTS

I wish to thank Karen Fisher and Sanda Erdelez for inviting me to present this paper, and Jenna Hartel for her very helpful suggestions on the manuscript.

REFERENCES

Ackerman, M.S. (2000). The intellectual challenge of CSCW: The gap between social requirements and technical feasibility. *Human-Computer Interaction, 15*(2-3), 179-203.

Allen, J.F. (1995). *Natural language understanding* (2nd ed.). Menlo Park, CA: Benjamin-Cummings.

American Heritage Dictionary of the English Language. (1969). Boston: Houghton-Mifflin.

Anderson, J.R. (1995). *Cognitive psychology and its implications* (4th ed.). New York: W.H. Freeman.

Appleby, J., Hunt, L., & Jacob, M. (1994). *Telling the truth about history.* New York: Norton.

Bar-Ilan, J., & Peritz, B.C. (2002). Informetric theories and methods for exploring the Internet: An analytical survey of recent research literature. *Library Trends, 50*(3), 371-392.

Barkow, J.H., Cosmides, L., & Tooby, J. (Eds.). (1992). *The adapted mind: Evolutionary psychology and the generation of culture.* New York: Oxford University Press.

Barzun, J., & Graff, H.F. (1992). *The modern researcher* (5th ed.). Fort Worth, TX: Harcourt Brace Jovanovich College Publishers.

Bates, J.A. (2004). Use of narrative interviewing in everyday information behavior research. *Library & Information Science Research, 26*(1), 15-28.

Bates, M.J. (1979). Information search tactics. *Journal of the American Society for Information Science, 30*(4), 205-214.

Bates, M.J. (1990). Where should the person stop and the information search interface start? *Information Processing & Management, 26*(5), 575-591.

Bates, M.J. (1994). The design of databases and other information resources for humanities scholars: The Getty Online Searching Project report no. 4. *Online & CDROM Review 18*(6), 331-340.

Bates, M.J. (1999). The invisible substrate of information science. *Journal of the American Society for Information Science, 50*(12), 1043–1050.

Bates, M.J. (2002). The cascade of interactions in the digital library interface. *Information Processing & Management, 38*(3), 381–400.

Bates, M.J. (2005). Information and knowledge: An evolutionary framework for information science. *Information Research, 10*(4), paper 239. Retrieved from http://InformationR.net/ir/10-4/paper239.html.

Bates, M.J. (2006). Fundamental forms of information. *Journal of the American Society for Information Science and Technology, 57*(8), 1033–1045.

Belkin, N.J. (1990). The cognitive viewpoint in information science. *Journal of Information Science, 16*(1), 11–15.

Belkin, N.J., Oddy, R.N., & Brooks, H.M. (1982). ASK for information retrieval: Part I. Background and theory. *Journal of Documentation, 38*(2), 61–71.

Berger, P.L., & Luckmann, T. (1990 reprint). *The social construction of reality: A treatise in the sociology of knowledge.* New York: Anchor Books.

Berkenkotter, C., & Huckin, T.N. (1993). Rethinking genre from a socio-cognitive perspective. *Written Communication, 10*(4), 475–509.

Bishop, A.P., & Star, S.L. (1996). Social informatics of digital library use and infrastructure. *Annual Review of Information Science and Technology, 31*, 301–401.

Blair, D.C. (2003). Information retrieval and the philosophy of language. *Annual Review of Information Science and Technology, 37*, 3–50.

Bowers, F.T. (1994). *Principles of bibliographical description.* New Castle, DE: Oak Knoll Press.

Bradford, S.C. (1948). *Documentation.* London: Crosby Lockwood.

Brookes, B.C. (1968). Derivation and applications of the Bradford-Zipf distribution. *Journal of Documentation, 24*(4), 247–265.

Budd, J.M., & Raber, D. (1996). Discourse analysis: Method and application in the study of information. *Information Processing & Management, 32*(2), 217–226.

Button, G. (1991). Introduction: Ethnomethodology and the foundational respecification of the human sciences. In G. Button (Ed.), *Ethnomethodology and the human sciences,* (pp. 1–9). Cambridge, England: Cambridge University Press.

Carmichael, J.V.J. (1998). *Daring to find our names: The search for lesbigay library history.* Westport, CT: Greenwood Press.

Carroll, J.M. (Ed.). (2002). *Human-computer interaction in the new millennium.* New York: ACM Press.

Case, D.O. (1991). Conceptual organization and retrieval of text by historians: The role of memory and metaphor. *Journal of the American Society for Information Science, 42*(9), 657–668.

Case, D.O. (2002). *Looking for information: A survey of research on information seeking, needs, and behavior.* New York: Academic Press.

Chatman, E.A. (1992). *The information world of retired women.* Westport, CT: Greenwood Press.

Cherry, C. (1966). *On human communication: A review, a survey, and a criticism* (2nd ed.). Cambridge, MA: MIT Press.

Chomsky, N. (1959). Review of B.F. Skinner, *Verbal behavior. Language, 35*(1), 26–57.

Chowdury, G.G. (2003). Natural language processing. *Annual Review of Information Science and Technology, 37*, 51–89.

Chu, C.M. (1999). Literacy practices of linguistic minorities: Sociolinguistic issues and implications for literacy services. *The Library Quarterly, 69*(3), 339–359.

Cool, C. (2001). The concept of situation in information science. *Annual Review of Information Science and Technology, 35*, 5–42.

Cooper, W.S. (1971). Definition of relevance for information retrieval. *Information Storage and Retrieval, 7*(1), 19–37.

Covi, L.M. (1999). Material mastery: Situating digital library use in university research practices. *Information Processing & Management, 35*(3), 293-316.

Crabtree, A., Nichols, D.M., O'Brien, J., Rouncefield, M., & Twidale, M.B. (2000). Ethnomethodologically informed ethnography and information system design. *Journal of the American Society for Information Science, 51*(7), 666-682.

Croft, W.B., & Thompson, R.H. (1987). I3R: A new approach to the design of document retrieval systems. *Journal of the American Society for Information Science, 38*(6) 389-404.

Dahlbom, B., Beckman, S., & Nilsson, G.B. (2002). *Artifacts and artificial science.* Stockholm: Almqvist & Wiksell International.

Day, R.E. (2001). *The modern invention of information.* Carbondale, IL: Southern Illinois University Press.

Dervin, B. (1983). Information as a user construct: The relevance of perceived information needs to synthesis and interpretation. In S.A.Ward & L.J.Reed (Eds.), *Knowledge structure and use: Implications for synthesis and interpretation* (pp. 155-183). Philadelphia: Temple University Press.

Dervin, B. (1997). Given a context by any other name: Methodological tools for taming the unruly beast. In Vakkari, P., Savolainen, R., & Dervin, B. (Eds.), *Information seeking in context. Proceedings of an international conference on research in information needs, seeking and use in different contexts* (pp. 13-38). London: Taylor Graham.

Dervin, B., (1999). On studying information seeking methodologically: The implications of connecting metatheory to method. *Information Processing & Management, 35*(6): 727-750.

Dervin, B. (2003). Human studies and user studies: A call for methodological inter-disciplinarity. *Information Research, 9*(1), paper 166. Retrieved March 12, 2004, from http://InformationR.net/ir/9-1/paper166.html.

Dewey, J. (1933). *How we think.* Lexington, MA: Heath.

Dewey, J. (1944). *Democracy and education.* New York: Macmillan.

Dick, A.L. (1995). Library and information science as a social science: Neutral and normative conceptions. *The Library Quarterly, 65*(2), 216-235.

Dick, A.L. (1999). Epistemological positions and library and information science. *The Library Quarterly, 69*(3), 305-323.

Dillon, A. (1994). *Designing usable electronic text: Ergonomic aspects of human information usage.* London: Taylor & Francis.

Dillon, A. (1995). Artifacts as theories: Convergence through user-centered design. *Proceedings of the ASIS Annual Meeting, 32*, 208-210.

Dretske, F.I. (1981). *Knowledge and the flow of information.* Cambridge, MA: MIT Press.

Ellis, D. (1989). A behavioural approach to information retrieval system design. *Journal of Documentation, 45*(3), 171-212.

Ellis, D. (1992). The physical and cognitive paradigms in information retrieval research. *Journal of Documentation, 48*(1), 45-64.

Ellis, D. (1993). Modeling the information-seeking patterns of academic researchers: A grounded theory approach. *The Library Quarterly, 63*(4), 469-486.

Ellis, D., & Haugan, M. (1997). Modeling the information seeking patterns of engineers and research scientists in an industrial environment. *Journal of Documentation, 53*(4), 384-403.

Fidel, R. (1993). Qualitative methods in information retrieval research. *Library & Information Science Research, 15*(3), 219-247.

Fielding, N. (1993). Ethnography. In N. Gilbert (Ed.), *Researching social life* (pp. 154-186). London: Sage.

Foucault, M. (1972). *The archaeology of knowledge.* London: Routledge.

Frohmann, B. (1994). Discourse analysis as a research method in library and information science. *Library & Information Science Research, 16*(2), 119-138.

Frohmann, B. (2001). Discourse and documentation: Some implications for pedagogy and research. *Journal of Education for Library and Information Science, 42*(1), 12–26.

Fuller, S. (2002). *Social epistemology* (2nd ed.). Bloomington, IN: Indiana University Press.

Gaines, B.R., Chen, L.L.J., & Shaw, M.L.G. (1997). Modeling the human factors of scholarly communities supported through the Internet and World Wide Web. *Journal of the American Society for Information Science, 48*(11), 987–1003.

Garfield, E. (1983). *Citation indexing: Its theory and application in science, technology, and humanities.* Philadelphia, PA: ISI Press.

Garfinkel, H. (1967). *Studies in ethnomethodology.* Englewood Cliffs, NJ: Prentice-Hall.

Gilovich, T., Griffin, D., & Kahneman, D. (Eds.). (2002). *Heuristics and biases: The psychology of intuitive judgment.* New York: Cambridge University Press.

Given, L.M., & Leckie, G.J. (2003). "Sweeping" the library: Mapping the social activity space of the public library. *Library & Information Science Research, 25*(4), 365–385.

Glaser, B., & Strauss, A. (1967). *The discovery of grounded theory: Strategies for qualitative research.* Chicago, IL: Aldine.

Harris, M.H. (1986). State, class and cultural reproduction: Toward a theory of library service in the United States. *Advances in Librarianship, 14*, 211–252.

Harris, R.M. (1992). *Librarianship: The erosion of a woman's profession.* Norwood, NJ: Ablex.

Hartel, J. (2003). The serious leisure frontier in library and information science: Hobby domains. *Knowledge Organization, 30*(3/4), 228–238.

Hayles, N.K. (1999). *How we became posthuman: Virtual bodies in cybernetics, literature, and informatics.* Chicago, IL: University of Chicago Press.

Hendry, D.G., & Harper, D.J. (1997). An informal information-seeking environment. *Journal of the American Society for Information Science, 48*(11), 1036–1048.

Hildenbrand, S., Ed. (1996). *Reclaiming the American library past: Writing the women in.* Norwood, NJ: Ablex.

Hildreth, C.R. (1989). *Intelligent interfaces and retrieval methods for subject searching in bibliographic retrieval systems.* Washington, DC: Library of Congress Cataloging Distribution Service.

Hjørland, B. (1998). Theory and metatheory of information science: A new interpretation. *Journal of Documentation, 54*(5), 606–621.

Hjørland, B. (2000). Library and information science: Practice, theory, and philosophical basis. *Information Processing & Management, 36*(3), 501–531.

Hjørland, B. (2002). Epistemology and the socio-cognitive perspective in information science. *Journal of the American Society for Information Science and Technology, 53*(4), 257–270.

Hjørland, B., & Albrechtsen, H. (1995). Toward a new horizon in information science: Domain-analysis. *Journal of the American Society for Information Science, 46*(6), 400–425.

Holquist, M. (2002). *Dialogism: Bakhtin and his world* (2nd ed.). New York: Routledge.

Ingwersen, P. (1992). *Information retrieval interaction.* London: Taylor Graham.

Ingwersen, P. (1999). Cognitive information retrieval. *Annual Review of Information Science and Technology, 34*, 3–52.

Jacob, E.K., & Shaw, D. (1998). Sociocognitive perspectives on representation. *Annual Review of Information Science and Technology, 33*, 131–185.

Kahneman, D., & Tversky, A. (Eds.). (2000). *Choices, values, and frames.* New York: Russell Sage Foundation.

Kelly, G.A. (1963). *A theory of personality: The psychology of personal constructs.* New York: W.W. Norton.

Kling, R., & McKim, G. (2000). Not just a matter of time: Field differences and the shaping of electronic media in supporting scientific communication. *Journal of the American Society for Information Science, 51*(14), 1306–1320.

Kraft, D.H., & Petry, F.E. (1997). Fuzzy information systems: Managing uncertainty in databases and information retrieval systems. *Fuzzy Sets & Systems, 90*(2), 183–191.

Kuhlthau, C.C. (1993). *Seeking meaning: A process approach to library and information services.* Norwood, NJ: Ablex.

Kuhn, T.S. (1996). *The structure of scientific revolutions* (3rd ed.). Chicago, IL: University of Chicago Press.

Kwasnik, B.H. (1991). The importance of factors that are not document attributes in the organisation of personal documents. *Journal of Documentation, 47*(4), 389–398.

Kwasnik, B.H. (1992). *Descriptive study of the functional components of browsing.* Paper presented at the Proceedings of the IFIP TC2\WG2.7 working conference on engineering for human-computer interaction, Elivuoi, Finland, August 10–14, 1992.

Lave, C.A., & March, J.G. (1975). An introduction to models in the social sciences. New York: Harper & Row.

Liddy, E.D., Jorgensen, C.L., Sibert, E.E., & Yu, E.S. (1993). A sublanguage approach to natural language processing for an expert system. *Information Processing & Management, 29*(5), 633–645.

Maack, M.N. (2000). "No philosophy carries so much conviction as the personal life": Mary Wright Plummer as an independent woman. *The Library Quarterly, 70*(1), 1–46.

Madden, A.D. (2004). Evolution and information. *Journal of Documentation, 60*(1), 9–23.

Marchionini, G. (1995). Information seeking in electronic environments. New York: Cambridge University Press.

Marchionini, G., & Komlodi, A. (1998). Design of interfaces for information seeking. *Annual Review of Information Science and Technology, 33*, 89–130.

McClure, C.R., & Hernon, P. (Eds.). (1991). Library and Information Science Research: Perspectives and strategies for improvement. Norwood, NJ: Ablex.

McKechnie, L.E.F. (2000). Ethnographic observation of preschool children. *Library & Information Science Research, 22*(1), 61–76.

Mellon, C.A. (1986). Library anxiety—A grounded theory and its development. *College & Research Libraries, 47*(2), 160–165.

Metoyer-Duran, C. (1991). Information-seeking behavior of gatekeepers in ethnolinguistic communities: Overview of a taxonomy. *Library & Information Science Research, 13*(4), 319–346.

Miller, G.A. (1951). *Language and communication.* New York: McGraw-Hill.

Minsky, M.L., Ed. (1968). Semantic information processing. Cambridge, MA: MIT Press.

Newell, A., & Simon, H.A. (1972). *Human problem solving.* Englewood Cliffs, NJ: Prentice-Hall.

Nielsen, J. (1993). *Usability engineering.* Boston, MA: AP Professional.

Norman, D. A. (1990). *The design of everyday things.* New York: Doubleday.

Orlikowski, W.J., & Yates, J. (1994). Genre repertoire: The structuring of communicative practices in organizations. *Administrative Science Quarterly, 39*(4), 541–574.

Over, P. (2001). The TREC interactive track: An annotated bibliography. *Information Processing & Management, 37*(3), 369–381.

Paisley, W.J. (1968). Information needs and uses. *Annual Review of Information Science and Technology, 3*, 1–30.

Palmer, C.L. (2001). *Work at the boundaries of science: Information and the interdisciplinary research process.* Dordrecht, Boston, MA: Kluwer Academic.

Pawley, C. (1998). Hegemony's handmaid—The library and information studies curriculum from a class perspective. *The Library Quarterly, 68*(2), 123–144.

Pettigrew, K.E. (2000). Lay information provision in community settings: How community health nurses disseminate human services information to the elderly. *The Library Quarterly, 70*(1), 47–85.

Pettigrew, K.E., & McKechnie, L. (2001). The use of theory in information science research. *Journal of the American Society for Information Science and Technology, 52*(1), 62–73.

Pettigrew, K.E., Fidel, R., & Bruce, H. (2001). Conceptual frameworks in information behavior. *Annual Review of Information Science and Technology, 35,* 43–78.

Pierce, J.R. (1961). *Symbols, signals, and noise: The nature and process of communication.* New York: Harper.

Poole, H. (1985). *Theories of the middle range.* Norwood, NJ: Ablex.

Powell, R.R. (1997). *Basic research methods for librarians* (3rd ed.). Greenwich, CT: Ablex.

Powell, R.R. (1999). Recent trends in research: A methodological essay. *Library & Information Science Research, 21*(1), 91–119.

Price, D.J. de Solla. (1986). *Little science, big science. . . and beyond.* New York: Columbia University Press.

Radford, G.P. (2003). Trapped in our own discursive formations: Toward an archaeology of library and information science. *The Library Quarterly, 73*(1), 1–18.

Rayward, W.B. (1996). The history and historiography of information science: Some reflections. *Information Processing & Management, 32*(1), 3–17.

Reynolds, P.D. (1971). *A primer in theory construction.* Indianapolis: Bobbs-Merrill.

Ritzer, G. (2000). *Modern sociological theory* (5th ed.). Boston, MA: McGraw-Hill.

Rogers, Y. (2004). New theoretical approaches for human-computer interaction. *Annual Review of Information Science and Technology, 38,* 87–143.

Ruesch, J., & Bateson, G. (1968). *Communication: The social matrix of psychiatry.* New York: Norton.

Russell, S.J., & Norvig, P. (1995). *Artificial intelligence: A modern approach.* Englewood Cliffs, NJ: Prentice-Hall.

Salton, G., & McGill, M.J. (1983). *Introduction to modern information retrieval.* New York: McGraw-Hill.

Sandstrom, A.R., & Sandstrom, P.E. (1995). The use and misuse of anthropological methods in library and information science research. *The Library Quarterly, 65*(2), 161–199.

Sandstrom, A.R., & Sandstrom, P.E. (1998). Science and nonscience in qualitative research: A response to Thomas and Nyce. *The Library Quarterly, 68*(2), 249–254.

Sandstrom, P.E. (1994). An optimal foraging approach to information seeking and use. *The Library Quarterly, 64*(4), 414–449.

Sandstrom, P.E. (1999). Scholars as subsistence foragers. *Bulletin of the American Society for Information Science, 25*(3), 17–20.

Schutz, A. (1967). *The phenomenology of the social world.* Evanston, IL: Northwestern University Press.

Shannon, C.E., & Weaver, W. (1975). *The mathematical theory of communication.* Urbana, IL: University of Illinois Press.

Simon, H.A. (1976). *Administrative behavior: A study of decision-making processes in administrative organization* (3rd ed.). New York: Free Press.

Simon, H.A. (1981). *The sciences of the artificial* (2nd ed.). Cambridge, MA: MIT Press.

Skinner, B.F. (1992). *Science and human behavior* (1992 ed.). New York: Classics of Psychiatry and Behavioral Sciences Library.

Small, H. (1999). A passage through science: Crossing disciplinary boundaries. *Library Trends, 48*(1), 72–108.

Sonnenwald, D.H., & Iivonen, M. (1999). An integrated human information behavior research framework for information studies. *Library & Information Science Research, 21*(4), 429–457.

Talja, S. (1999). Analyzing qualitative interview data: The discourse analytic method. *Library & Information Science Research, 21*(4), 459–477.

Talja, S. (2001). *Music, culture, and the library: An analysis of discourses.* Lanham, MD: Scarecrow Press.

Talja, S., Tuominen, K., & Savolainen, R. (2005). "Isms" in information science: Constructivism, collectivism and constructionism. *Journal of Documentation, 61*(1), 79–101.

Thomas, N.P., & Nyce, J.M. (1998). Qualitative research in LIS—redux: A response to a (re)turn to positivistic ethnography. *The Library Quarterly, 68*(1), 108–113.

Trosow, S.E. (2001). Standpoint epistemology as an alternative methodology for library and information science. *The Library Quarterly, 71*(3), 360–382.

Updike, D.B. (2001). *Printing types: Their history, forms, and use* (4th ed. expanded). New Castle, DE: Oak Knoll Press.

Vakkari, P. (1997). Information seeking in context: A challenging metatheory. In P. Vakkari, R. Savolainen, & B. Dervin (Eds.), *Information seeking in context,* Proceedings of an international conference on research in information needs, seeking and use in different contexts, (pp. 451–464). London: Taylor Graham.

Vaughan, M.W., & Dillon, A. (1998). The role of genre in shaping our understanding of digital documents. *Proceedings of the ASIS Annual Meeting, 35,* 559–566.

Vygotsky, L. (1978). *Mind in society: The development of higher psychological processes.* Cambridge, MA: Harvard University Press.

Wagner, D.G., & Berger, J. (1985). Do sociological theories grow? *American Journal of Sociology, 90*(4), 697–728.

Wang, P.L. (1999). Methodologies and methods for user behavioral research. *Annual Review of Information Science and Technology, 34,* 53–99.

Wang, P.L., & White, M.D. (1999). A cognitive model of document use during a research project. Study II. Decisions at the reading and citing stages. *Journal of the American Society for Information Science, 50*(2), 98–114.

Westbrook, L. (1994). Qualitative research methods: A review of major stages, data analysis techniques, and quality controls. *Library & Information Science Research, 16*(3), 241–254.

White, H.D., & McCain, K.W. (1998). Visualizing a discipline: An author co-citation analysis of information science, 1972–1995. *Journal of the American Society for Information Science, 49*(4), 327–355.

Wiegand, W.A. (1999). Tunnel vision and blind spots: What the past tells us about the present: Reflections on the twentieth-century history of American librarianship. *The Library Quarterly, 69*(1), 1–32.

Wiegand, W.A., & Davis, D.G. (Eds.). (1994). *Encyclopedia of library history.* New York: Garland.

Wiener, N. (1961). *Cybernetics: Or, control and communication in the animal and the machine* (2nd ed.). Cambridge, MA: MIT Press.

Wilson, P. (1968). *Two kinds of power: An essay on bibliographical control.* Berkeley, CA: University of California Press.

Wilson, P. (1977). *Public knowledge, private ignorance: Toward a library and information policy.* Westport, CT: Greenwood Press.

Wilson, P. (1983). *Second-hand knowledge: An inquiry into cognitive authority.* Westport, CT: Greenwood Press.

Wilson, T.D. (1999). Models in information behaviour research. *Journal of Documentation, 55*(3), 249–270.

Wilson, T.D., & Streatfield, D.R. (1981). Structured observation in the investigation of information needs. *Social Science Information Studies, 1*(3), 173–184.

Wright, R. (1994). *The moral animal: The new science of evolutionary psychology.* New York: Vintage Books.

Zipf, G.K. (1949). *Human behavior and the principle of least effort: An introduction to human ecology.* Cambridge, MA: Addison-Wesley.

8

The role of the Ph.D. in a professional field

Questions regarding the role of the Ph.D. degree in our professional field arise, in my experience, quite frequently and in a variety of environments. Among my academic colleagues at the university, there seems to be an unwritten assumption, which sometimes emerges in very explicit form, that the letters and sciences are not only the heart of the university, but are also somehow on a higher plane than professional fields. Research in those areas is "pure," in professions "applied," and, therefore, lesser.

On the other hand, among at least some library practitioners, one can find a quite pronounced view that we who are teaching in professional programs are foggy-headed researchers—hopelessly impractical, incapable of preparing our students for the real world of practice. More specifically, our research is often seen to be so unconnected to real-world situations that it can offer little of use to the working professional.

So what is it that we are doing when we get a doctorate in this profession? Doing "merely" applied research, which our academic colleagues dismiss, while, on the other hand, the practitioners view as so impractical that it is of no use anyway?

On the positive side, almost all universities recognize that researchers in professional fields can earn the Ph.D., that is, a doctor of philosophy in a profession. In other words, they recognize that even professions can

First presented as Bates, M. J. (Oct. 16,1999). *The role of the Ph.D. in a professional field* (Srygley Lecture). Florida State University, Tallahassee, FL.

do research that taps basic questions, and are not only applied fields. By the way, in our field, there have only been one or two cases that I know of where a School has both a DLS, or Doctor of Library Science, and a Ph.D. In most places, even where the student earns a DLS, such as at the now-closed school at Columbia University, the degree is understood to be a Ph.D., de facto. Only where both degrees exist is a distinction made between them.

So just what is the nature of a Ph.D. in a professional field—specifically, in this one? Let's look first at what a profession is. A profession develops a body of knowledge around roles and activities that a society values. Professional skill then involves mastering that body of knowledge and developing experience applying it to situations where the knowledge is relevant.

Professional knowledge is not algorithmic; that is, it does not involve following a rule-based sequence of actions that are applicable in every common situation. Rather, a profession always requires developed judgment. There are complexities, subtleties, and variations from situation to situation that require experienced judgment in order to apply the professional knowledge properly.

On the other hand, the knowledge a profession develops always has some application to real situations in life. Knowledge developed in a research discipline may be abstruse and complex, and certainly not algorithmic, but if it has no application to some identifiable real situations, then it is, in the time-worn phrase, "purely academic," and not professional knowledge.

Professions develop out of social need and out of social consciousness. In the nineteenth century, when the size of library collections reached a point where older ad hoc and idiosyncratic approaches would no longer suffice, standards, training, and a self-conscious world view toward library and information services began developing. At the same time, a conceptualization of the very idea of a profession was also ripening and maturing. Medicine, teaching, nursing, the law had all been around for centuries, but their conscious formulation as professions is a fairly recent development.

Indeed, there are many social forces shaping the modern idea of a profession, and our profession in particular. But today, I want to concentrate on the question I started with—what is the role of the Ph.D. in a professional field? And I want to address that question with respect to the intellectual content of Ph.D.-level research and its relation to the profession of library and information science—or, as it is becoming known, information studies.

To cut to the chase, the fundamental point I want to make is that doctoral level research is the seed-bed for ideas and practices of the profession. AND it achieves that role by doing research that is as fundamental as any done in the letters and sciences.

I have argued elsewhere that information studies is fundamentally socio-technical. Our professional functions are derived from social needs and circumstances, and, to be successful, must serve to meet those needs. Information studies meets those social needs by drawing on a wide range of intellectual and physical technologies. Achieving the objectives of our profession depends on the skillful melding of an understanding of social, political, economic, and psychological factors related to information needs and uses with a mastery of both theoretical and practical issues in information organization and system design.

There is something of a culture war going on these days between those in the field who see the institution of the library and all it has meant to our society as the defining core of our field and those who employ the concept of information to define the field. To the "library" people, the information perspective is coldly technological and indifferent to important social values and a rich history of service to a wide range of people, many of whom were socially and economically enfranchised by the education they received at the library. To these "library" partisans, the information perspective means pure computer technology, carrying with it implicit authoritarian, even fascist, adoration of the power of technology, to the detriment of precious human values.

I confess that I nonetheless come down on the "information" side of this argument, and that, needless to say, I have a different perspective on the information view of our field. Though I do not believe that most "information" partisans carry such an adoring view of technology, to whatever extent they do, I side with the "library" partisans. Librarianship has precious values of service to society, and I would like to see those values carry over to any conceptualization of our field as information studies.

However, I do think the "information studies" approach is the right one for now in our field. To many people, the term "information" is a cold and neutral one, and is limited in their minds to pieces of data, like the population of Turkey or the distance to the sun. In this field, however, we can and should use the term much more broadly, and apply it to all experiences that can or do impact the human understanding and memory. Such a concept is indisputably rooted then not only in the single mind, but also, necessarily, in human society, as the mind is never wholly independent of the social matrix in which it was nurtured and continues to act daily.

Some would then answer, well, in that case, "information" means everything—all knowledge and experience. No one field can possibly encompass all of that.

I agree, and I don't make such a claim. We do address all of knowledge, but in a very particular way. Information studies is a meta-field, like education or communication. Those fields, likewise, deal with all knowledge, but each does something distinctive with that world of knowledge and experience that is unique to each field.

Our field organizes information systems and services to offer that knowledge and experience to information seekers, education shapes curricula and uses teaching skills to convey that knowledge to learners, and communication/journalism uses reporting and writing skills to convey discovered news to others. Each field holds a particular rhetorical stance toward all knowledge, a stance that enables it to understand the human relationship to information in its own distinctive way, and in service to a larger social purpose that is unique to each field.

Once we see information studies in this manner, it becomes clear that the activities traditionally engaged in by librarians in libraries bear many similarities to the activities of archivists, of professionals in technical documentation, paper and electronic records, digital databases and libraries, and so on. The shift to information studies is not, should not, involve any rejection of libraries or librarianship; rather it should broaden our perspective to all the contexts within which these meta-field activities take place.

However this debate in the field over our fundamental world view and philosophy is resolved, the fact that there are such debates in the first place should provide a clue to the depth of thinking and theory that lies at the heart of our work. In truly algorithmic work, such as that of a restaurant cashier, it is doubtful that there are raging debates about the philosophy of cashiering.

I would like now to provide an extended example of how our work draws upon, and needs to draw upon, research that is as fundamental as any in the letters and sciences. The example involves the design of information retrieval systems. Now, surely this is at the technical, applied end of the business if anything is, right? Well, let's see.

Over the last two or three years I have come to see that the design of information systems is vastly more complex than it was in the early days of automated information systems. Early online database systems were crude and clunky. We didn't mind, however, because it was so exciting that we could search online for information in real time. People mastered arcane techniques and worked around the limitations of these systems as they needed to.

By now, however, the vast storage capabilities, sophisticated computer programming languages, the varieties of search engines, the speed of networks, and the subtlety of database design have all opened up a vast array of new design possibilities and complexities. Furthermore, these technical elements I've mentioned by no means exhaust the design elements that must be taken into consideration in creating the contemporary information system. The selection of the information content itself, of systems of metadata and indexing, design of search capabilities, selection or design of various kinds of access front-ends, as well as interface design, must all go into the design of an automated information system.

However, there remain still two more huge factors in this system design that are essential to its success. First, the needs and characteristics of anticipated users of the system must be taken into account with respect to every decision made on each of these technical and informational elements. Second, all the design elements or layers need to be taken into account *simultaneously and in relation to each other.*

Still, however, despite the complexity, all this might still be viewed by some as a merely technical or applied problem—inviting the conventional dismissal. Now, in fact, design of all kinds, including information system design, requires as much creativity and originality as anything—but that's another speech. Here, I want to address the more conventional, recognized forms of academic research.

So let us look deeper. To make the following points, I am drawing on an extensive study done by the Getty Information Institute, which I participated in and wrote six articles about. In the study, Visiting Scholars in the humanities at the Getty were given the opportunity to be trained to search databases provided by the vendor DIALOG, and then do as much searching as they wished, for free, during their year as Visiting Scholars. In exchange, they would participate in interviews and agree to have their entire searches recorded automatically by computer.

In such a case, when we examine the users' needs, that examination must be conducted as the very subtlest of original, basic social science research. To design an information system that people will actually use, we must understand, in a profound way, how people use and relate to the information in their domains of interest. In this case, we were dealing with humanities scholars who had had, sometimes, decades in which to develop research styles and in which to shape their perspectives on the world of recorded information that they dealt with so intensively as scholars.

Here is where the term "information" must be used in the broader and deeper sense I was arguing for earlier. In particular, in the Getty study it became clear to me that we could not understand how the scholars used their searching opportunity, and could not suggest features needed for them in information systems, without understanding something of the entire social and educational system of practices and values within which they operated. This is not simply a matter of gathering statistics on what percentage of the scholars' resources were books, what percentage journals, and the like.

Rather, it involved developing a deep and subtle understanding of the *culture* of humanities scholarship. It was necessary to understand how the scholars conceptualized their research intellectually, and how they conceptualized the process of going about locating and using research materials. It became clear that many assumptions in information science about methods of research and desirable system design characteristics were rooted in the research practices of the sciences and were inapplicable to the humanities.

For example:

- We learned that the selection of resources for the typical data-base—with a heavy emphasis on current journal articles—was much less useful for the scholars than for scientists. Currency is very important in the sciences, while historical range is more important in the humanities. This may seem obvious, but this basic reality was seldom understood and acted upon by the data-base producers, who were creating products on the science model.

- Second, fundamental differences in the way scholars conduct research and in how they attempt to find new information require at least a partial reconceptualization of information retrieval. Again, the paradigm of information retrieval research has been based on the science model. Some of the basics of that model do not make sense for the humanities scholar.

One such assumption is that the user of an information system is unfamiliar with the majority of the records in the system relating to their topic of interest. We found, however, for reasons that are based in the way scholars see their own relationship to their literature, compared to the way that scientists do, that humanities scholars almost always knew most of the literature in their field already. Thus, they saw their task as scholars to be to achieve mastery of a large number of particular works by other scholars.

For scientists, on the other hand, knowledge of particular publications is of less interest than knowledge of work going on in other laboratories. Scientific researchers on the cutting edge will know of discoveries long before they reach print in any form—thus the print version generally serves only an archival function.

In humanities scholarship, the particular insights achieved by the scholar are so important that the texts of publications must be read closely and absorbed, then reread. I am generalizing, of course, but there were nonetheless marked trends that can be seen to distinguish the two groups of researchers. Most of the time, for scholars, a database search functioned only as a way to catch any stray materials they were not already familiar with. If such is the typical situation of the humanities information system user, then we may need to design information systems in ingenious new ways that the science model did not suggest. The fourth and fifth reports in the Getty series explain these points in more detail, and suggest further design implications.

- Third, we found that the nature of humanities queries differed from those in the sciences not only in the obvious respect of content, but also in conceptual structure. I have argued that one consequence of this systematic difference is that humanities materials should probably be indexed by faceted vocabularies, not by conventional alphabetical thesauri in the scientific model. Furthermore, given this presumptive mode of indexing, then information system search capabilities and interface character-istics should be designed quite differently to accommodate these inherent information and system differences.

In fact, virtually every layer in the information system design would be affected dramatically if these fundamental social characteristics of humanities research and researchers were taken into account.

I've described social factors that go very deep into the information seeking and using behavior of humanities scholars. Along the same lines, we could describe ergonomic, psychological, and perceptual characteristics of people that must be taken into account, for system design, based on basic research in the cognitive sciences. Further, we could identify—and this is unique to the information sciences—social, statistical, and economic characteristics of the humanities literatures themselves that need to be understood to optimize information system design.

If we think of the natural sciences as studying the physical world as their universe, the social sciences studying the social world, and the arts and humanities studying the creative products of human beings, then we can think of the information sciences as taking the universe of recorded information—and human beings' interactions with it—as our domain of study. It is uniquely our challenge to identify the statistical, social, economic, and design properties of the world of recorded information.

At every point, research results of basic research must be drawn upon to create optimal information systems. Frequently, the only people doing the kind of basic research we need in this field, are people in the field.

I could make analogous arguments about the need for basic research as a grounding to developing information policies, or to developing the best institutional bases for information services. In other words, in every area, this profession must draw on original, basic research to achieve the best professional service.

This brings us to the other problem I mentioned earlier, however. If those of us with Ph.D.'s in information studies are busy doing basic research, or developing experimental systems and services that draw on basic research, how does this effort affect the working practitioner? There have been many lamentations, at conferences and in print, that researchers keep doing things that the practitioner cannot directly make use of. Is the availability of the Ph.D. degree creating a group of irrelevant people producing results no one will use?

The first thing we need to understand in relation to this question is that this is a near-universal problem, not one specific to our field. It is a problem in the letters and sciences and a problem in the professions. The bench chemist in a pharmaceutical firm no more uses the latest chemistry results, as a rule, than the working librarian applies the latest results published in *Library Quarterly* or the *Journal of the American Society for Information Science*.

The problem arises for reasons very similar to the ones mentioned earlier regarding the differences between scholars and scientists. Scholars and scientists have different cultures, values, and ways of operating because the nature of their work demands it. The same holds true for researchers and practitioners generally. Different native talents, training, and experience all add up to create people—professors and practitioners—who work and produce differently, and have different cultures and outlooks. Distinct types of minds and personalities will be drawn to these distinct activities.

I have spent just two years as a manager—as the Chair of my Department. And at the end of the two years I was a walking nervous breakdown. I have boundless admiration for Dean Robbins and other academic leaders who skillfully navigate the dangerous waters of large egos, small budgets, and killer work schedules. People who have a good understanding of research, and who are simultaneously good at the practice of academic management are rare, and universities are perennially, and often vainly, hunting for them.

It is in the nature of basic research that it often does not have immediate applications. But even where there are immediate applications, the work produced by the researcher often cannot be translated readily by the practitioner into practice. At this point the criticisms arise.

It needs to be understood by all involved that the ability to translate between different intellectual and practice cultures is a very special talent, and not all that common in any field. Most researchers cannot do it and most practitioners cannot. Those who can, should be treasured.

Unfortunately, the rewards are few to the people who engage in this translation process. The librarian is unlikely to have the time to devote to the extensive literature review and analysis needed to prepare a research-into-practice article—and may not have the research training to understand all the literature. The researcher, on the other hand, gets scant recognition or merit raises for producing what are generally dismissed as "service publications." Consequently, those of us in academia seldom write such boundary-spanning works.

To sum up, I believe that the Ph.D. in a profession is viewed as impure or applied precisely because it must hold a broader view of its subject matter. The holder of a Ph.D. in a professional field must be able to do basic research AND select research questions that may contribute—if not this century then the next—to the practice of that socially useful activity that is a profession. I'm hinting here, that doing good research in the professions may actually require more talent than the work in the traditional disciplines. At the very least, it requires a considerable breadth of vision, a flexibility, and an ability to see connections others may miss.

Few are the researchers who can take it still a step further and extract implications for practice, and present those implications in a way that practitioners can make immediate use of. That is quite a challenge. Where that ability does exist, however, it deserves to be richly rewarded.

9

A doctorate in
library/information science

Going for a doctorate can be an exciting prospect, and as is the case with other career decisions, it is vital to have all the information you can get going in. I have served as chair of the committee that admits doctoral students and supervises the doctoral program at both the University of Maryland and at UCLA. I have also worked with doctoral students in other capacities in these two programs for many years.

Applicants and incoming students often carry a number of common misconceptions about doctoral work. In this article I hope to set those misunderstandings aright and thus possibly help you in making more satisfying decisions in this area.

There are two main questions to address: 1) Should I go for a doctorate? 2) If so, where should I apply?

Should I go for a doctorate?

Many people find the one-year MLS program frustratingly brief. Just as soon as the basics of cataloging, reference, etc. are mastered, plus a few electives, wham! it is all over. You may be left wanting to take a dozen other courses, feeling that you need more to improve your general preparation

First published as Bates, M. J. (1986). A doctorate in library/information science. *The Library Journal, 111*(14), 157–159.
Note: Superseded information has been removed from this article.

for the field. Alternatively, out on the job, you may conclude that further work in one particular area of specialization would be particularly valuable. The thought then naturally occurs: Should I go back to school to work on a further degree?

If these are the only reasons for working on a doctorate, then I would not recommend applying for a doctoral program. There are other, better, ways to get the further work you need.

Academic or professional

Advanced degrees are structured in a different manner in a professional field like librarianship than they are in conventional academia. In English literature or sociology the student majors in the subject as an undergraduate, gets more training at the master's level, and still more at the doctoral level. The subject is "academic" throughout, that is, a BA in English provides no preparation for a professional career unless one also pursues professional studies in such areas as education or journalism.

In contrast, in a professional field, whether law, medicine, or librarianship, there is a single "terminal" professional degree—the JD, MD, or MLS. That degree fully qualifies one for any position in the field, whether lowly beginner or exalted top administrator.

In a professional field, the doctorate is not intended to make one more qualified to work as a practitioner. Rather, the doctorate prepares you to enter a new field: university teaching and research. Since the teaching is in the area of librarianship, you must, of course, have the professional master's as well as other advanced course work in the field, but the point of getting a doctorate is to prepare you for a new career, not to make you more qualified for the career you have.

Obviously you will use your doctoral degree any way you wish, including such possibilities as library administration, work in the information industry or in private research organizations, and so on. The training you get, however, will be strongly oriented to a career in scholarship.

To supplement the MLS

If your desire is solely to supplement the professional learning you got in master's training, then there are several good ways to pursue that desire. Many library schools offer certificate programs, some a year long.

If a year is more than you want, many schools allow you to register for individual courses under a variety of arrangements. Some have special miniprograms in the school aimed at the working professional. These programs often offer courses at night or at distant locations to accommodate those who cannot come to the established library school.

In other cases you may sign up for regular courses as a special student and so not be required to work for a degree. Sometimes university extension programs offer librarianship courses. Generally, with extension programs it is not necessary to go the whole involved route of application, transcripts, etc., yet continuing education credit may nonetheless be available.

The nature of doctoral studies

Doctoral work is different in several important respects from master's work in library/information science. Understanding those differences can help you sort out whether doctoral work is for you.

The purpose of a doctorate is to prepare you to become a scholar. So while you are a student you are viewed as an apprentice scholar and given the kind of training a scholar needs. Whereas master's programs tend to be highly structured, you are expected to show much self-direction at this level, because that is what you will need on the job as a professor.

You learn to identify and shape research problems, you learn research methods, you may practice/teach courses, and you study in great depth what is already known in a wide range of subject areas.

The standards at the doctoral level are the highest anywhere in academia and it requires a great deal of self-direction and perseverance to make it through virtually any doctoral program.

Perhaps the single most widespread misconception about advanced work is found in the phrase "working on a Ph.D. on the side." The image here is of one taking a course now and then, gradually building up the credits over the years. After a while, one takes exams—then comes the "easy" part, working on the dissertation. The dissertation is thought to be easy because one can work on it "on your own" instead of having to attend classes.

Well, you cannot get a doctorate by stacking up credits. The material must all be in mind at once for qualifying exams—instead of forgotten at the end of a course five years back. You take as many courses, for credit or not, as are needed to master the material for the exams and to be prepared to do a dissertation on your chosen topic. If that requires ten credits over the required minimum number of units, so be it.

In fact, our program at UCLA has no required number of units at all—the point is whether you get the knowledge, not the units. And the dissertation is the hardest part of all. It not only comes at the end of a long haul, but it is also a major research project which you must mount yourself and carry through from beginning to end. You do not get a C+ on a dissertation and graduate the way you might with a term paper for a course. The dissertation is redone and redone until it meets the highest standards of scholarship—or no degree.

Full time is better

Is it practical, then, to work for a doctorate part-time? Because of the time, effort, and expense involved, most people working on a doctorate go part-time during at least some of their program. But I would strongly advise that you arrange to have at least a half-year to a year (and the more the better) for full-time work sometime during the time you are working on the degree.

There are two reasons for this recommendation. First, the drop-out rate, particularly at the point of doing the dissertation, is fairly high, and often arises out of the sheer exhaustion of having studied for years on nights and weekends.

Second, if you give yourself time to really concentrate on your studies when your energies are not distracted by the politics and demands of work, then you can explore ideas, study new areas, and make connections between bodies of research that you would not have the time and energy to notice otherwise.

The ideas you get for research while a student will constitute a fund that you can draw upon for the rest of your career. Remember that after you get the degree, the next hurdle will be to get tenure in your teaching job—and that requires evidence of significant research beyond the dissertation.

An individualized program

The plus side of the hard work you do on the doctorate is that you are seen much more as an individual at this level. Working with your advisor you shape a truly individualized course of study—in the end you may be the only person in the country with your particular mix of knowledge. Your exams are tailored specifically to you and your work throughout with faculty as an individual—more like a junior professor than a student.

The person you work with most is your advisor—who is more a mentor than an advisor in the traditional sense. The selection of an advisor is critical to your doctoral work and your success in it. Once having accepted you as a student, the advisor is committed to help you get through the program. You have to hold up your end of the bargain, of course, but the advisor has an obligation to see that you make it through if you possibly can. Since you are entering a different career—university teaching—your advisor and the other professors you work with can also give you valuable advice on how to survive under the different game rules of academia.

Where should I apply?

Essentially, *you select the school by selecting the people you want to work with.* Good libraries and facilities are important, and so is being housed in a good university, but far and away the most important factor in the quality and suitability of a school for your needs is the faculty.

You want good people and people with research interests in the areas that interest you. Every library school in the country can teach you something about reference, for example, at the master's level, but not every school with a doctoral program can give you a really good start in that area *at the doctoral level.* That is, not every school has people who are interested in research and theory of reference. If that is what you want to do your dissertation on, and later teach, then you will have to find a school that has good people in that subject.

Since doctoral work is individualized and specialized, you go where you know you can work with specific individuals.

How can you make that choice? There are a lot of ways to get information. First of all, find candidate schools by reading the *Educational Media Yearbook* (Libraries Unlimited) which lists the schools that offer doctoral work in library/information science, their major areas of specialization, who to write, etc.

To find the individuals, go back to *Library Literature* and/or *Information Science Abstracts* and find current articles in the subject areas that interest you. Following up the articles you find there, whose work in the field excites you? Check out the doctoral programs where they are teaching.

Each year the *Journal of Education for Library and Information Science* publishes a directory issue which lists all the library schools, their faculty, and each professor's areas of specialization. You can get a sense of the overall strength of a school by seeing who is teaching there.

Another valuable source of information is recent doctoral graduates of the schools that interest you. Always be sure, however, to get the most current information. Professors leave or retire, programs are reorganized—even last year's information may be out of date.

Choosing a specialty

What if you do not know what you want to specialize in yet? That is all right; you do not have to have interests cast in stone before entering doctoral work. But presumably, there are some areas that interest you more than others, that seem to be candidates for specialization. Be sure to go to a school where you can pursue one or more of those interests.

And what do you do if you do not feel free to go to the best program for you? Suppose you must apply at the local university because you can get in-state fees, or your spouse's job keeps you there?

If the problem is money, many universities have fee waivers, fellowships, and research assistantships that make it possible for you to attend. The library field also offers financial support of various kinds. When seeking financial help, the crucial requirement is to apply well in advance. Many of the best fellowships have lead times of six months to a year.

What if you feel tied to a particular city for reasons other than finances? In that case, you must make some hard choices. If the local program is weak, or none of the faculty interests match yours, then you must nonetheless either find a way to attend a more suitable program elsewhere, or you have to get your degree outside of your area of interest. It is better to understand this before entering that local program, than to realize it a disappointed two years later, after frustrations have built up because you are not getting what you want.

Program quality

There is another factor in this matter of choice of schools that deserves more emphasis: quality. Because you work so closely with your advisor and a few other faculty, particularly at the dissertation stage, your thinking will be molded, for good or ill, by your interactions with these individuals, especially the advisor. I have seen dull, unimaginative dissertations produced by the students of dull, unimaginative professors, methodologically defective dissertations produced by the students of faculty who never quite mastered research design, etc.

If you want to do good work, look for someone who does good work now. If you feel you do not yet have the qualifications to evaluate the research and writing of your potential professors, there are two aids you can draw upon. The major journals in the field operate on the principle of "peer review." Experts in subjects serve as referees for incoming manuscripts, and articles are not published unless evaluated as acceptable by (usually) two or more such peers.

The Library Quarterly, Journal of the American Society for Information Science, College and Research Libraries, Library and Information Science Research, and *Information Processing and Management* are examples of some of the top journals in our field that operate by peer review. Check out the publications of the faculty in your candidate schools. How often does their work appear in top journals such as these? If your candidate faculty members have published books, what do reviewers in library/information science have to say about their work?

The second aid you can use is citation patterns. Look up these individuals in the *Social Sciences Citation Index.* Is their work being cited, i.e., do they have *impact* in the field? Since our field is relatively small, anyone who is being cited more than half a dozen times a year is having considerable impact.

By the time you have finished your investigations, you will know a lot about library education in general as well as where you can get the best preparation for your new career in teaching and research.

10

Future directions for the library profession and its education

[Our profession is presently] experiencing its most profound transformation and challenge since the invention of printing in the fifteenth century. . . . Whatever changes history might bring, the fundamental mission of librarianship—the service of society through meeting its information needs—will remain a fixed and constant necessity.

—*Kevin Starr, California State Librarian*

BACKGROUND

Born out of a concern over the fate of library school education in California, as well as a keen awareness of rapid technological advances which promised to change the traditional face of librarianship, the Task Force on the Future of the Library Profession in California was appointed by California Library Association (CLA) President Joy Thomas in March 1994. Its members, representing academic, public and school libraries, included practicing librarians Betty Blackman, Dale Bublotz, Colleen Foster, and Jim Schmidt, library educators Marcia Bates, Stuart Sutton, and Nancy Van House, library school student Sheri Irvin, and State Librarian Kevin Starr. Linda Crowe and Luis Herrera were co-chairs.

First published as California Library Association (1996). Future directions for the library profession and its education. Sacramento, CA: California Library Association Task Force on the Future of Librarianship.

THE REPORT

This Task Force report, issued in November 1995, includes several elements that serve as the framework for further discussion within the California library community:

A mission statement for the California library profession, defining:

- leadership in meeting information needs in a changing social, technological and economic environment

- unique mix of knowledge and information management

- strong professional advocacy

- ongoing commitment to education and training, shared knowledge and an exchange of ideas

The core areas of professional expertise and practice that define the profession of librarianship and present some forecasts of coming changes in the field:

- knowledge of information resources and organization

- understanding user needs, patterns of information use, and the matching of user needs to information

- information technology

FOLLOW-UP ACTIVITY

Based on the Task Force's recommendations, this report is being disseminated statewide to members of CLA, the California School Library Association, and the California Association of Research Librarians, as well as local chapters of the Special Library Association and Medical Library Association, in order to seek reactions and stimulate mutual interest in professional development. It is also being made available to national professional associations and publications. Over the next several months, forums will be held to discuss this report and the steps California librarians must take to ensure meeting the challenges outlined herein. CLA officials have already instigated an effort to collaborate with library school programs to provide continuing education opportunities throughout the state.

Recommendations on implementing a course of action to meet the challenges outlined in this report will be presented at the CLA Assembly meeting in May, 1997, to Mary Jo Levy, chair of the CLA Task Force on Implementation of the Library Education Report, c/o the California Library Association, 717 K Street, Sacramento, CA 95814-3477 or info@cla-net.org. The CLA newsletter, California Libraries, will inform association members of all developments related to this matter.

CALIFORNIA LIBRARY ASSOCIATION
Task Force on the Implementation of the Library Education Report
August 1996

PROPOSED MISSION STATEMENT OF CALIFORNIA LIBRARIANSHIP

The State of California is a dynamic and diverse community with critical information needs. California librarians are committed to a leadership role in meeting these needs in a changing social, technological and economic environment. Librarians apply a unique mix of knowledge and information management to empower Californians through strong professional advocacy that supports the value of information, intellectual diversity and freedom, equity of access, and innovative information system design and service delivery. Our values embrace an ongoing commitment to education and training, shared knowledge and an exchange of ideas.

Future directions for the library profession and its education

Introduction

The purpose of this document is to reflect on the core areas of professional expertise and practice that define librarianship and to present some of our forecasts of coming changes in the field as a basis for discussion in the profession.

Librarianship is in a period of rapid change and of competition for its professional domain. The explosive growth in the capacity, number, and power of computers (information machines), telecommunications channels (carrying the Information Superhighway), and new media have made the entire society information-conscious to a degree inconceivable a few years ago. Ironically, that growth has led to unprecedented interest in our professional domain on the part of segments of society previously indifferent to it.

Our profession now faces competitive challenges to which we must respond effectively. We will best be able to do so if we understand well our unique expertise, apply that expertise appropriately to the newly developing realities of the Information Society, and stake a very public, high-profile claim to our professional territory, or niche. Employment opportunities are both continuing in the public sector as well as developing explosively in the private sector in contexts ranging from companies that provide value-added information products to the public to corporate strategic intelligence. Library and information education in California must respond to the full range of these needs.

Given the radical change confronting our whole society, the institutions of our field, including both libraries and library education institutions, must be flexible and effectively adaptive to change. Individuals, likewise, must be able to adapt to the times. What beginning librarians need to know today and what they needed twenty years ago are not the same. Lifelong continuing education needs to be embraced by our field as a whole and by the individuals within it. Greater emphasis on the offering of continuing education opportunities is imperative for educational institutions in the field.

The nature of the profession

For a body of knowledge to be professional, as opposed to technical, it must be composed of some general knowledge (a body of theory, principles and practices). A practitioner of a profession then makes specific applications of this knowledge to particular situations. The application is not routine or automatic, but rather is of a nature that requires judgment and experience. To put it another way, the application is not algorithmic, like a computer program; rather, it has unpredictable, variable, and ambiguous elements. It is at those points where professional judgment and experience come into play.

Professions are distinguished from one another by the societal problems they address and the tools, methods, and service models they use to address those problems. For example, the legal profession deals with the social problems of conflict avoidance and resolution. Over the centuries, lawyers (along with the judicial system) have developed procedural rules, methods and skill sets for interpreting and applying law in service to this social problem. However, the legal profession is only one of a number of professions that have carved out niches from the same societal problem. For example, family and marriage counselors, therapists of all sorts, and arbitrators have niches that overlap and frequently compete with lawyers. Thus the societal problem defines a domain of practice that may be shared by more than one profession—each approaching resolution of the problem from a set of core areas of expertise.

Implicit in the above, and vital to successful professional activity, is the existence of a population of individuals and institutions that draw upon a field's professional skills. Currently, changes in the larger society have implications both for the nature of the practice of librarianship and for the nature of our relationship to individual and institutional users of our services. One of the pivotal dynamics facing librarianship today is competition for at least parts of its traditional niche. To focus too closely on the institution of the library in which librarianship has largely been practiced is to miss the social problems that it addresses and the fact that the tools, methods, and values that define librarianship will continue to do so as its institutional base changes. For our profession to compete successfully, it must identify and assert its professional niche on the larger societal stage.

A hierarchy of professional expertise

In what follows, we do not attempt a definitive description of the library profession. Instead, we set out the parameters of the professional niche of librarianship. Those parameters can be represented at various levels of abstraction and detail. At the lowest level of abstraction, professional competencies could be addressed (but will not be) in terms of skill sets and the craft of execution. And, at the highest level of abstraction, we can address core areas of professional expertise. What we are calling the dimensions of practice occupy a middle ground between these two.

- Core Areas of Professional Expertise

- Dimensions of Practice

- Professional Competencies

Movement upward through this hierarchy represents increasing abstraction and stability, as we move toward a somewhat immutable core that defines the profession and distinguishes it from other adjacent or overlapping (and competing) professions.

It is not the objective of the CLA Task Force on the Future of the Library Profession to attempt to define the myriad, shifting skill sets and competencies of the craft of librarianship, but rather to develop an outline of the core areas of professional expertise and dimensions of practice which can assist us in framing the identity of our profession in a rapidly changing information universe.

Core areas of professional expertise

The core professional expertise of librarianship is at the confluence of three areas of information, information technology, and users. It is the knowledge of these three areas that uniquely identify the domain of librarianship and distinguishes it from other professions. Information professionals should also understand their profession's role in relation to the larger social context in which information, information technology and users interact.

Information

Knowledge of information resources and organization has historically been the central expertise of librarians. While other disciplines or professions are concerned with the information content of their areas of expertise, only librarianship has been concerned with management of the full-range of recorded knowledge, frequently quite independent of the particular content or context of that record. We know about information acquisition (collection development and management), organization (cataloging, indexing, database design), storage (physical management of the resources), and retrieval. We know how to facilitate access to information.

Users

Librarians are concerned about the users and uses of information: understanding user needs, patterns of information use, and the matching of user needs to the contents of the information store. Many of the values of librarianship, such as intellectual freedom, come from a concern for users and their access to information.

Information technology

Information technology provides tools with which librarians perform their work. Librarians must understand those technologies to use them effectively. Technology is not an end in itself, but a tool evaluated by its utility to information management and use.

Our unique blend

What is unique about librarianship is the intersection of these three areas. While other fields—including computer science, communications, and cognitive science—address some part of these three areas, no other field addresses all three, and no other field pulls them all together in research and practice with information seekers and users. Effective information retrieval and service requires the unique professional mix of knowledge of information, users, and information technology. These three broad areas of knowledge and expertise and their intersection provide the framework for the niche librarianship needs to stake out.

Most of us in this field have been drawn to it precisely because we had a wider range of interests than usual, and can handle—indeed enjoy—dealing with many sorts of knowledge and activity in our work. It is this strength we should build on in staking out our professional niche. While individual librarians may specialize—some more technical, some more information organization oriented, and some more people-oriented—all need a strong grounding in all three areas to function effectively. That, in turn, requires the personality and intellectual disposition to encompass this broad range of skills and talents in our work.

Dimensions of practice

These three highly abstract core areas of professional expertise—information, users, and information technology—cut across four dimensions of practice and inform how the profession developed in the past and will evolve in the future. Like the core areas of expertise, these dimensions are relatively immutable and serve to define the scope of the niche of librarianship. These dimensions provide a framework for a more precise definition of information work. From these dimensions, new and highly variable skill sets are derived.

These dimensions are:

1. tool making,

2. information management,

3. service, and

4. management of information organizations.

While the professional must possess knowledge in all three of the core areas of expertise, specialization along one or more of these dimensions is common, and, most likely desirable.

Tool making dimension

As a profession, we create tools and service models to solve society's information problems. As the emerging information environment grows more complex and competitive, and as computing and telecommunications make an increasing variety of tools possible, this tool making function will become ever more critical. The changes in how information is created, distributed, and used in the emerging arena of distributed digital information and the library without walls requires new tools to help people store, find, and retrieve information in these new forms and contexts.

These developments will require not only the traditional principles and bodies of knowledge of librarianship, but also an understanding both of: (1) how these principles and knowledge can be applied to new contexts, and (2) new interdisciplinary knowledge incorporating, for example, knowledge from cognitive science, computer science, linguistics, communications, and educational psychology.

The survival of the profession (to say nothing of its playing a significant role in the emerging information universe) will require adaptive solutions for these new contexts. The profession must pursue those solutions quickly and aggressively since it is in the arena of digital information that the profession is most vulnerable to the emerging (and intensifying) competition. The continued contribution of librarianship will rest largely with its ability to establish a vigorous tool-making presence in the eyes of the public, the competition, and the framers of national information policy.

Information management dimension

The second dimension relates directly to the application of the tools of the profession to the management of information from its acquisition to its dissemination. Librarians have traditionally been the experts on the tools used for information storage, organization and retrieval. We have both used those tools on behalf of users and advised users on their use. As increasingly abundant digital media accelerates the transition of library as "place" to library as "logical entity," many traditional functions (e.g., collection development) will be performed in dramatically new ways using equally dramatic new tools. One fact is clear, in this emerging environment of geographically distributed digital information, the function of information management will become increasingly critical to the processes of information access, filtering, analysis and use.

This fact was recently reaffirmed by Paul Saffo when he said of the digital future:

It is this plethora of content that will make context the scarce resource. Consumers will pay serious money for anything that helps them sift and sort and gather the morsels that satisfy their fickle media hungers. The future belongs neither to the conduit or content players, but those who control the filtering, searching, and sense-making tools we will rely on to navigate through the expanses of cyberspace.

Due to rapid changes in the information environment, professionals must be prepared to perform the information management functions along the full continuum from the traditional library defined by its resident collection to the library without walls defined by its logical collection of geographically distributed information.

All evidence indicates that the rate of these changes will accelerate. To prepare the large base of practicing professionals to meet the emerging demands, we will need to develop structured mechanisms for retooling. As a result, the vitality of the profession will hinge substantially on a professional commitment to life-long learning and its aggressive pursuit by individual professionals.

Service dimension

Librarians have traditionally provided users with two kinds of direct service: help in using the tools to find information, and training in the use of the tools.

In helping users find information, librarians have traditionally provided a service that, while highly useful, is somewhat divorced from the information need. We have generally provided an array of sources containing potential solutions to the user's need while avoiding the responsibility for evaluation of sources (and the information they contain) in terms of specific needs and the choice of an authoritative answer. In other words, we often direct the user toward sources of information without evaluating those sources in light of the specific need.

What we see in the emerging information universe is: (1) users relying less on the librarian for direction to resources as they rely more on machine intermediation, (2) more reliance on the librarian for training, and (3) the librarian as agent or partner in the information enterprise by way of information analysis, synthesis and the presentation of an information product (that is, the librarian as the information expert in an information problem-solving team with the consumer).

This consultative or partnership model, while relatively common in many special libraries and information centers, is rare in other library contexts. However, we foresee librarians functioning as the information expert in a multidisciplinary problem-solving team in an increasing variety of contexts. The trend toward the consultative model will increase as information structures grow in both complexity (e.g., multi- and hypermedia) and abundance. The services provided by professional librarians will range along a continuum from a traditional resource provision model, in which the librarian provides that user with resources that potentially contain the answer, to a consultative model in which the information professional directly solves the information need of the user or actively collaborates with the user in reaching that solution.

In the past, with the exception of some courses in information brokering and special libraries, instruction in the schools of library and information science has focused primarily on the resource provision model. Professionals must receive a balanced education that permits them to function effectively along the continuum from a resource provision model to the consultative model.

We also foresee more emphasis on user training as tools are increasingly designed for direct use by end-users and become more complex and changeable. In addition, the need for another kind of training will grow critical: aggressive information literacy education. The abundance of information and the complexity of new information structures will require that people be information literate. Full citizen participation in the benefits promised by the National Information Infrastructure will require knowledge of the four sub-literacies of information literacy—literacy in reading and writing, computing literacy (managing the enabling technologies and productivity software), media literacy (creating, manipulating and integrating various media including images, sound and video), and network literacy (finding and retrieving globally distributed information). Librarians should be prepared to engage in such education—first by being information literate, and second by mastering the processes required to create instructional resources and to teach the four sub-literacies to a largely unprepared populace.

Management of information organizations dimension

The issues of management are not unique to librarianship. Throughout our society all sorts of organizations—government, educational, and business—are being challenged by economic pressures to modernize, downsize, and generally make management practices more streamlined and sophisticated. Public organizations must now attract financial support through grants and gifts in a major way; public funding is no longer enough.

These pressures, plus the fusion of computing and telecommunications, are leading to a flattening of organizational structures and the emergence of adhocracy. These trends will affect libraries as well as the other organizations within which libraries function.

These pressures toward efficiency, effectiveness, and accountability will require that librarians have expertise in new processes of management and organizational behavior. While not unique to librarianship, the need for this expertise is critical, because without sufficient resources (and effective use of available resources), libraries and other information organizations or units will be unable to deliver the services needed by society and will ultimately lose their legitimacy as solvers of society's information problems.

Conclusion

The message that we want to convey is three-fold. First, the library profession is facing continuing, unprecedented change. What was adequate preparation for professional practice in the past is no longer. What was once adequate performance from librarians, libraries, and other information organizations is no longer.

Second, uncertainty will be an enduring aspect of the professional environment. No one knows what will happen other than, whatever it is, it will be unexpected and unpredictable. The invention of the World Wide Web, for example, drastically changed the information environment in unpredictable ways.

Third, the survival of the library profession, and therefore of its values and principles, is at stake. Other professions also see opportunities in the burgeoning information world. This competition is both disquieting and energizing. Developments in information technology and in the knowledge base of the library profession have vastly increased the variety, quality, and level of information services possible. Competition from other professions should spur us to find more creative and enterprising ways to apply our knowledge base and dimensions of practice.

Each component of the library world must respond to these challenges. Libraries and other information organizations have to look to new ways of providing services and products, and to continually improve and expand operations and services to meet new opportunities despite resource constraints if they are to continue to exist. We have to be proactive about inventing new generations of tools and services, and continually improving our management methods and tools.

Librarians have to look for new arenas and methods for applying our unique knowledge base. Jobs are changing in all parts of the economy; the old jobs, for librarians and others, may simply not exist in the future. New jobs are emerging in other parts of the information environment. If librarians don't stake a claim (based on our knowledge base and dimensions of practice), others will.

Librarians must continually update their skills to fit the changing information world. This is the responsibility not only of the employer but of the individual professional as well. Without continuous education, the professional will lack the ability to stay current and will be left behind by accelerating change. Library education programs are one possible source of continuing education opportunities.

Library education programs are likewise responsible for dynamically responding to these challenges brought about by changing needs and opportunities. The library education program of twenty or ten or even five years ago no longer meets the needs of today's professionals.

The California Library Association and other library and information associations must also take responsibility for ensuring that the profession and its institutions remain responsive to the changes in the information world by looking for new ways to apply our professional expertise as we reshape our current domain and explore new ones.

This is a challenging time for the library profession. And it's not temporary; we expect that the development of new technology and opportunities, and the competition and uncertainty, are now permanent features of information's environmental landscape. We have the choice of responding with energy and forthrightness, or seeing our profession—our values, principles, knowledge base, dimensions of practice, and jobs—disappear. We are confident that the profession will embrace this challenge.

11

A criterion citation rate for information scientists

ABSTRACT

Citation rates, as a measure of the quality of a researcher's works, are influenced not only by the inherent value of the research, but also by the size of the pool of available citers in a given field. Proper evaluation of a researcher's work should be in relation to a *criterion rate of citation*, the citation rate of the top researchers in that field. This paper 1) provides a criterion citation rate for information science (found to be 13.5 citations per year), 2) presents the names of the top researchers in the field by citation rate, and 3) compares these results to those of four earlier studies on information science researchers.

Introduction and background

The rate, or frequency, of citation to a researcher's works is coming increasingly to be used as a basis for evaluating the impact and importance of that person's work in a field. (See discussion in Broadus, 1977, p. 310–313.) But citation rates are a function not only of the importance or value of the works, but also of the size of the pool of available citers and their publication and citing behavior. A high citation rate for one field might be a low one for another field. Zuckerman found, for example, that Nobel prize winners in science averaged 222 citations a year just prior to receipt

First published as Bates, M. J. (1980). A criterion citation rate for information scientists. *Proceedings of the 43rd ASIS Annual Meeting, 17,* 276–278.

of the prize (Zuckerman, 1977, p. 187). Because of our smaller pool, such a rate of citation in information science would be almost inconceivable.

Thus, it would appear that proper citation rate evaluation of a researcher's work should be in relation to a criterion rate of citation as a basis for comparison within a field. Formally, the *criterion rate of citation* is defined as the median score of the top 20 researchers' mean annual citation rates (computed on a 10-year base) in the top journals in a field. "Top journals" are those covered in *Science Citation Index (SCI)*.

The main purpose of this paper is to present a criterion citation rate for information science. The secondary purposes are to present a list of the top information scientists and to compare the list to those produced by earlier studies—keeping in mind the differences in method among various studies.

Others have looked at citation rates and/or attempted to determine who the top researchers were in information science. Cuadra (1964) sought to identify a core body of literature in the field in 1963. He did this several ways, asking knowledgeable colleagues for names, looking at citations in textbooks, and looking for frequently listed works in bibliographies of the field. A consensus list of top names produced by the first of these methods (names receiving two or more mentions by the four colleagues) will be used later in comparison with other studies. Salton (1973) used the seven-year cumulative index of the *Annual Review of Information Science and Technology (ARIST)* and Dahlberg's *Literature zu den Informationswissenschaften* (Information Sciences Literature) to determine who the most-mentioned people in the field were in 1973. He selected people who had more than four lines of mentions in these sources. His names from ARIST will be used later in the comparisons. Note: Citations in *ARIST* indexes will be called "mentions" in this paper to distinguish them from citations to an author's work from the field as a whole. Usually, a given work receives only one mention in *ARIST* the year after it is published, while it may receive none to indefinitely many citations over many years from the research literature as a whole, depending on its importance to the field.

Cawkell (1974) responded to Salton's article by producing a list of information scientists who received 10 or more citations during 1972 in the *SCI*. He did not say how he found the names to look up in the *SCI*. Finally, Brace (1976) produced a list of most-cited authors in library and information science dissertations for the years 1961–70. According to Brace, only 11% of the dissertations were in information science. None of these four studies was concerned with anything like a criterion rate of citation, nor did any use the method of identifying top researchers to be described below.

Method

A major, respected annual review such as *ARIST* can be presumed to identify the conceptual territory and the corresponding researchers of a field. Thus, those who are most frequently mentioned in the 10-year index to *ARIST* (1966–75) are presumably of great importance in the field. If we look up these high-mention names in the *SCI*, we should be able to determine the criterion citation rate.

But number of mentions in *ARIST* is to some extent a function of number, rather than impact of publications. By the nature of annual reviews, the author of many small-but-worth-mentioning papers will receive more mentions than the author of the occasional major work. So rather than looking up the 20 researchers with the most *ARIST* index mentions in the *SCI*, a much larger number with top mentions were looked up, in order to produce a pool from which the 20 actually *cited* most (as opposed to most *mentioned*) might be expected to emerge. As it happened, 99 information scientists had received 16 or more mentions, so these were made the base.

The top 20 by mean annual frequency of citation in *SCI* from among the 99 names are considered, for purposes of this paper, to be the top people in the field. The criterion citation rate for information scientists is the median of these 20 people's mean annual citation rates for a ten year period, 1965–74, corresponding to the two five-year cumulations of *SCI*.

The approach of taking a larger pool proved to be valuable. Members of the top 20 in citations ranked as low as 57th in number of *ARIST* mentions (20 mentions). On the other hand, since none of the 29 people receiving 16 to 19 mentions appear in the top 20, odds are good that few other highly cited people would be picked up by starting with a base larger than 99. (Fourteen people had 20 mentions, thus accounting for the discrepancy between 57 and 29 in adding up to 99.)

In actually counting citations in *SCI*, both practical and substantive problems arise. Works by people with the same names are interfiled and must be distinguished. Further, if the intention is to look at citation rates within information science, then two additional problems need handling: 1) A researcher starts life in, say, psychology, and then moves into information science—a not uncommon occurrence in this young field. Earlier works by this researcher will continue to draw citations from the huge pool of psychologists even after he/she has come into information science. The information science criterion rate will be raised by citations that have nothing to do with the field. 2) The work of some researchers bridges two fields, information science and another. Citations to information-science-relevant works from outside information science should be counted.

After all, we do not want to penalize the researcher for having impact beyond our field. But at the same time, we do not want to count citations to all those other non-information-science-relevant publications that the researcher is also writing. The huge pool of his/her other field will again produce disproportionately many citations.

All three of these problems were dealt with by the following rule in counting citations: If the *cited* item looks like it is in information science (uses words such as "information," "documentation," etc.) *or* at least one of the *citing* journals is on the list of library/information science journals covered by *SCI* for the period of that cumulation, then count all of the citations to a particular item for that author. If neither of these qualifications is met, count none of the citations to that item. The reasoning here was that if an item is neither cited by a single information science journal in five years, nor sounds like it is itself in the field, then it is reasonable to say that the item and its citations are outside the field. In this way we also automatically eliminated citations to other researchers with the same names.

Caveats

ARIST mentions are affected by when a person comes to prominence. Even a stellar performer who came to prominence in only the last two or three years of the period covered would probably not get enough mentions to be in the 99 studied. To bring the study forward, the same 99 names were looked up in the annual *SCI* volumes for 1975–78. Since the same, earlier, pool of people is used as a base, newly prominent people will not appear in the counts for these years either. Finally, since the counting rule with the 1975–78 volumes had to be based on citation by a library/information science journal in a given year, rather than in a five-year period, the count for 1975–78 is on a slightly more stringent basis.

Results and discussion

The top 20 people by mean annual citations for the period 1965–74 had a range of 10.4 to 97.6 (mean) citations per year. The median of these 20 scores, and thus the criterion citation rate for information science, is 13.5. So if an information scientist averages ten or more citations per year, he or she is in the very top ranks of the field, as measured by citation rates. This is about 1/20 the rate cited earlier for top scientists in Nobel fields. Only 45 people averaged even 5 or more citations per year. Even the top figure, 97.6

TABLE 1. *Mean citation rates,* ARIST *mentions, and ranks of information scientists averaging five or more citations per year*

RANK BY MEAN ANNUAL CITATION RATE	NAME	MEAN ANNUAL CITATION RATE	NUMBER OF MENTIONS IN *ARIST* 1966–75	RANK BY NUMBER OF MENTIONS IN *ARIST* 1966–75 INDEX
1	Chomsky, N.	97.6	21	53
2	Salton, G.	39.8	107	1
3	Price, D.J.	29.2	23	44
4	Cleverdon, C.W.	26.8	33	18
5	Lancaster, F.W.	21.6	82	2
6	Borko, H.	19.7	49	4
7	Vickery, B.C.	19.6	26	29
8	Cuadra, C.A.	16.8	44	6
9	Allen, T.J.	14.1	40	11
10	Kilgour, F.C.	13.7	44	6
11	Menzel, H.	13.3	20	57
12	Fairthorne, R.A.	13.0	22	46
13	Kent, A.K.	12.8	27	27
14	Bourne, C.P.	12.5	29	22
15	Rees, A.M.	12.3	22	46
16	Swanson, D.R.	12.0	26	29
17	Orr, R.H.	11.8	22	46
18	Avram, H.	11.4	43	9
19	Garvey, W.D.	10.7	37	13
20	Becker, J.D.	10.4	40	11
21	Kochen, M.	9.7	30	21
22	Maron, M.E.	9.2	18	77
23	Simmons, R.F.	8.7	32	20
24	Freeman, R.R.	8.5	25	32
24	Herner, S.	8.5	20	57
26	Farradane, J.L.	8.0	16	91
27	Stevens, M.E.	7.9	19	71
28	Licklider, J.R.	7.4	26	29
29	Line, M.B.	7.2	19	71
30	O'Connor, J.J.	7.1	16	91
31	Gardin, J.-C.	6.5	16	91
32	Hayes, R.M.	6.1	36	14
33	Dougherty, R.M.	5.6	20	57
34	Atherton, P.	5.5	44	6
34	Sparck Jones, K.	5.5	41	10
36	Giuliano, V.E.	5.4	17	81
36	Leimkuhler, G.	5.4	20	57
38	Hillman, D.J.	5.3	28	23
38	Lipetz, B.-A.	5.3	24	36
38	Veaner, A.B.	5.3	28	23
41	Crane, D.	5.2	20	57
41	Taylor, R.S.	5.2	22	46
43	Artandi, S.	5.1	35	15
43	Paisley, W.J.	5.1	34	16
43	Pizer, I.H.	5.1	17	81

TABLE 2. *Top information scientists, various studies, 1963–80*

1963 CUADRA	1973 SALTON	1974 CAWKELL	1976 BRACE	1980 BATES	1980 BATES	1980 BATES
1963 DATA ALPHABETICAL MENTIONS	1966–72 DATA ALPHABETICAL MENTIONS	1972 DATA ALPHABETICAL CITATIONS	1961–70 DATA RANKED CITATIONS	1965–74 DATA RANKED CITATIONS	1966–75 DATA RANKED MENTIONS	1975–78 DATA RANKED CITATIONS
Bar-Hillel	Atherton	Bar-Hillel	Shera	Chomsky	Salton	Chomsky
Cleverdon	Avram	Carnap	Wheeler	Salton	Lancaster	Salton
Kent	Becker	Cherry	Wilson, L.	Price, D.J.	Parker	Lancaster
Luhn	Bobrow	Garfield	Downs	Cleverdon	Borko	Line
Maron	Borko	King	Ranganathan	Lancaster	King	Vickery
Mooers	Cuadra	Lancaster	Tauber	Borko	Atherton	Price, D.J.
Perry	Hayes	Licklider	Metcalf, K.	Vickery	Cuadra	Williams, M.
Stiles	Kilgour	MacKay	Martin, L.	Cuadra	Kilgour	Lynch, M.F.
Swanson	King	Perry	Joeckel	Allen	Avram	Allen
Taube	Lancaster	Price, D.J.	Evans, C.	Kilgour	Sparck Jones	Buckland
	Parker	Salton	Gaver	Menzel	Allen	Cuadra
	Salton	Shannon	McMurtrie	Fairthorne	Becker	Slamecka
	Simmons	Taube	Price, D.J.	Kent	Garvey	Sparck Jones
		Vickery	Danton	Bourne	Hayes	Cleverdon

citations per year, is a special case: it is for Noam Chomsky. The counting rule eliminated about one-third of the citations to Chomsky but still left many in from other fields. His work is very important to information science (he was 53rd in number of mentions in the *ARIST* index), but also in many other fields. Chomsky, in effect, belongs to all of science and would probably be a top cite-getter in a number of fields.

Table 1 summarizes the data for the top 45 people. The table includes name, mean annual *SCI* citation rate for 1965–74, number of mentions in the *ARIST* 1966–75 index, and ranks for these last two.

Even though the counting rule eliminated a lot of work done by individuals outside of information science, several of the top-listed information scientists have other audiences for their work besides information science, which may be responsible in part for their high citation rates. Chomsky has already been mentioned. Salton has an audience in computer science, Price in the history of science, Allen in management, Menzel in sociology, Garvey in psychology, and so on. One suspects that for those whose work has been entirely in information science, the criterion citation rate is even lower.

The top-ranking woman on the list is Henriette Avram at rank 18. There are only six women in the top 45 ranked researchers, or 13%. It appears that information science research, for the period covered, at least, is dominated by men.

Finally, in Table 2 the top-ranked names in the various studies mentioned in the background section are compared with the results of this study. Corporate authors are deleted from Brace's (1976) librarianship-oriented list, which is included mainly for contrast.

As can be seen in Table 2, there is little overlap between lists. This change in names may suggest, as Salton (1973, p. 220) discussed, rapid shifts in dominance in the field. But differences in method also no doubt account for some part, perhaps a large part, of these differences. A solid comparison through time will have to await a study using the same method for all periods.

ACKNOWLEDGMENT

I wish to thank my assistant, Lili Angel, for her hard work on the more tedious portions of this project.

REFERENCES

Brace, W. (1976). Frequently cited authors and periodicals in library and information science dissertations, 1961–1970. *Journal of Library & Information Science, [Taiwan]*, 2(1), 16–34.

Broadus, R.N. (1977). The applications of citation analyses to library collection building. In M.J. Voigt, & M. Harris (Eds.), *Advances in Librarianship, 7*, 299–335. New York: Academic Press.

Cawkell, A.E. (1974). [Letter]. *Journal of the American Society for Information Science*, 25(5), 340.

Cuadra, C.A. (1964). Identifying key contributions to information science. *American Documentation, 15*(4), 289–295.

Salton, G. (1973). On the development of information science. *Journal of the American Society for Information Science, 24*(3), 218–220.

Zuckerman, H. (1977). *Scientific elite: Nobel laureates in the United States*. New York: Free Press.

12

Information science at the University of California at Berkeley in the 1960s: A memoir of student days

ABSTRACT

The author's experiences as a master's and doctoral student at the University of California at Berkeley School of Library and Information Studies during a formative period in the history of information science, 1966–71, are described. The relationship between documentation and information science as experienced in that program is discussed, as well as the various influences, both social and intellectual, that shaped the author's understanding of information science at that time.

Introduction

I am writing this article not to claim myself as a pioneer of information science but rather to describe what it was like to be a student in the pioneering days of information science. There is much discussion nowadays of the history of information science, and in some instances it is argued that the early twentieth-century documentation theories of Paul Otlet (1990) and Suzanne Briet (n.d./1951) were the intellectual antecedents of information science. I am not a historian of the field, and I make no claim one way or the other about its historical roots. The understanding

First published as Bates, M. J. (2004). Information science at the University of California at Berkeley in the 1960s: A memoir of student days. Library Trends, 52(4), 683–701.

I developed of information science as a doctoral student in the 1960s at the University of California at Berkeley (U.C. Berkeley), however, had little to do with Otlet, Briet, or documentation in general. We saw information science as something brand new that was drawing on a range of earlier ideas, to be sure, but those sources were from realms very different from documentation. I believe that some of these sources are being lost sight of in the current discussion of the history of information science. In what follows, I present a memoir of my experience as a student at a formative moment in information science and describe the effort, as I saw it, to develop the field as a meaningful, distinctive discipline in one large doctoral program in a major university.

Becoming a Master in Library Science student

Sometimes in life we fall into where we were meant to be all along. Upon returning from Peace Corps service in Thailand, where I had taught English as a foreign language in two Thai high schools, I was confronted with the same question I had had as a fresh college graduate before I left: What should I do with the rest of my life? While sorting this out, I went to live with my parents in Lafayette, California, just over the hills from the University of California at Berkeley. I went down to Berkeley to take aptitude tests and get career counseling. It must be remembered that this was 1965, and women did not routinely get career guidance. Most female life-models for me in those years were homemakers. In fact, despite having attended Pomona College, one of the top liberal arts colleges in the country, I did not personally know a female Ph.D. in a tenure-track position until a high school girlfriend got her doctorate and an academic position.

The woman counselor was blunt: With my Phi Beta Kappa key and B.A. from a good college, about all that was available to me was to "type or teach." In those days, that meant working as a secretary or teaching in elementary or high school. In fact, at that time, the University of California required its secretaries to have bachelor's degrees. Other, more remunerative jobs for B.A.'s at the university went to men. The counselor said I would have to do graduate work of some kind if I wanted an interesting job of any other type. The high school children I had taught in Thailand had been far better behaved than typical American high school students, yet I disliked the little disciplining I had had to do there. I knew I did not want to teach in U.S. high schools.

I sensed that this was the first straight talk I had ever heard about careers and knew that she was right. The trouble was that I did not feel like going back to school. My undergraduate schooling had been very intense, and I wanted to play for awhile. What was worse, in the aptitude tests I scored right down the middle on everything—interested in everything and nothing. What was I to do? I went to the career information center and looked through the brochures for graduate programs. I looked for the program with the shortest time to attain a degree—maybe if the schooling did not last too long, I could stand it. The library program at Berkeley took just one calendar year. I applied.

The admissions officer at the library school asked me if I had ever read anything by Theodore Dreiser. I had not, which worried me a bit, but I was admitted anyway. I needed money, however, as the arrangement with my parents had always been that they would support me through college but not beyond. I applied at the newly founded Institute for Library Research (ILR), which was associated with the library school, and was hired as a graduate assistant by Ralph Shoffner.

In the first semester in the master in library science (M.L.S.) program, I studied book selection, cataloging, reference services, and the history of the book. Leroy Merritt, who later founded the short-lived library program at the University of Oregon in Eugene, taught book selection with a strong academic library orientation. I became expert at consulting book auction catalogs for out-of-print books. Roger Levinson, a fine printer by trade, conveyed his deep love of books as physical objects as he expounded on them in class in the Rare Book Room of the library.

The library school had its quarters on the top floor of the main Doe Library. Desks for students ran in alcoves next to the windows around the outer wall of the floor. The main library was walled off from the student quarters, and the library for the school was carved out of a portion of the main library stacks. (Later, the school moved into South Hall, the oldest building on campus, which had been renovated for its occupancy.) I was greatly relieved that, as a graduate student, I had direct access to the general stacks of the main library; undergraduates and visitors were not allowed in the stacks and had to handwrite request cards for every book they wanted to examine.

The atmosphere at the ILR made an interesting contrast with the more humanities-oriented world of the library program. The ILR was housed on the second floor of an old "temporary" building on campus that was supposed to be torn down at the end of World War II. The building was painted a pale institutional green and looked like those quickly constructed wooden military units seen in World War II movies. The ILR was directed by Robert Hayes at the University of California at Los Angeles (UCLA), but Ralph Shoffner ran the institute on a day-to-day basis at Berkeley; I reported to him or to others he directed. Shoffner has long since gone on to found his own consulting firm in Oregon, but at the time he was not many years from a very intense engineering education with an emphasis on operations research at the Massachusetts Institute of Technology (MIT). He had a driven quality, a fierce grin, and a wry sense of humor. He lived and breathed systems analysis, and every one of the ILR's projects was approached in a system-analytic way. Systems analysis itself was not so old then; in fact, one of its pioneers, Wes Churchman, taught at U.C. Berkeley.

Everything I was learning while a graduate assistant was new to me; I was unsure of myself and asked questions till I must have driven Shoffner crazy. Trained in the discursive language of the humanities, I found this new way of thinking utterly different, absorbing, and interesting. In the course of the first year I worked there, this way of thinking literally transformed how my mind worked. Gradually, I realized that I had an aptitude for this particular type of analytical thought. I worked on a project to speed up interlibrary loan processes among the University of California campuses by using fax machines to communicate between campuses; I also worked on a project to get the contents of the catalog of the California State Library in Sacramento into machine-readable form. In those days, there was no "online"; the machine-readable records were used to produce printed book catalogs. I soon formed a plan to become a library systems analyst with my master's degree.

Computers were still a relatively new phenomenon then. Though most librarians favored their use, there were debates in the library literature about whether computers were a good thing for libraries and, if they were used, what they should be used for. I wrote my own FORTRAN programs to do basic statistical analyses on some data for a small research project at the ILR. Such work would be done with standard statistical programs today, of course. At that time, all computer processing was done by feeding punched cards into big mainframe computers. In fact, because sciatica in my hip made it painful for me to walk during one term, I dropped out of a programming course because I could not walk up the hill over and over again to where the computers were in order to pick up my paper printouts.

My triumph as a student assistant in the ILR came one day after I and two others had been sent in a university car to Sacramento to draw a sample from the State Library's card catalog. I soon realized that the sampling method we were using was seriously biasing the results. Upon our return, it took me forty-five minutes to persuade my supervisor, who was just a rung above me in the institute hierarchy, that what we were doing was not right. He was finally convinced, and we retook the sample.

Going for the doctorate

One day at the ILR, Shoffner said to me, "So when are you going to apply to the doctoral program?" I had not seriously entertained this thought before but, when asked in this way, it seemed like quite a natural thing to do. At about this time other events took place that also made going on for a doctorate seem like an exciting thing to do. The federal government was dramatically expanding support for doctoral students in library and information science (LIS), in the form of what were known as Title II-B grants. At the same time, the Berkeley program launched a new information science emphasis with the hiring of M. E. (Bill) Maron and, a while later, William Cooper, Victor Rosenberg, and Michael Buckland. The new direction was exciting and felt like a natural follow-up to my interest in library systems analysis. I applied for and was admitted to the program and also received a three-year Title II-B fellowship. Fellow students entering the program within a year or two of my entrance included the following (for those who went on to teach, their university affiliations are given in parentheses): Hilary Burton, Michael Cooper (University of California at Berkeley), Ruth Gordon, Theodora Hodges (University of California at Berkeley), Caryl McAllister, Edmond Mignon (University of Washington), Jerry Nelson (University of Washington), Barbara Nozick, Ruth Patrick, Ralph Shoffner, Keith Stirling (Brigham Young University), Irene Travis (University of Maryland), Diana Thomas (UCLA), Howard White (Drexel University), and Harriet Zais.

Within weeks of Maron's arrival, I and nine other doctoral students had signed up to be his advisees. That was, I believe, a majority of the doctoral students in the early stages of the program, even with the boost of the Title II-B grants, and gives an indication of the enthusiasm and excitement surrounding the new initiative. Maron taught a course entitled Introduction to Information Science, which drew many students and, in effect, defined what information science was for the Berkeley program.

I have been unable to find my notes from Maron's class; however, a couple of years later, I was invited to teach the same introductory course as an acting instructor.[1] The introduction in my notes for the class states that I retained the content of Maron's course largely intact. The principal changes were my additions of a section on user studies and some material by Marshall McLuhan. Here is the main sequence of content of the quarter-length course syllabus as I presented it in spring 1970. Each indented line represents one class day; the class met three days a week for one hour.

---

1 Currently enrolled doctoral students are not now permitted to teach graduate courses as sole instructor in their own department in the University of California; presumably, the rules were different then.

Librarianship 240, Spring 1970
Introduction to the Information Sciences

I. Introductory
   The Information Explosion

II. The Organization of Information for Access
   What Is "Access"?
   Some Indexing Systems–I
   Some Indexing Systems–II
   The Descriptive Continuum

III. Automatic Procedures
   Set Theory
   Computers–I
   Computers–II
   Artificial Intelligence
   Automatic Indexing and Abstracting–I
   Automatic Indexing and Abstracting–II
   Associative Indexing
   Search Strategy
   Question-Answering Systems
   Field Trip
   Midterm

IV. Analysis and Evaluation
   Systems Analysis (guest speaker)
   "The Scientific Method"–I[2]
   "The Scientific Method"–II
   Statistical Procedures–I
   Statistical Procedures–II
   The User in the System–I
   The User in the System–II
   The User in the System–III
   Evaluating Information Systems–I
   Evaluating Information Systems–II
   Computers and Privacy
   Overflow

---

2 "The Scientific Method" was put into quotation marks because there are many such methods. The purpose of this section was to provide a simplified, general conception of how scientific research is conducted.

For a while, Maron was the only information science faculty member in the program, and the question quickly arose for those of us in the area of what other courses to take to prepare for the field. Maron later developed a follow-up information science course, which I took, and offered a seminar. I took other courses in the library school as well, which I will describe shortly. One person does not a discipline make, however. There was generally a feeling, supported and promoted by Maron, that information science was developing out of a number of disciplines, and a full education in the field required gaining that knowledge from outside the program as well as within. That often meant taking a course only part of which was of interest for my purposes. In the end, with the help of that wonderful fellowship, I took three full years of classes, culminating in qualifying exams in March of 1970.

Partly on Maron's advice, and partly based on my own interests, I took or audited the courses listed below, which were offered outside the library school. (I have made up the titles, as the Berkeley transcripts are quite cryptic.) Home departments for the courses are listed in parentheses; these are quarter, rather than semester-long, courses.

> Introduction to statistical inference (Statistics)
> Probability theory (Statistics)
> Cost/benefit analysis (School of Public Health)
> Linear algebra (Mathematics)
> Reading course in communication research (Psychology)
> Psycholinguistics (Psychology; took one quarter, audited second quarter)
> Artificial intelligence seminar on automatic game-playing[3] (Psychology)
> Propositional and first-order logic (Philosophy)

In the end, it was the social science work, rather than the mathematical, that most appealed to me, but the math enabled me to understand reasonably well the formulas and theory behind Gerard Salton's work (1968) and Maron's own work (Maron, 1961; Maron & Kuhns, 1960) on the design of automatic indexing systems.[4] One of my two chosen doctoral exam specialization areas was then known as "intellectual access"; it would be called "information retrieval" in most schools today.

---

3 These were not what are currently known as computer games; rather, we addressed established nonautomated games with known rules for play. As computer processing power was limited, software had to be based on strategic heuristics rather than on brute force computation of all options, and there was much interest at the time in such heuristics.

4 Wherever possible in this article, the cited book editions are the ones that I would have seen at the time rather than the latest edition available now.

In a seminar with Maron, I wrote a lengthy paper analyzing and comparing eleven efforts that had been made to that time to come up with formulas for effective automatic indexing. The paper covered the work of H. P. Luhn, Don Swanson, Fred Damerau, Harold Borko, and Paul Switzer, among others. It also covered the work of three women: Phyllis Baxendale, Myrna Bernick (Borko's coauthor) and Mary E. Stevens. Stevens wrote a widely cited review of literature on automatic indexing (1965), which I relied on a great deal in my studies. Many of these indexing approaches were re-invented in the 1990s in the early days of Web retrieval. In the 1960s the emphasis was on automatic indexing, rather than automatic retrieval, but the thinking was essentially the same. In another seminar, I wrote about the history and applications of citation indexing.

The social context of the times

Before I discuss other intellectual influences, something should be said about the general context of the times and its impact on this particular student. I started the M.L.S. degree in February of 1966, the last semester before Berkeley switched to the quarter system (they switched back to semesters again after I left). I started the Ph.D. program in the spring quarter of 1967 and left to take a teaching position at the University of Maryland in January of 1972, finishing the doctorate in December 1972. Those years that I was at Berkeley, 1966–71, encompassed most of the years of the 1960s revolution, of which it might fairly be said that Berkeley and San Francisco were the national headquarters. Those years were a time of almost continual ferment—there were movements for black and female equality, for sexual liberalization and general relaxing of rigid social constraints, and in opposition to the Vietnam War.

One cannot understand how liberating the 1960s were without understanding how oppressive the 1950s were for anyone raised during that decade. Our parents' generation, which had had a long hard slog through the 1930s Depression and World War II, just wanted peace and quiet, and they enforced that desire with an imposed conformism that was frightening in its intensity. (I am speaking about society in general here; my parents were not particularly strict.) Young people nowadays who wish they could have lived in that time would, in fact, be horrified at the almost Victorian constriction of 1950s life.

Not surprisingly, the prospect of equal rights for women in the 1960s had particularly intense meaning for me. I read Betty Friedan's *The Feminine Mystique* (1963) early on. When I heard that an organization called the

National Organization for Women was being founded, I made the necessary contacts to be involved in the creation of the first West Coast chapter of NOW in San Francisco.[5] It was organized by a middle-aged businesswoman named Inka O'Hanrahan. I was one of the youngest women at the founding meeting, and I am proud that I was the recording secretary for the meeting.

For a time I was part of a conscious-raising group (as they were then called) that we billed as a group for women who had already had their consciousnesses raised. Oh, were we naïve! I believed that once men realized that they had been discriminating against women, they would be happy to change things to make them more fair. I was baffled when they seemed to be angry that we wanted equal rights. After all, we were the ones who had been discriminated against! It was a long time before I understood that equality for women and for men of color caused some white men to feel they were losing their former privileges.

The pervasive inequality of women in society was certainly reflected in the university as well, including the library school. Despite the fact that about 30 percent of the doctorates in librarianship had gone to women at that time (based on a count I made at the time in Cohen, Denison, & Boehlert, 1963), there were no women in tenure-track positions in the school. (The one exception, Anne Markley, had only a master's degree and had been tenured and promoted to associate professor many years earlier, when that was still possible without a doctorate.) There was, in fact, a kind of upstairs/downstairs culture at the school, with the professors having all the privileges of tenure-track faculty and the lecturers and cataloging revisers constituting the downstairs, with much less pay and security. Most of the latter were women. In fact, this culture was so established, accepted, and out of consciousness that it was not until close to the time I graduated that I finally noticed that the work of cataloging instructors, such as Grete Frugé, was also about "intellectual access," and I wondered for the first time why there was not more connection between her work and Maron's.

Throughout this time, I participated in many marches against the Vietnam War. The movement climaxed in May of 1970, at the time of the U.S. invasion of Cambodia, which seemed a particularly egregious violation of the rights of a country that was not a party to the war, though Vietnamese troops were in Cambodia. The last several weeks of the school term were lost to rallies, marches, and organizational meetings. Students in the school, in line with our training, developed a clearinghouse of information on the war. This activity led to my first publication in the field (Bates, 1970).

---

5 I heard later that there is some dispute between the Los Angeles and San Francisco chapters regarding which chapter was actually founded first.

Altogether, I taught the Introduction to Information Science course three times at Berkeley, in the spring and fall of 1969 and spring of 1970. Because of riots or other disruptions, I was not able to complete the entire ten weeks of the quarter any one of those three times. I was reluctant to cancel classes for the sake of those students who wanted to continue, but, for a variety of reasons, it was sometimes just not possible to hold a class. In the spring of 1970, the university cancelled the last several weeks of classes for safety reasons.

Other influences

Another major influence during my years at Berkeley was William Paisley. He was a professor in the Communication Department at Stanford and was invited to teach a course in information needs and uses at the library school. Paisley's original background was social psychology, and his move into communication research had enabled him to have a broader understanding of and appreciation for the commonalities among normally distinct academic disciplines. He and Edwin Parker were the faculty members at Stanford who were looking at the information aspects of communication. They represented a small salient away from the conventional orientation of communication research toward the study of mass media.

Paisley taught a second course in the user area as well. I have found the two syllabi in my papers. One course was entitled Behavioral Study of Scientific Information Flow, and the other was entitled The Flow of Information to the Public. It is the first of these courses, taught in 1968, that I want to describe in part. The first week introduced "the information systems of science in their historical context." The next four weeks were devoted to "behavioral research methods useful in information studies." Assigned readings for these weeks drew heavily from a classic social sciences methods text, Kerlinger's *Foundations of Behavioral Research* (1964). Each week Paisley took a different broad class of research methods and illustrated it with relevant studies from information science, communication, and related fields. After an introduction to behavioral research methods in week two, he addressed, in succession during the next three weeks, the following: "the logic of nonexperimental descriptive research," "the logic of experimental explanatory research," and "the logic of nonexperimental explanatory research." I quote these exactly to demonstrate the language used at the time. In effect, he was supporting the legitimacy of what is now called quantitative and qualitative research. The terms "experimental," "quasi-experimental," and "nonexperimental" were widely used. I do not recall anyone ever calling such research "quantitative" or "qualitative."

In the second five weeks of the term, Paisley addressed "the information systems of science in their social context," including the "effects of the cultural and political systems on information flow," followed by the "effects of the professional association," "effects of the 'invisible college,'" "effects of the employing organization and work team," and, finally, "information inputs and cognitive processes." The reader may recognize these various contexts from Paisley's 1968 review chapter on information needs and uses in the *Annual Review of Information Science and Technology* (Paisley, 1968). "Information seeking in context" has become a popular byword in modern LIS research and has even generated a separate conference by that name, which began in the 1990s. But such thinking was already well launched thirty years earlier.

It is popular these days to speak of information-seeking behavior research and theory as though it only truly came around to a user-centered orientation in the 1980s. Before then, it is said, user research was system oriented. Paisley used the word "system" repeatedly in his class, as evident from the above, but the scientist is a very real actor in these systems, not a helpless pawn. These are not technological systems but rather human social systems. Throughout his course and the research he drew upon, there was very much a sense of the scientists both creating and being influenced by these several social contexts.

Many writers, however, cite Dervin and Nilan's 1986 literature review on information seeking as marking, essentially, the beginning of a modern user-centered orientation to information-seeking research. I have long been puzzled at this apparent blanking out of the rich body of information behavior research by Paisley and many other excellent researchers with social science research training prior to 1986. In reviewing Dervin and Nilan's paper, I note that their remit was to review the literature from 1978 forward, as 1978 was the date of the last preceding review of the topic. The following is the first paragraph of a section entitled "Call for a Paradigm Shift" (capitalization of author names was the standard format for the *Annual Review* at the time):

Since 1978 some scholars have focused their primary efforts on identifying the underlying premises and assumptions that they see as having guided information needs and uses research. They call for developing an alternative set of premises and assumptions—in essence, for the introduction of an alternative paradigm. Notable among these are: BELKIN (1978), BROOKES, DERVIN (1977; 1983b), HAMMARBERG, JÄRVELIN & REPO, LEVITAN, MARKEY, MICK ET AL., NEILL, RUDD, AND THOMAS D. WILSON (1981; 1984). (Dervin & Nilan, 1986, p. 12)

They then go on to summarize what they consider to be the differences between the user orientation represented by the above authors and a systems orientation. Thus, it would appear that subsequent generations of information behavior researchers have read this section and assumed that modern, user-sensitive research on information users began only about 1978 and that Dervin and Nilan were the first to capture this new move in their article. Yet in 1965, Paisley and Parker wrote an article entitled "Information Retrieval as a Receiver-Controlled Communication System" (Paisley & Parker, 1965). Colin Mick, cited above, was a doctoral student of Paisley's.[6] Further, the extensive research cited by Paisley in another, much longer review (1965) and by Herbert Menzel in a 1960 review is overwhelmingly user-centered; these are not studies of information system performance with standardized relevance assessments or collections of studies on library circulation.

I subsequently took User Studies as the second of my two doctoral examination areas for the Ph.D. As the first student to request that area, I typed a giant binder full of notes on all the research of note that I had found to that date and wrote a sixty-page literature review of the essential studies, in effect, to educate the faculty. I subsequently submitted the review to ERIC (1971). I should have developed the work into a book, but I lacked the confidence at the time. I now very much wish I had done so, because it appears that much of our field currently thinks that user-centered research began in 1978!

I subsequently founded the first courses in information-seeking behavior at the University of Maryland, the University of Washington, and UCLA, in, respectively, about 1975, 1977, and 1981. I have always tried to teach the full historical arc of information-seeking research rather than only the latest work. In ignoring that earlier literature, we collectively have failed to benefit from a rich body of findings that were often based on top-quality research designs that were supported by abundant funding. The 1960s were a golden era where federal social science research funding

---

6 Three other advisees of Paisley's in the Communication Department at Stanford have been influential in information studies: Christine L. Borgman, Donald O. Case, and Ronald E. Rice. Case has recently published a comprehensive book on information seeking behavior (Case, 2002).

was concerned; we have not seen the equal since. These studies were not all soulless statistical monstrosities, as so often caricatured in the current world of qualitative research theory. For example, the thirteen information seeking studies that appear in the 1959 *International Conference on Scientific Information*, which I studied closely as a doctoral student, employ a wide range of methods, most quite sensitive to a user perspective. Indeed, Menzel's research (1959) in that volume on the ways scientists serendipitously discovered new information of value to them could be reported today in a modern journal as a qualitative study, and, except for changes in the technologies used, the results are still of value—because *people and the social system of science* change much more slowly than does the technology.

My 1970 Information Science course (discussed above) lists the "User in the System" as a topic because I was a doctoral student attempting to bring a user orientation to a course that had been entirely devoted to a systems approach. I felt I had to relate this interest of mine to the main content of the course, thus I entitled the section "User in the System." In the meantime, however, I had already taken Paisley's and other courses that drew me to a social science research paradigm and an interest in both information seekers and system design. I was particularly interested in access vocabularies that were oriented to users and designed from their needs.

When I took Paisley's course, I was absolutely fascinated. I went to see him in his office one day. We talked for forty-five minutes, and it felt like coming home. There was an intellectual "just-right-ness" about how he thought about things and what his interests were. I had found my preferred intellectual style and content, and, ultimately, my mentor. (He could not formally be my advisor because he was at Stanford.) I was to learn that he had tremendous personal and professional integrity as well. He took me absolutely seriously as a researcher in training—something that was not always easy to come by for a young woman in the era of miniskirts. I could not have picked a better person to look to for guidance and as a model. (Paisley's wife, Matilda Butler, might have served as a female model; however, she was beautiful and had her life so well organized that I could not imagine ever being like her.)

Not long after that, I left Maron as an advisee. My increasing interest in social science approaches was not a good match with his mathemetico-logical theoretical orientation. Victor Rosenberg, who had joined the Berkeley faculty in the meantime, became my new advisor. Rosenberg had done an early and important study on information seeking for professional needs (Rosenberg, 1967). That study was one of the best-known sources of evidence for the Principle of Least Effort in information seeking at that time.

While working with Rosenberg, I had the opportunity to attend, as a student "gofer," an exciting conference in Palo Alto on the new technology of online database searching, which was just appearing on the horizon in its most primitive form. I did the bibliography for the book that appeared out of that conference (Walker, 1971) and got to meet a number of leading lights of the field, including Pauline Atherton (now Cochrane), Douglas Engelbart, Margaret Fisher, Robert Katter, Frederick Kilgour, Robert Landau, Davis McCarn, Edwin Parker, Mary Stevens, and Roger Summit, the founder of the DIALOG search service.

Rosenberg and I got along well in the advisor relationship, but he left for the University of Michigan before my work was completed. So I switched, finally, to Ray Swank, the dean of the library school, whose interests were less well linked to mine but were sufficiently close to complete the dissertation. Swank was a thoughtful and supportive advisor. My dissertation was entitled "Factors Affecting Subject Catalog Search Success," a topic that nicely melded my two interests in intellectual access and users (Bates, 1972; published in Bates, 1977a and b).

Another influential course was a seminar taught in the library school by James Dolby. Dolby was a professor of mathematics at San Jose State University and was working at that time with Harold Resnikoff on a grant to study information storage and access, especially in library catalogs (Resnikoff & Dolby, 1972). Dolby's course was immensely important to me in ways I did not fully recognize at the time. The seemingly disparate topics he raised and discussed in the class all had in common a deep understanding on his part of the ways in which we can think about information independently of content and still discover wonderful and valuable things about it. It is popular nowadays to be somewhat dismissive of the fascination with information that characterized the 1950s and 1960s. The work from that time is often viewed as reflecting a naïve assumption that information is an objective entity to be transferred from a sender to a recipient and has an identical meaning to both parties in the transaction (Dervin, 1983, Tuominen et al., 2002). I believe that this view misreads how sophisticated at least some of the writers were at that time. But, more importantly, this view also fails to see the positive benefits that arose, and can still arise, from the study of information as an entity distinct from its meaning content. Information can be an indicator of social processes, and it can be considered as a phenomenon of interest in and of itself in a variety of senses.

The final influential course to be mentioned was a seminar given by Patrick Wilson. I took it in my first quarter in the doctoral program, in the spring of 1967, when Wilson was in the final throes of writing his first book,

Two Kinds of Power: An Essay on Bibliographical Control (1968). Wilson was trained as a philosopher, and he brought philosophical rigor to the discussion in class. We essentially worked through the ideas in his book during the course of the quarter. The class taught me that our discipline can be as intellectually demanding and as exciting as any body of thought. Though I had had a couple of philosophy courses in college, I had not understood the game of philosophy as it is played by its theoreticians. Wilson's course piqued my interest in philosophy in a more sophisticated sense. Some ten years later, while teaching at the University of Washington, I audited a couple of philosophy courses and frequently went to philosophy colloquia. This background has enriched my understanding ever since.

During this entire time as a student I was also taking reading courses and reading on my own, always with a sense of exploration in a new world and in an effort to pull together a coherent sense for myself of what information science was and where it could go. There had not been many books written within the field yet, but these four formed my understanding of the then existing core: Joseph Becker and Robert Hayes's *Information Storage and Retrieval* (1963), F. W. Lancaster's *Information Retrieval Systems* (1968), Charles Meadow's *The Analysis of Information Systems* (1967), and Manfred Kochen's collection, *The Growth of Knowledge* (1967). For me, Becker and Hayes's work was the canonical description of information science as it began in the 1960s. The book was a rare mixture of the key elements of a science named for information: it covered the management, both physical and intellectual, of information, the structure of retrieval systems, and the theoretical background. Lancaster's book provided a very insightful conceptualization of indexing theory. Meadow's book, though subtitled "A Programmer's Introduction to Information Retrieval," was useful to me because it presented a database management perspective. On the contents page of Kochen's book, there are checkmarks indicating that I had read over half the articles in it, but the article that influenced me the most by far was Derek de Solla Price's "Networks of Scientific Papers" (1967). Along with Price's two short books on the bibliometrics of scientific communication (1961, 1963), this work demonstrated how powerfully the seemingly trivial barebones statistics of information could tell stories of great interest from a sociological and historical perspective. I also read Thomas Kuhn's (1964) and James Watson's (1968) books, both of which, in different ways, shattered some standard assumptions about the way science works, and, implicitly, how science information flows.

Because of the newness of the subdiscipline of information-seeking behavior, there were few books on it, hence my extensive use, as noted above, of my own and others' literature reviews to identify a wide range of partially or wholly relevant resources. As for sources outside of information science, I read Shannon and Weaver's popularization of Shannon's information theory (1963), which was an important part of Maron's courses. In addition, as the theory drew on mathematics, engineering, and physics, which I found difficult, I also gave close attention to expositions on information and communication theory by Colin Cherry (1966), J. R. Pierce (1961), and Jagjit Singh (1966). I was particularly fascinated with Norbert Wiener and his book on cybernetics (1961). I also read his autobiography, *I Am a Mathematician* (1956), which was rooted in a classic child-prodigy tale. Though cybernetic theory itself has been marginalized subsequently, the core idea that some systems are governed by information fed back from the environment was a breakthrough of enormous significance at the time. We use terms like "feedback" so casually today that we do not realize how fundamentally such ideas shook up science and human understanding generally in the 1940s and 1950s. Wiener's work reinforced for me the idea that there is great power in understanding the role of information at a systems level. After I had held Wiener as a hero for many years, I worked briefly on a consulting job with Joseph Becker (co-author of the above-mentioned book by Becker & Hayes) in 1989. He told me that he had met Wiener and that Wiener had dismissed our field as "sorting things into jam pots." So much for hero worship.

More harmonious with my native abilities and cognitive style were materials I read in psychology and linguistics. The psycholinguistics course I took was taught by Dan Slobin, and it represented the cutting edge work of the day. We read Noam Chomsky's brilliant dissection (1959) of B. F. Skinner's book, *Verbal Behavior* (1957). Chomsky's review was one of several forces propelling a movement to restore the validity of studying the mind to the discipline of psychology, in contrast to the mandate to study only observable behavior, which had been the position of the behaviorist paradigm of Skinner and others. In a bibliometric study I did in 1980 of our field, covering somewhat earlier literature, I found that Chomsky was the most-cited person in our field at the time (Bates, 1980). Books such as George Miller's *Language and Communication* (1951), which analyzed language from the standpoint of Shannon's information theory, and Miller, Galanter, and Pribram's *Plans and the Structure of Behavior* (1960), informed my thinking and reinforced the value of understanding life from its pattern and structure, from its information, in addition to its meaning.

Final of influences were the guest speakers in classes or in the regular colloquia that were held in the school or in other departments. I kept notes on these talks. Apart from faculty in the school, such as William Cooper and J. Periam Danton, the speakers whose talks I attended included (in no particular order): Robert Hayes, Paul Wasserman, Lotfi Zadeh, John Bennett, Robert Sommer, Robert Katter, Donald Kraft, Carlos Cuadra, Warren McCulloch, and Michael Lesk. These speakers represented a mixture of the social, engineering, and information sciences. The one woman speaker I recall and have notes on was Christine Montgomery. She presented a creditable, professional talk on "Content Representation and Information Processing." It was a novel sensation to hear a speaker with whom I could identify more directly than I was accustomed to with the male speakers.

Through all these various influences, I developed a sense of information science as being, actually, about information. For a long time, I took my own understanding for granted, as representing a general way of thinking about the subject in the field. Finally, however, as more and more new influences entered the field, many of them powerful and interesting as well, it seemed more and more important to try to articulate just how our discipline can carve out its own particular territory among the many disciplines competing for some of the same intellectual turf. In a 1987 conference paper, entitled "Information: The Last Variable," I argued for more attention to the discovery of the variables that are unique to the study of information. In 1999, in "The Invisible Substrate of Information Science," I presented much more extensively my view of what uniquely distinguishes information science from other disciplines. See, especially, the section titled "Information Science Theory" in that article. These ideas, developed over thirty years in the field, had their roots in my experiences at U.C. Berkeley in the 1960s.

Discussion and conclusions

This article began with a question regarding the role of the history of documentation in the development of information science at the University of California at Berkeley. I have reviewed a wide range of influences that chiefly formed my thinking as a doctoral student in the school at the time. These influences are all about scientific, engineering, logical, social, and psychological thinking that formed early thinking in information science as we experienced it at U.C. Berkeley rather than about documentation. It is fitting at this point, however, to refer to the tiny role that documentation did play in my studies there.

When I arrived at the library school, there was still a course on the books titled Documentation, and my recollection is that it was Dean Swank whom I asked about it. He said that it had not been given in several years, and the subject had been a precursor to information science. As far as I am aware, the course was not given again. In reviewing the notes from my schooling, however, I found a lecture on documentation. In a course numbered Librarianship 212A—unfortunately with no title, but I recall it as a course in advanced reference sources—taught by Ray Held in the winter quarter of 1967, the first lecture of the course was on documentation. Perhaps Held had taught it previously? In the notes I took on that lecture, I wrote that documentation largely overlaps librarianship but has slightly different concerns. Documentalists were said to be more interested than librarians were in dissemination; were more likely to focus on new systems, theories, and technologies; and worked most often in science and technology disciplines. This description sounds a lot like the work of special librarians. In that lecture, I got no sense, or at least retained no sense, of the long history behind the idea of documentation. For me, at Berkeley, information science was something new under the sun, drawing on theory and research from a number of fields, none of them being documentation.

It is no doubt information science's loss that we did not develop a better linkage with the larger theoretical history of our field while students at Berkeley. At the same time, a wide range of deeply developed thinking in the social and engineering sciences *did* enormously enrich our understanding. In the recent enthusiasm for reconnecting with the earlier history of our field, it seems to me that the middle-term history, that of the 1950s and 1960s that I have described herein, is being rather ignored, and the full richness of understanding that is available to our field thus is not integrated.

Of the nine of us who went into academia from the Berkeley doctoral group listed earlier, all but two have taught mainly in the West. Thus the vision of information science developed at Berkeley may not have penetrated much beyond the Rocky Mountains. Considering the standing today of the subject matter that we covered, the subfield of information retrieval has certainly thrived subsequently. Gerard Salton's work at Cornell University in New York, however, surely had a great deal to do with the subsequent success of that subfield as well. Patrick Wilson's sophisticated philosophical analyses of access and information seeking in his three books (1968, 1977, 1983), as well as in the book written by Howard White, Wilson, and myself in 1992, and Paisley's legacy in information seeking (see also Paisley, 1980; Paisley & Parker, 1965, 1968; Parker & Paisley, 1966; Rees & Paisley, 1968, among others),[7] seem to have been much less recognized subsequently—much to the loss of the field, I believe. Whatever the reasons, perhaps now with publications such as this issue of *Library Trends* we are at last developing a

---

7 Both Paisley and Parker subsequently left Stanford to establish their own information and communication industry businesses.

sufficient sense of ourselves as a discipline to bring together all of the rich
sources from which we draw and to create an intellectual edifice worthy
of the exciting questions we study.

REFERENCES

Bates, M.J. (1970). A campus information clearinghouse. *California Librarian, 31*(3), 171–172.
Bates, M.J. (1971). *User studies: A review for librarians and information scientists.* (ERIC ED 047 738).
Bates, M.J. (1972). *Factors affecting subject catalog search success.* (Unpublished doctoral dissertation). University of California, Berkeley.
Bates, M.J. (1977a). Factors affecting subject catalog search success. *Journal of the American Society for Information Science, 28*(3), 161–169.
Bates, M.J. (1977b). System meets user: Problems in matching subject search terms. *Information Processing & Management, 13*(6), 367–375.
Bates, M.J. (1980). A criterion citation rate for information scientists. *Proceedings of the American Society for Information Science Annual Meeting, 17,* 276–278.
Bates, M.J. (1987). Information: The last variable. *Proceedings of the 50th ASIS Annual Meeting, 24,* 6–10.
Bates, M.J. (1999). The invisible substrate of information science. *Journal of the American Society for Information Science, 50*(12), 1043–1050.
Becker, J., & Hayes, R.M. (1963). *Information storage and retrieval: Tools, elements, theories.* New York: Wiley.
Briet, S. (n.d.). Qu'est-ce que la documentation? [What is documentation?], (R.E. Day & L. Martinet, Trans.). Paris: Éditions documentaires, industrielles et techniques, 1951. Retrieved April 18, 2004, from http://www.lisp.wayne.edu/~ai2398/briet.htm.
Case, D.O. (2002). *Looking for information: A survey of research on information seeking, needs, and behavior.* New York: Academic Press.
Cherry, C. (1966). *On human communication: A review, a survey, and a criticism* (2nd ed.). Cambridge, MA: MIT Press.
Chomsky, N. (1959). Review of B.F. Skinner, *Verbal behavior. Language, 35*(1), 26–58.
Cohen, N.M., Denison, B., & Boehlert, J.C. (1963). *Library science dissertations: 1925–1960: An annotated bibliography of doctoral studies.* Washington, DC: United States G.P.O.
Dervin, B. (1983). Information as a user construct: The relevance of perceived information needs to synthesis and interpretation. In S.A. Ward & L.J. Reed (Eds.), *Knowledge structures and use: Implications for synthesis and interpretation* (pp. 153–184). Philadelphia: Temple University Press.
Dervin, B., & Nilan, M. (1986). Information needs and uses. *Annual Review of Information Science and Technology, 21,* 3–33.
Friedan, B. (1963). *The feminine mystique.* New York: Norton.
International Conference on Scientific Information. (1959). *Proceedings,* (2 Vols.). Washington, DC: National Academy of Sciences, National Research Council.
Kerlinger, F.N. (1964). *Foundations of behavioral research: Educational and psychological inquiry.* New York: Holt, Rinehart, & Winston.
Kochen, M. (Ed.). (1967). *The growth of knowledge: Readings on organization and retrieval of information.* New York: Wiley.
Kuhn, T.S. (1964). *The structure of scientific revolutions.* Chicago: University of Chicago Press.

Lancaster, F.W. (1968). *Information retrieval systems: Characteristics, testing, and evaluation.* New York: Wiley.

Maron, M.E. (1961). Automatic indexing: An experimental inquiry. *Journal of the Association for Computing Machinery, 8*(3), 404–417.

Maron, M.E., & Kuhns, J.L. (1960). On relevance, probabilistic indexing, and information retrieval. *Journal of the Association for Computing Machinery, 7*(3), 216–244.

Meadow, C.T. (1967). *The analysis of information systems: A programmer's introduction to information retrieval.* New York: Wiley.

Menzel, H. (1959). Planned and unplanned scientific communication. In *Proceedings of the International Conference on Scientific Information* (Vol. 1, pp. 199–243). Washington, DC: National Academy of Sciences, National Research Council.

Menzel, H. (1960). *Review of studies in the flow of information among scientists* (2 vols). New York: Columbia University, Bureau of Applied Social Research.

Miller, G.A. (1951). *Language and communication.* New York: McGraw-Hill.

Miller, G.A., Galanter, E., & Pribram, K.H. (1960). *Plans and the structure of behavior.* New York: Holt.

Otlet, P. (1990). *International organisation and dissemination of knowledge: Selected essays of Paul Otlet* (W. Boyd Rayward, Ed. & Trans.). New York: Elsevier.

Paisley, W.J. (1965). *The flow of (behavioral) science information: A review of the research literature.* Stanford, CA: Stanford University, Institute for Communication Research.

Paisley, W.J. (1968). Information needs and uses. *Annual Review of Information Science and Technology, 3*, 1–30.

Paisley, W.J. (1980). Information and work. *Progress in Communication Sciences, 2*, 113–165.

Paisley, W.J., & Parker, E.B. (1965). Information retrieval as a receiver-controlled communication system. In L.B. Heilprin, B.E. Markuson, & F.L. Goodman (Eds.), *Symposium on education for information science. Proceedings* (pp. 23–31). Washington, DC: Spartan Books.

Paisley, W.J., & Parker, E.B. (1968). The AAPOR conference as a communication medium. *Public Opinion Quarterly, 32*(1), 65–73.

Parker, E.B., & Paisley, W.J. (1966). Research for psychologists at the interface of the scientist and his information system. *American Psychologist, 21*(11), 1061–1071.

Pierce, J.R. (1961). *Symbols, signals, and noise: The nature and process of communication.* New York: Harper.

Price, D.J. de Solla. (1961). *Science since Babylon.* New Haven, CT: Yale University Press.

Price, D.J. de Solla. (1963). *Little science, big science.* New York: Columbia University Press.

Price, D.J. de Solla. (1967). Networks of scientific papers. In M. Kochen (Ed.), *The growth of knowledge: Readings on organization and retrieval of information* (pp. 145–155). New York: Wiley.

Rees, M.B., & Paisley, W.J. (1968). Social and psychological predictors of adult information seeking and media use. *Adult Education Quarterly, 19*(1), 11–29.

Resnikoff, H.L., & Dolby, J.L. (1972). *Access: A study of information storage and retrieval with emphasis on library information systems.* Washington, DC: U.S. Department of Education. (ERIC ED 060 921).

Rosenberg, V. (1967). Factors affecting the preferences of industrial personnel for information gathering methods. *Information Storage and Retrieval, 3*(3), 119–127.

Salton, G. (1968). *Automatic information organization and retrieval.* New York: McGraw-Hill.

Shannon, C.E., & Weaver, W. (1963). *The mathematical theory of communication.* Urbana, IL: University of Illinois Press.

Singh, J. (1966). *Great ideas in information theory, language, and cybernetics.* New York: Dover.

Skinner, B.F. (1957). *Verbal behavior.* London: Methuen.

Stevens, M.E. (1965). *Automatic indexing: A state-of-the-art report,* (NBS monograph no. 91). Washington, DC: United States G.P.O.

Tuominen, K., Talja, S., & Savolainen, R. (2002). Discourse, cognition, and reality: Toward a social constructionist metatheory for library and information science. In H. Bruce, R. Fidel, P. Ingwersen, & P. Vakkari (Eds.), *Emerging frameworks and methods: Proceedings of the Fourth International Conference on Conceptions of Library and Information Science* (CoLIS4), (pp. 271–283). Greenwood Village, CO: Libraries Unlimited.

Walker, D.E. (Ed.). (1971). *Interactive bibliographic search: The user-computer interface.* Montvale, NJ: AFIPS Press.

Watson, J.D. (1968). *The double helix: A personal account of the discovery of the structure of DNA.* New York: Atheneum.

White, H.D., Bates, M.J., & Wilson, P. (1992). *For information specialists: Interpretations of reference and bibliographic work.* Norwood, NJ: Ablex.

Wiener, N. (1956). *I am a mathematician.* Garden City, NY: Doubleday.

Wiener, N. (1961). *Cybernetics: Or control and communication in the animal and the machine* (2nd ed.). Cambridge, MA: MIT Press.

Wilson, P. (1968). *Two kinds of power: An essay on bibliographical control.* Berkeley, CA: University of California Press.

Wilson, P. (1977). *Public knowledge, private ignorance: Toward a library and information policy.* Westport, CT: Greenwood Press.

Wilson, P. (1983). *Second-hand knowledge: An inquiry into cognitive authority.* Westport, CT: Greenwood Press.

13

Acceptance speech for ASIST Award of Merit

I stand before you as Llewellyn C. Puppybreath the 3rd's second ex-wife. It's great to get some recognition and not have it all go to Llewellyn.

But, seriously, I cannot tell you how much this award means to me. It represents this community's recognition of my work and participation in the field. This high honor caps my career and fulfills my lifetime's work. I want to thank the nominators and the jury and all of you for selecting me to receive the Award of Merit.

As this is such a special honor, I want to use the time I have to talk about the distinctive experience I've had as a woman of a particular generation, who set out to teach and do research in her discipline. The poet Muriel Rukeyser has said: "What would happen if one woman told the truth about her life? The world would split open."

Opportunities in university teaching for women in generations prior to mine were virtually non-existent. Women faculty in earlier generations were the exception that proved the rule. They offered no serious threat to male dominance.

When women did begin to come into university teaching in large numbers in the 1970's, mine was the first generation large enough to be perceived as a serious threat to the comfortable boys' club. And the boys were not happy. My generation paid a high price for our opportunities. I had naively thought that men would be embarrassed to realize how much women had been discriminated against and would quickly act to correct things once the unfair rules of the game came to the light of day in the 1960's. Au contraire.

First presented as Bates, M. J. (Nov. 1, 2005). *Acceptance speech for the American Society for Information Science and Technology Award of Merit.* Annual Conference of the American Society for Information Science and Technology, Charlotte, NC.

I saw this in my own life when my effort to be promoted to full professor was turned down in the late 1980's, despite an excellent record. Finally, after ten years without a merit raise of any kind, when I was considering going up for professor again in 1991, I looked back at the cohort of doctoral students who had gone through Berkeley's program with me and then gone on to teach. There were four men and four women in regular positions. Two of the men were now full professors, and two were associate professors. Two of the women, including myself, were associate professors, and two had failed to make tenure. In other words, the guys were exactly one promotion further along. I hadn't realized that being a woman dropped my odds of tenure to 50 percent!

That's scary enough, but the work of the two women who didn't make tenure and left academia continued to be cited more than the work of the male associate professors who did get tenure and continued to have the opportunity to do research and write. That's the operational moment when discrimination becomes visible. That's when the rubber meets the road. This award is in part for those two women colleagues and all the others who didn't get a fair shake.

The personal lives of women of my generation were not easy either. The cost of reconciling two different careers was overwhelmingly borne by the woman. A doctoral student I met in another department had gotten a Ph.D. in English literature and received a coveted offer of a tenure-track position. But her husband, also a doctoral student, couldn't find a position in the same city. So she was now working on her *second doctorate*, in the hopes that they could still find a job together.

But I don't want to speak only of men. I learned another big lesson over these years, and that is that we are ALL sexist. Almost anyone old enough to be in this room is old enough to have been raised in a different era. Most of us act in sexist ways quite out of awareness. We were steeped in it in our own families. We saw it in our friends' families and on a hundred sitcoms, TV dramas, and movies.

Daddy's job was important, and Mother's job wasn't. So when junior gets sick at school, it's Mother who leaves work to pick him up, not Daddy. Mother always took up the slack, whatever it was, while Daddy did the important things. As for the work around the house, the relative contributions of men and women were summarized along these lines by a male respondent in a research study not long ago: "I take care of the garage and the dog, and my wife takes care of the house and the children." He wasn't joking.

Faculties are not unlike large families. Need someone for a high-demand committee? Appoint a woman, even though she has had an overload of committee work for the last three years. Ask it because this is what women do; they take up the slack. I've seen men excused from high enrollment required courses because the subject matter is outside their specialty. A woman, however, gets told, "If you don't know the subject matter, learn it!"

Indeed, in the approximately 25 classes I taught during the four and a half years I was an assistant professor at the University of Maryland, 23 of the 25 were sections of the three required courses that the school offered. And it was a woman dean who asked this of me. It never crossed her mind that this was unfair.

It wouldn't be so bad if all this were in the past, but I've heard a number of stories just like these at this very conference.

In our culture, the sacrifices of women are invisible in a way that the sacrifices of men are not. The husband can't get a job? The *wife* gets a second doctorate. Not the husband, the wife! We don't, as a culture, acknowledge and value these sacrifices. We just take them, and take them for granted.

A study published in *Science* a few years ago demonstrated that in all the different contexts in which scientists work—business, government, academia, etc.—women were paid 72 to 83 percent as much as men with comparable backgrounds [v. 252, 24 May 1991, employment sector percentage calculations are mine]. This isn't a particular university's problem, or an individual's problem; it's not those people's problem over there. The second-class citizenship of women has something universal in it. Let's turn that pay figure around. A woman making 72 percent as much as a man would need a 39 percent raise to make as much as he does.

Virginia Valian, in a book titled *Why So Slow*, studied why women in academia advance more slowly and are paid less. Men feel entitled to a variety of forms of support in the faculty family. Out of habit, women don't expect that support, and even when they realize they should have it, they feel guilty asking for it. Further, when they do ask, they may be treated as though they were selfish and demanding, both by men and by other women.

Finally, when it comes down to it, there's a big difference between going through life with the wind at your back, and going through life leaning into the wind. I retired at 61 not because I really wanted to, but because I was worn out.

Fair treatment of women can happen only when we ALL self-consciously ask ourselves what we are doing every time we apportion work and rewards to men and women. Fair treatment does not happen without a self-conscious effort to change.

That's why THIS recognition, the Award of Merit, is so very important to me, and I value it so highly. After all, I'm only the 8th woman to have received this honor, in the 40-plus years it has been awarded. Thank you.

14

A Note from Marcia J. Bates

(Papers in honor of Pauline Atherton Cochrane)

Pauline Atherton Cochrane has been a pioneer in many ways in her life. She had a significant role in developing the foundational theory and practice of scientific indexing early in her career, then of online catalogs and, later, online information retrieval. The writers in this volume will celebrate her research through their own contributions. It happened that I was unable to contribute a chapter to this Festschrift at the time Will Wheeler was organizing it, but I do wish to write a brief, more personal, note about Pauline.

When I started out as a doctoral student in information science, Pauline Atherton (now Cochrane) was one of few women researchers in the field. Her friendliness and support were evident from the very beginning, and I always felt I could talk with her.

It is hard to remember nowadays how difficult it was in the late 1960s and early 1970s for substantial numbers of women to enter academia and other research environments. Graduate admissions were frequently biased against women—a professor I talked with in one (non-librarianship) program told me I would have to be "twice as smart" as any male applicant to be admitted to graduate school—and hiring and promoting of women in faculty positions was rare.

First published as Bates, M. J. (2000). A note from Marcia J. Bates. In W. J. Wheeler (Ed.), *Saving the time of the library user through subject access innovation: Papers in honor of Pauline Atherton Cochrane* (pp. 5-6). Champaign, IL: Graduate School of Library and Information Science.

I went through undergraduate education at a private college and a master's and doctorate in librarianship at the University of California at Berkeley without once being taught by a woman holding a doctorate in a tenure-track position. In fact, the first woman I knew personally with a doctorate was a high school girlfriend who received hers before I got mine.

In such an environment, I hungered for professional women in the field who could be models for me as a researcher. In the 1950s and 1960s, women who worked were commonly portrayed in the media as mannish unhappy creatures, obvious failures because they were not at home taking care of husband and children. We did not have mothers or friends of our mothers as examples to see how one should act in the professional world. Neither of my parents had a college education, so it was all new to me.

Some women who did make it under such harsh conditions turned on young newcomers to their fields, sometimes being tougher on these women than even the men were. Somehow Pauline avoided all those dangers. She had a natural professional manner, and made it seem easy to earn the respect of the largely male worlds she worked in. She had become one of only two women presidents of the American Society for Information Science during its first forty years of existence.

At several key points in my career, Pauline provided the support and encouragement I needed to move forward. When I was starting out, she found grant money to hire me to develop some ideas that appeared in several later publications of mine. Another time she helped me get an article published that had been turned down, it seemed, because of the new ideas in it. There is much in common in our research interests, and we have talked on numerous occasions. I value deeply our professional relationship.

The research and academic worlds can be very competitive. Pauline Cochrane is one researcher who has never forgotten that research should be exciting, fun, and collegial. Whether I was a junior person just coming on, or a more mature and confident scholar, she has always been a true colleague and friend to me, and I will treasure that always.

MARCIA J. BATES
Department of Information Studies
University of California at Los Angeles

Marcia J. Bates

CURRICULUM VITAE

Marcia Jeanne Bates
Curriculum Vitae

Department of Information Studies
GSEIS Building
University of California (UCLA)
Los Angeles, CA 90095-1520

mjbates@ucla.edu

https://pages.gseis.ucla.edu/faculty/bates/

RANK

Professor VI, U.S. Citizenship

DATE OF BIRTH

July 30, 1942, Terre Haute, Indiana, USA

NATURE OF APPOINTMENT

Full time. Regular faculty. Tenured.

EDUCATIONAL BACKGROUND

Pomona College, 1959–1963, Claremont, CA, B.A. German, 1963.

University of California, Berkeley, 1966–1967, M.L.S., Librarianship, 1967.

University of California, Berkeley, 1967–1972, Ph.D., Librarianship, 1972.

AREAS OF SPECIAL COMPETENCY IN INFORMATION STUDIES EDUCATION

Information search strategy; Information seeking behavior; User-centered design of information retrieval systems and interfaces; Subject access in manual and online systems; Science and technology information services.

TEACHING AND ADMINSTRATIVE EXPERIENCE

Peace Corps Volunteer, taught English as a Foreign Language, Thailand, 1963-1965.

Acting Instructor, School of Information Studies, University of California, Berkeley, CA, 1969-1970.

Assistant Professor, College of Library and Information Services, University of Maryland, College Park, MD, 1972-1976.

Assistant Professor, 1976-1980; Associate Professor, 1980-1981, Graduate School of Library and Information Science, University of Washington, Seattle, WA.

Associate Professor, Graduate School of Library and Information Science, University of California, Los Angeles, CA, 1981-1991.

Professor, Graduate School of Library and Information Science, University of California, Los Angeles, CA, 1991-, Professor Step VI, 2000-.

Associate Dean, Graduate School of Library and Information Science, UCLA, 1992-1993.

Department Chair, Department of Library and Information Science, UCLA, 1993-1995.

CONSULTING EXPERIENCE

Institute of Library Research, University of California, Berkeley, CA, October 1972-February 1973.

U.S. National Commission on Libraries and Information Science, March 1973-October 1974.

American Chemical Society, October 1973.

Academic Press, May 1976, December 1978.

For Prof. P. Atherton, Syracuse University School of Information Studies, project funded by federal grant, December 1978.

Atari Inc., Sunnyvale, CA, November 1982.

U.S. Library of Congress, Cataloging Division, July 1986, May 1991.

John Wiley and Sons, Publishers, July 1986.

Washington Consulting Group, Washington, DC, August 1987.

Getty Art History Information Program, Santa Monica, CA. December 1988, December 1989-January 1990, April-August 1990, October 1990-June 1991.

Bancroft Group, Los Angeles, CA, January-March 1989.

California System Design, Inc., Los Angeles, CA, July-December 1989, July 1990-January 1991.

Litton Guidance and Control Systems, Woodland Hills, CA, November 1989.

OCLC Online Computer Library Center, June 1990.

Stone and Webster Engineering Corp., Los Angeles, CA, 1991-1996.

Information Access Co., Foster City, CA, April 1992-June 1993. (Infotrac and a variety of business and computer databases.)

Amgen, Inc., Thousand Oaks, CA, May 1995.

Ensemble, Inc. (then ensemble.com), San Rafael, CA, June 1997.

Getty Information Institute, Santa Monica, CA, June–August 1997.

Getty Research Institute, Los Angeles, CA, February 1998–March 1999.

Electric Schoolhouse (lightspan.com), Pasadena, CA, November 1998–February 1999.

Centro de Investigaciones Regionales de Mesoamerica (cirma.net), Antigua, Guatemala, August 2000.

Library of Congress, Cataloging Directorate, Washington, DC, July 2002–January 2004.

Southern California Edison, Rosemead, CA, May–June 2010.

PUBLICATIONS

Books

Cuadra, CA., & Bates, M.J. (Eds.) (1974). *Library and information service needs of the nation: Proceedings of a conference on the needs of occupational, ethnic, and other groups in the United States.* Sponsored by the National Commission on Libraries and Information Science. Washington, DC: U.S.G.P.O. [ERIC #ED 101 716]. 314 pp.

White, H.D., Bates, M.J., & Wilson, P. (1992). *For information specialists: Interpretations of reference and bibliographic work.* Norwood, NJ: Ablex.310 pp.

Rudell-Betts, L., & Bates, M.J. (1992). *LADWP department vocabulary (Thesaurus)* (print version, 2nd ed.). Los Angeles, CA: Los Angeles Department of Water and Power. 450 + 424 pp.

Bates, M.J., & Maack, M.N. (Eds.) (2010). *Encyclopedia of library and information sciences* (3rd ed.). Boca Raton, FL: CRC Press. 7 vols., 5742 pp.

Bates, M.J. (Ed.). (2012). *Understanding information retrieval systems: Management, types, and standards.* Boca Raton, FL: CRC Press. 732 pp.

Bates, M.J., (Comp.). (2016). *Information and the information professions: Selected works of Marcia J. Bates,* vol. 1. Berkeley, CA: Ketchikan Press. 394 pp.

Bates, M.J., (Comp.). (2016). *Information searching theory and practice: Selected works of Marcia J. Bates,* vol. 2. Berkeley, CA: Ketchikan Press. 414 pp.

Bates, M.J., (Comp.). (2016). *Information users and information system design: Selected works of Marcia J. Bates,* vol. 3. Berkeley, CA: Ketchikan Press. 396 pp.

Memoir

Bates, M.J. (2014). *Oral History* (M. Buckland, interviewer). 129 pp. http://infoscileaders.libsci.sc.edu/asis/wp-content/uploads/2014/04/Bates-Transcript.pdf

Book Chapters

(Anon.) (1974). Analysis of the user needs descriptions. In C.A. Cuadra & M.J. Bates (Eds.), *Library and information service needs of the nation,* (pp. 249–263). Washington, DC: U.S.G.P.O. [ERIC #ED 101 716].

(Anon.) (1974). Implications of the conference. In C.A. Cuadra & M.J. Bates (Eds.), *Library and information service needs of the nation* (pp. 265–279). Washington, DC: U.S.G.P.O. [ERIC #ED 101 716].

Bates, M.J. (1974). Library and information services for women, homemakers and parents. In C.A. Cuadra & M.J. Bates (Eds.), *Library and information service needs of the nation* (pp. 129–141). Washington, DC: U.S.G.P.O. [ERIC #ED 101 716].

Bates, M.J. (1974). Speculations on the socio-cultural context of public information provision in the seventies and beyond. In C.A. Cuadra & M.J. Bates (Eds.), *Library and information service needs of the nation* (pp. 51–76). Washington, DC: U.S.G.P.O. [ERIC #ED 101 716].

Bates, M.J. (1991). The berrypicking search: User interface design. In M. Dillon (Ed.), *Interfaces for information retrieval and online systems: The state of the art* (pp. 55–61). Westport, CT: Greenwood Press.

Bates, M.J. (2000). World Wide Web opportunities in subject cataloging and access. In T.H. Connell & R.L. Maxwell (Eds.), *The future of cataloging: Insights from the Lubetzky symposium* (pp. 143–148). Chicago, IL: American Library Association.

Bates, M.J. (2005). An Introduction to metatheories, theories, and models. In K.E. Fisher, S. Erdelez, & L. McKechnie (Eds.), *Theories of information behavior* (pp. 1–24). Medford, NJ: Information Today.

Bates, M.J. (2005). Berrypicking. In K.E. Fisher, S. Erdelez, & L. McKechnie (Eds.), *Theories of information behavior* (pp. 58–62). Medford, NJ: Information Today.

Bates, M.J. (Ed.) (2008). *Klondike trek: Jim Hinkle's life in the gold rush of 1898* (Preface). Paso Robles, CA: Ketchikan Press.

Bates, M.J. (2010). Information. In M.J. Bates & M.N. Maack (Eds.), *Encyclopedia of library and information sciences, 3rd Ed.* (vol. 3, pp. 2347–2360). Boca Raton, FL: CRC Press. [http://www.gseis.ucla.edu/faculty/bates/articles/information.html].

Bates, M.J. (2010) Information behavior. In M.J. Bates & M.N. Maack (Eds.), *Encyclopedia of library and information sciences, 3rd Ed.* (vol. 3, pp. 2381–2391). Boca Raton, FL: CRC Press. [http://www.gseis.ucla.edu/faculty/bates/articles/information-behavior.html].

Bates, M.J. (2016). Many paths to theory: The creative process in the information sciences. In D. H. Sonnenwald (Ed.), *Theory development in the information sciences* (pp. 22–49). Austin, TX: University of Texas Press.

Articles

Bates, M.J. (1970). A campus information clearinghouse. *California Librarian* *31*(3), 171–172.

Bates, M.J. (1974). SIG cabinet news. *Bulletin of the American Society for Information Science 1*(1), 28, 35–36.

Soergel, D., & Bates, M.J. (1974). A modern integrated introduction to library and information services. *SIG Newsletter* ED-1, 7–11.

Bates, M.J. (1974). SIG cabinet report. *Bulletin of the American Society for Information Science 1*(4), 24–25.

Articles, (cont'd.)

Bates, M.J. (1976). Rigorous systematic bibliography. *RQ 16*(1), 7–26.

Bates, M.J. (1977). Factors affecting subject catalog search success. *Journal of the American Society for Information Science 28*(3), 161–169. *Refereed.*

Bates, M.J. (1977). System meets user: Problems in matching subject search terms. *Information Processing & Management 13*(6), 367–375. *Refereed.*

Bates, M.J. (1979). An exercise in research evaluation: The work of L.C. Puppybreath. *Journal of Education for Librarianship 19*(4), 339–342.

Bates, M.J. (1979). Information search tactics. *Journal of the American Society for Information Science 30*(4), 205–214. *Refereed.*

> *Also:* Reprinted in B. Katz & A. Clifford (Eds.), *Reference and information services: A new reader* (pp. 275–299). 1982, Metuchen, NJ: Scarecrow.

> *Also:* Reprinted in A.H. Jackson (Ed.), *Training and education for online* (pp. 96–105). 1989, London: Taylor Graham.

Bates, M.J. (1979). Idea tactics. *Journal of the American Society for Information Science 30*(5), 280–289. *Refereed.*

> *Also:* Excerpted and reprinted in *IEEE Transactions on Professional Communication* (1980) PC-23 (2), 95–100.

> *Also:* Reprinted in A.H. Jackson (Ed.), *Training and education for online,* (pp. 106–115). 1989, London: Taylor Graham.

Bates, M.J. (1981). Search techniques. *Annual Review of Information Science and Technology 16*, 139–169. *Refereed.*

Bates, M.J. (1984). Locating elusive science information: Some search techniques. *Special Libraries 75*(2), 114–120. *Refereed.*

> *Also:* Reprinted in B. Katz (Ed.), *Reference and information services: A reader for today,* pp. 307–315. 1986, Metuchen, NJ: Scarecrow Press.

Bates, M.J. (1984). The fallacy of the perfect thirty-item online search. *RQ 24*(1), 43–50. *Refereed.*

> *Also:* Reprinted in E. Auster (Ed.), *Managing online reference services* (pp. 238–250). 1986, New York: Neal-Schuman.

Bates, M.J. (1986). A doctorate in library/information science: Advice for librarians considering Ph.D. studies. *The Library Journal 111*(14), 157–159.

Bates, M.J. (1986). Subject access in online catalogs: A design model. *Journal of the American Society for Information Science, 37*(6), 357–376. *Refereed.*

Bates, M.J. (1986). What is a reference book? A theoretical and empirical analysis. *RQ 26*(1), 37–57. *Refereed.*

Bates, M.J. (1987). How to use information search tactics online. *Online 11*(3), 47–54.

> *Also:* Reprinted in *Database search strategies and tips: Reprints from the best of Online/Database* (pp. 21–28). n.d., Weston, CT: Online, Inc.

Bates, M.J. (1988). How to use controlled vocabularies more effectively in online searching. *Online, 12*(6), 45–56.

Bates, M.J. (1989). Rethinking subject cataloging in the online environment. *Library Resources & Technical Services, 33*(4), 400–412. *Refereed.*

Bates, M.J. (1989). The design of browsing and berrypicking techniques for the online search interface. *Online Review, 13*(5), 407-424. *Refereed.*

Bates, M.J. (1990). Where should the person stop and the information search interface start? *Information Processing & Management 26*(5), 575-591. *Refereed.*

Bates, M.J., Wilde, D.N., & Siegfried, S. (1993). An analysis of search terminology used by humanities scholars: The Getty Online Searching Project report no. 1. *The Library Quarterly, 63*(1), 1-39. *Refereed.*

Siegfried, S., Bates, M.J., & Wilde, D.N. (1993). A profile of end-user searching behavior by humanities scholars: The Getty Online Searching Project report no. 2. *Journal of the American Society for Information Science 44*(5), 273-291. *Refereed.*

Bates, M.J. (1994). The design of databases and other information resources for humanities scholars: The Getty Online Searching Project report no. 4. *Online & CDROM Review 18*(6), 331-340. *Refereed.*

Bates, M.J., Wilde, D.N., & Siegfried, S. (1995). Research practices of humanities scholars in an online environment: The Getty Online Searching Project report no. 3. *Library and Information Science Research 17*(1), 5-40. *Refereed.*

Bates, M.J. (1995). Nevertheless. *Journal of the American Society for Information Science* (p. 32; special issue: 20 Years of SIG/CON).

Bates, M.J. (1996). Learning about the information seeking of interdisciplinary scholars and students. *Library Trends 45*(2), 155-164. *Invited Paper.*

Bates, M.J. (1996). Document familiarity in relation to relevance, information retrieval theory, and Bradford's law: The Getty Online Searching Project report no. 5. *Information Processing & Management 32*(6), 697-707. *Refereed.*

Bates, M.J. (1996). The Getty end-user Online Searching Project in the humanities: report no. 6: Overview and conclusions. *College & Research Libraries 57*(6), 514-523. *Refereed.*

Kafai, Y., & Bates, M.J. (1997). Internet web-searching instruction in the elementary classroom: Building a foundation for information literacy. *School Library Media Quarterly 25*(2), 103-111. *Refereed.*

Bates, M.J., and Lu, S. (1997). An exploratory profile of personal home pages: Content, design, metaphors. *Online & CDROM Review 21*(6), 331-340. *Refereed.*

Bates, M.J. (1998). The role of publication type in the evaluation of LIS Programs. *Library & Information Science Research 20*(2), 187-198. *Refereed.*

Bates, M.J. (1998). Response to "The academic elite in library science. . .". *College & Research Libraries 59*(3), 275-280. *Refereed.*

Bates, M.J. (1998). Indexing and access for digital libraries and the internet: Human, database, and domain factors. *Journal of the American Society for Information Science 49*(13), 1185-1205. *Refereed.*

Bates, M.J. (1999, July 15). Another information system fails—why? (OpEd article) *Los Angeles Times*, B9.

Bates, M.J. (1999). Guest editor, 50th Anniversary Issues, numbers 11 and 12, *Journal of the American Society for Information Science 50*(11, 12), 958-1162. 30 refereed articles.

Bates, M.J. (1999). The 50th Anniversary of the *Journal of the American Society for Information Science:* Guest editor introduction. *Journal of the American Society for Information Science 50*(11), 958-964.

Articles, (cont'd.)

Bates, M.J. (1999). A tour of information science through the pages of *JASIS*. *Journal of the American Society for Information Science 50*(11), 975–993.

Bates, M.J. (1999). The invisible substrate of information science. *Journal of the American Society for Information Science 50*(12), 1043–1050. *Refereed.*

Bates, M.J. (2002). The cascade of interactions in the digital library interface. *Information Processing & Management 38*(3), 381–400. *Refereed.*

Bates, M.J., Hulsy, C., & Jost, G. (2002). Multimedia research support for visiting scholars in museums, libraries, and universities. *Information Technology and Libraries 21*(2), 73–81. *Refereed.*

Bates, M.J. (2002). After the dot-bomb: Getting web information retrieval right this time. *First Monday 7*(7). http://firstmonday.org/ojs/index.php/fm/article/view/971. *Refereed.*

Bates, M.J. (2002). Toward an integrated model of information seeking and searching. *New Review of Information Behaviour Research 3*, 1–15.

Bates, M.J. (2004). Information science at the University of California at Berkeley in the 1960s: A memoir of student days. *Library Trends, 52*(4), 683–701.

Bates, M.J. (2005). Information and knowledge: An evolutionary framework for information science. *Information Research, 10*(4) paper 239. [Available at http://InformationR.net/ir/10-4/paper239.html].

Bates, M.J. (2006). Fundamental forms of information, *Journal of the American Society for Information Science and Technology, 57*(8), 1033–1045.

Bates, M.J. (2007). Defining the information disciplines in encyclopedia development. *Information Research, 12*(4), paper colis29. [Available at http://InformationR.net/ir/12-4/colis/colis29.html]

Bates, M.J. (2007). What is browsing—really? A model drawing from behavioural science research. *Information Research, 12*(4), paper 330. [Available at http://InformationR.net/ir/12-4/paper330.html]

Bates, M.J. (2008). Hjørland's critique of Bates' work on defining information. *Journal of the American Society for Information Science and Technology, 59*(5), 842–844.

Bates, M.J. (2011). Birger Hjørland's Manichean misconstruction of Marcia Bates' work. *Journal of the American Society for Information Science and Technology, 62*(10), 2038–2044.

Mizrachi, D. & Bates, M.J. (2013). Undergraduates' personal academic information management and the consideration of time and task-urgency. *Journal of the American Society for Information Science and Technology, 64*(8), 1590–1607.

Bates, M.J. (2015). The information professions: Knowledge, memory, heritage. *Information Research, 20*(1), paper 655. Retrieved from http://InformationR.net/ir/20-1/paper655.html (Archived by WebCite® at http://www.webcitation.org/6WyNray1L)

Bates, M.J. (1978). The testing of information search tactics. *Proceedings of the American Society for Information Science Annual Meeting 15*, 25-27. *Refereed.*

Bates, M.J. (1980). A criterion citation rate for information scientists. *Proceedings of the American Society for Information Science Annual Meeting 17*, 276-278. *Refereed.*

Bates, M.J. (1984). Some design ideas for subject access in online systems. In M.E. Powell (Ed.), *First Conference on Computer Interfaces and Intermediaries for Information Retrieval* (pp. 1-10). Williamsburg, VA, October 3-6. Alexandria, VA: Defense Technical Information Center, Office of Information Systems and Technology. [NTIS #AD A167 700]. *Invited Paper.*

Bates, M.J. (1985). An exploratory paradigm for online information retrieval. In B. C. Brookes (Ed.), *Intelligent information systems for the information society. Proceedings of the Sixth International Research Forum in Information Science (IRFIS 6)*, Frascati, Italy, September 16-18. (pp. 91-99). Amsterdam: North-Holland. *Refereed.*

Bates, M.J. (1986). Terminological assistance for the online subject searcher. In Defense Technical Information Center and M.I.T., *Proceedings of the Second Conference on Computer Interfaces and Intermediaries for Information Retrieval*, Boston, MA, May 28-31, (pp. 285-293). Alexandria, VA: Defense Technical Information Center, Office of Information Systems and Technology. [Report No. NTIS/TR-86/5]. *Invited Paper.*

Bates, M.J. (1987). Optimal use of controlled vocabularies in online searching. *Online '87 Conference Proceedings, Part I* (pp. 14-19). Weston, CT: Online, Inc.

Bates, M.J. (1987). Information: The last variable. *Proceedings of the 50th ASIS Annual Meeting 24*, 6-10. *Refereed.*

Bates, M.J. (1989). Designing online catalog subject access to meet user needs. In *Fifty-fifth IFLA Council and General Conference*, Paris, France, August 19-26, (pp. 40-24-40.26). Division of Bibliographic Control, Section on Classification and Indexing. The Hague, Netherlands: IFLA. *Invited Paper.*

Bates, M.J. (1990). Design for a subject search interface and online thesaurus for a very large records management database. *Proceedings of the 53rd ASIS Annual Meeting 27*, 20-28. *Refereed.*

Bates, M.J. (1991). OPAC use and users: Breaking out of the assumptions. *Think tank on the present and future of the online catalog: Proceedings*, ALA Midwinter Meeting, Chicago, IL, January 11-12 (pp. 49-58). Chicago, IL: American Library Association, Reference and Adult Services Division. [RASD Occasional Papers, No.9]. *Invited Paper.*

Bates, M.J. (1992). Implications of the subject subdivisions conference: The shift in outline catalog design. In M.O. Conway (Ed.), *The future of subdivisions in the Library of Congress Subject Headings system: Report from the Subject Subdivisions Conference* (pp. 92-98). Washington, DC: Library of Congress Cataloging Distribution Service. *Invited Paper.*

Bates, M.J. (1996) Learning about your users' information needs: A key to effective service. In A. Cohen (Ed.), *PIALA '95: Preservation of Culture Through Archives & Libraries*, (pp. 5-12). Colonia, Yap, Federated States of Micronesia: Pacific Islands Association of Libraries and Archives. *Invited Paper.*

Conference Proceedings, (cont'd.)

Bates, M.J. (1999). Applying user research directly to information system design, (Panel abstract; panel organizer and speaker). *Proceedings of SIGIR '99*, (p. 263). Berkeley, CA. *Refereed Panel.*

Bhavnani, S., & Bates, M.J. (2002). Separating the knowledge layers: Cognitive analysis of search knowledge through hierarchical goal decompositions. *Proceedings of the 64ᵗʰ ASIST Annual Meeting 39*, 204–213. *Refereed.*

Bates, M.J. (2002). Speculations on browsing, directed searching, and linking in relation to the Bradford Distribution. In H. Bruce, R. Fidel, P. Ingwersen, & P. Vakkari, (Eds.), *Emerging Frameworks and Methods: Proceedings of the Fourth International Conference on Conceptions of Library and Information Science (CoLIS4)*, (pp. 137–150). Greenwood Village, CO: Libraries Unlimited. *Refereed.*

Bates, M.J. (2010). An operational definition of the information disciplines. *Proceedings of the iConference, February 3–6*, (pp. 19–25). Available at https://www.ideals.illinois.edu/handle/2142/14900

Research Reports

Bates, M.J. (1971). *User studies: A review for librarians and information scientists.* [ERIC #ED 047 738].

Bates, M.J. (1971). The assignment of index terms. In R. Shoffner & J.L. Cunningham (Eds.), *The organization and search of bibliographic records component studies* (pp. 23–62). Berkeley, CA: University of California, Institute of Library Research. [ERIC #ED 061 975].

Bates, M.J. (1973). Review of literature relating to the identification of user groups and their needs. In C.P. Bourne, V. Rosenberg, M.J. Bates, & G.R. Perolman (Eds.), *Preliminary investigation of present and potential library and information service needs: Final report* (pp. 36–68). Washington, DC: U.S. National Commission on Libraries and Information Science. [ERIC #ED 073 786].

Bates, M.J. (1994). *Expanded entry vocabulary for the Library of Congress subject headings: Final report.* Washington, DC: Council on Library Resources. [CLR Grant #891].

Kafai, Y., Bates, M.J., Braxton, P.D., Childs, D., Ender, P., Lo, H.H., Martin, M., Rose, K., & Yarnall, L. (1996). *Building a foundation for information literacy: Creating an annotated WWW index by children for children* (Apple Technical Report No. 154). Cupertino, CA: Apple Computer Advanced Technology Group.

Kafai, Y.B., Bates M.J., Ender, P., et al. (1997). *SNAPshots of presidential elections: Providing collaborative WWW-information resources and exchanges for elementary and middle school classrooms.* Los Angeles, CA: UCLA-Kids Interactive Design Studios Report No. 97-3.

Bates, M.J. (1987). *Interaction in information systems: A review of research from document retrieval to knowledge-based systems*, by N.J. Belkin & A. Vickery. *Information Processing & Management 23*(1), 65–66.

Bates, M.J. (1987). *Four indications of current North American library and information doctoral degree programs*, by W.H. Reid. *Library and Information Science Research 9*, 236–237.

Bates, M.J. (1988). *Success in answering reference questions: Two studies*, by F. Benham & R.R. Powell. *The Library Quarterly 58*(4), 405–406.

Bates, M.J. (1989). *Manual of online search strategies*, edited by C.J. Armstrong and J.A. Large. *Information Processing and Management 25*(2), 215–216.

Bates, M.J. (1993). *Matching OPAC user interfaces to user needs*, by F.J. Murphy, A.S. Pollitt, & P.R. White. *Journal of Documentation 49*(1), 72–74.

Bates, M.J. (1994). *Seeking meaning: A process approach to library and information services*, by C.C. Kuhlthau. *The Library Quarterly 64*(4), 473–475.

Bates, M.J. (1997). *Networking in the humanities*, by S. Kenna and S. Ross (Eds.). *The Library Quarterly 67*(1), 90–92.

Bates, M.J. (1999). *Information seeking and subject representation: An activity-theoretical approach to information science*, by B. Hjørland. *The Library Quarterly 69*(1), 112–113.

Bates, M.J. (1999). *Information seeking in context*, by P. Vakkari (Ed.). *The Library Quarterly 69*(3), 378–380.

Bates, M.J. (2015). *The discipline of organizing*, by R.J. Glushko (Ed.). *Journal of the Association for Information Science and Technology, 66*(2), 432–436.

Other

Bates, M.J. (1972). *Factors affecting subject catalog search success* (Ph.D. dissertation). University of California, Berkeley.

Bates, M.J., Kalbarczyk, S., & Beckman, C. (1974). *Information facilities*. [Slide/tape, 125 slides.] College Park, MD: University of Maryland, College of Library and Information Services.

Soergel, D., & Bates, M.J. (1974). *A modern integrated introduction to library and information services*. Presented at the 37th Annual Meeting of the American Society for Information Science, Atlanta, GA, October 15.

Bates, M.J. (1984). Response. In Cochrane, P.A., Modern subject access in the online age (p. 252). *American Libraries 15*(4), 250–255.

Bates, M.J. (1984). Foreword. In K. Markey, *Subject searching in library catalogs* (p. xiii). Dublin, OH: OCLC Online Computer Library Center.

Bates, M.J. (1986). Disastrous fire at central library. [Letter to the editor]. *Los Angeles Times*, 11 May.

Bates, M.J. (1987). *Fast library research: Techniques of expert researchers and librarians*. Los Angeles, CA: UCLA, Graduate School of Library and Information Science.

Anon. (1992, November 1). UCLA Graduate School of Library and Information Science. Self Review 1992/93 to the Graduate Council of the UCLA Academic Senate.

Other, (cont'd.)

CLA Task Force on the Future of Librarianship (member). (1996). *Future directions for the library profession and its education.* California Library Association.

Bates, M.J. (1996). Berrypicking. [Letter]. *Journal of Documentation 52*(3), 350–351.

Bates, M.J. (1998). Response to the academic elite in library science. . . . *College & Research Libraries 59*(3), 275–280.

Bates, M.J. (1998). Response to Bair and Barrons' response. *College & Research Libraries 59*(5), 481.

Bates, M.J. (1999). Statement from the UCLA Dept. of Information Studies on LIS Curricula for the ALA Congress on Professional Education.

Bates, M.J. (2000). Proper citations. [Letter]. *Journal of the American Society for Information Science and Technology, 51*(9), 882.

Bates, M.J. (2011). [Letter to the editor]. *Journal of Documentation, 67*(6), np.

RESEARCH GRANTS

Principal Investigator, University of Maryland, General Research Board Grant, 2 months' salary, Summer 1975.

Principal Investigator, University of Washington Graduate School Research Fund Grant, 1 month salary, Summer 1978.

Principal Investigator, Library Career Training Program Fellowships, Higher Education Act, Title II-B, U.S. Dept. of Education. Separate grants awarded annually:

1985–86	$28,000
1986–87	$40,000
1987–88	$56,000
1989–90	$10,800
1994–95	$44,000

Principal Investigator, with James Trent, Graduate School of Education, Co-Principal Investigator; "Evaluation of New Approach to Undergraduate Bibliographic Instruction," UCLA Academic Senate Committee on Research: $3,291, 1987–88.

Principal Investigator, "Review of Issues Associated with the Provision of Expanded Entry Vocabulary for the *Library of Congress Subject Headings.*" Council on Library Resources contract, $6,000, 1992–93.

Principal Investigator, "Sources of Error in Reference Interviews," UCLA Academic Senate Committee on Research.

1993–94	$2,100
1994–95	$1,600
1995–96	$1,600

Co-Principal Investigator, "Social Aspects of Digital Libraries" Conference, February 15–17, 1996, Department of Library and Information Science, UCLA. Funded by National Science Foundation, $36,027.

Principal Investigator, "Faceted Classification Techniques for Indexing Digital Libraries," UCLA Academic Senate Committee on Research, $1,360, 1997–98.

Principal Investigator, with Mary N. Maack, Co-Principal Investigator, "Surfing and Searching: Accessing and Using Print, Electronic, and Multimedia Information," UCLA Office of Instructional Development Instructional Improvement Grant, $4,320 plus .17 Teaching Assistant release time, 1998–99.

HONORS, AWARDS, ETC.

Phi Beta Kappa, 1963.

Graduated Pomona College cum laude, 1963.

Pomona College Scholarships, 1959–1961.

California State Scholarships, 1961–1963.

Doctoral Fellowships, Title II-B, Higher Education Act of 1965, U.S. Office of Education, 1967–1968, 1968–1969, 1969–1970.

"Best *Journal of the American Society of Information Science* Article for 1979." Awarded by American Society for Information Science, October 1980, for "Information Search Tactics."

Fellow, American Association for the Advancement of Science, elected February, 1990.

Distinguished Lectureship Award, New Jersey, ASIS (American Society for Information Science), October 10, 1991.

Listed, *Who's Who in Science and Engineering*, 1992– editions.

Visiting Scholar, School of Information and Library Science, University of North Carolina, Chapel Hill, NC, week of September 13, 1993.

"Contributions to Information Science" Award, Annual Awards Dinner of the Los Angeles Chapter of the American Society for Information Science, October 25, 1995.

Listed, *Who's Who in America*, 1997– editions.

"Top Twenty Articles on Library Instruction" Award, Library Instruction Round Table, American Library Association, 1997, for "Internet Web-Searching Instruction in the Elementary Classroom: Building a Foundation for Information Literacy."

Research in Information Science Award, American Society for Information Science, 1998.

"Interviews with ASIS Winners," Marcia Bates, 1998 Award for Research in Information Science, in *OASIS* 36 (Winter 1999): 7 +10 +14.

Srygley Lecture (endowed), School of Information Studies, Florida State University, Tallahassee. October 16, 1999.

Lazerow Lecture (endowed), School of Library and Information Science, University of Kentucky, Lexington, KY, October 13, 2000.

"Best *Journal of the American Society of Information Science* Article for 1999." Awarded by American Society for Information Science, November 2000, for "The Invisible Substrate of Information Science."

Frederick G. Kilgour Award for Research in Library and Information Technology, 2001. Library and Information Technology Association/American Library Association.

"Professional Contribution to Library and Information Science Education" Award, Association for Library and Information Science Education, 2005.

Award of Merit, American Society for Information Science and Technology, 2005.

Best Information Science Book Award, 2010, American Society for Information Science and Technology, for *Encyclopedia of Library and Information Sciences, 3rd Ed.*

Distinguished Lecture in Library and Information Science, University of Wisconsin-Milwaukee, September 11, 2012.

American Association for the Advancement of Science, member, 1968–
> AAAS Section on Information, Computing, and Communication, Electorate Nominating Committee, member, 1980–1984.
> Chair, AAAS Section on Information, Computing, and Communication, Electorate Nominating Committee, 1983–1984.
> Fellow, AAAS, 1990–
> Member-at-Large, Section Committee (governing board) of Section on Information, Computing, and Communication, 2000–2004.

American Library Association, member, various years.
> Committee on Accreditation Evaluation Team for University of North Carolina School of Information and Library Science, Fall, 1999.
> Committee on Accreditation Evaluation Team for Syracuse University School of Information Studies, Fall, 2001.

American Society for Information Science, member, 1968–
> ASIS Classification Research Special Interest Group Cabinet Representative, 1972–1973.
> ASIS Council Executive Committee member, 1973–1974.
> Special Interest Group Cabinet Councilor (i.e., member, ASIS Board of Directors), 1973–1974.
> Publication Awards Committee to select best information science book and best *Journal of the American Society for Information Science* article of the year, member, 1978.
> ASIS Jury to select best submitted dissertation proposal to receive Institute for Scientific Information scholarship award, member, 1982, 1983, 1999.
> Chair of Jury to select "Best *Journal of the American Society for Information Science* Article of the Year Award," 1988.
> Research Committee, member, 1988–89; Chair, Research Award Jury, 2001.
> Moderator, Session of Special Interest Group on Human-Computer Interaction "User Online Interaction Retrospective." American Society for Information Science Midyear Meeting, San Diego, CA, May 23, 1989.
> Member of the Jury to select "Best *Journal of the American Society for Information Science* Article of the Year Award," 1993.
> Member of the Jury to select UMI Doctoral Dissertation Award, 1999.
> Mentor for doctoral student mentoring program, ASIS Conference, November 1999; November 2002.
> Member, Annual Conference Technical Program Committee, 2001.
> Chair of Annual Conference Technical Program Committee, Long Beach, CA, October 2003.
> Regular attendance and frequent speaker at ASIS conferences.

Association for Academic Women, UCLA, member, 1983–1988.
> Co-Organizer of AAW monthly program, "Women in 'Women's Professions': Second Class Citizens Or...?" UCLA Faculty Center, March 18, 1986.

Association for Computing Machinery, 1996–
> Technical Papers Committee, ACM Digital Libraries '97 Conference, member, 1996–97.
> Final "Meta-reviewer" panel for annual conference papers, ACM Special Interest Group on Information Retrieval (SIGIR), 1999, 2000.

Association for Library and Information Science Education, member, 1988–
> Program Planning Committee, member, 1995–1996.
> Research Committee, member, 1996–1998.
> Research Committee awards jury, member, 1995–1997, 1998–1999.

Association of Records Managers and Administrators, member, 1991–2001.
California Cataloging Colloquium, member, 1983–1993.
> Program Organizer and Chair of California Cataloging Colloquium, February 15–16, 1986.
> Program Co-Organizer and Chair of California Cataloging Colloquium, April 3–5, 1992.
> Regular speaker at CCC.

California Library Association, member, 1993–2000.
> CLA Task Force on the Future of the Library Profession, member, 1994–1996.

International Communication Association, member, 1977–1989.
Phi Beta Kappa, member,1963–
Stanford Professional Women, member, 1985–
UCLA Faculty Seminar on Women, Culture, and Theory, member, 1986–1990.
Women Studies Research Group, University of Washington, member, 1977–1981.

GENERAL PROFESSIONAL ACTIVITIES

Grant and Fellowship Reviewing

National Science Foundation, research grants:
> Division of Information Science and Technology, 1978–1986.
> Political Science Program, 1982.
> Division of International Programs, 1984.
> Databases, Software Development Program, 1992.

National Endowment for the Humanities, research grants, 1986, 1996.
Council on Library Resources, research grant, 1986.
Jacob K. Javits Fellows Selection Panel, U.S. Dept. of Education, Office of Higher Education Programs, member, 1987.

Journal Article Refereeing and Editorial Board Service

Journal of the American Society of Information Science, 1978–
Member, Editorial Board, *Journal of the American Society for Information Science*, 1989–2008.
Library and Information Science Research, 1982–
Information Processing & Management, 1985–
Library Resources & Technical Services, 1986–
Proceedings of the American Society for Information Science—Annual Meeting, 1989, 1990
Library Quarterly, 1989–
Member, Editorial Board, *Library Quarterly*, 1993–2000.
Online, 1990–
Canadian Journal of Information Science, 1991–
Annual Review of Information Science and Technology, 1991–
Member, Advisory Committee, *Annual Review of Information Science and Technology*, 1994–96.
International Information and Library Review, 1995–
Computers and the Humanities, 1996–
Member, Editorial Board, *Encyclopedia of Library and Information Science*, 1999–
Information Retrieval, 2001–
Knowledge Organization, 2001–
Journal of Digital Information, http://jodi.ecs.soton.ac.uk/, 2002–
Editor-in-Chief, *Encyclopedia of Library and Information Sciences, 3rd Ed.*, 2005–2010.

Workshop Design (hired as consultant to design and present)

"Science and Technology Literature" for approximately 70 librarians of the Santiago Library System in Orange County, CA, February 10, 1983.

"Enhancing Online Search Skills." Two workshops for a total of 70 librarians. Sponsored by Continuing Library Education Project—Oregon (project funded by the U.S. Library Services and Construction Act). Beaverton, OR, May 1, 1986, and Eugene, OR, May 2, 1986.

"Perspectives in Indexing and Searching." With Ev Brenner. All-day workshop with 23 attending. National Online Meeting, New York, April 30, 1990.

Invited Conference Participant (speaker or discussant)

Library and Information Service Needs of the Nation. Sponsored by the U.S. National Commission on Libraries and Information Science, Denver, CO, May 1973. (24 attendees)

A Library and Information Science Research Agenda for the 1980s. Sponsored by the Office of Libraries and Learning Technologies, U.S. Dept. of Education, Airlie House, VA, July 1981. (26 attendees)

Users Group Conference. Sponsored by Institute for Scientific Information. Philadelphia, PA, November 13–15, 1985. (25 attendees)

Classification as Subject Enhancement in Online Catalogs. Sponsored by OCLC and The Council on Library Resources, Dublin, OH, January 27-28, 1986. (30 attendees)

Think Tank on the Present and Future of the Online Catalog. Sponsored by Reference and Adult Services Division, Catalog Use Committee, American Library Association. Chicago, IL, January 11-12, 1991. (10 attendees)

Subject Subdivisions Conference. Sponsored by U.S. Library of Congress. Airlie House, VA, May 9–12, 1991. (45 attendees)

User Interfaces for Online Public Access Catalogs. Sponsored by the University of Maryland Human-Computer Interaction Laboratory and the Library of Congress. Washington, DC, November 1, 1991. (35 attendees)

"Four Challenges to the Field." Conference on the Future of Library Education, University of Michigan, Ann Arbor, April 3, 1995. (40 attendees)

Social Aspects of Digital Libraries Conference. Sponsored by UCLA Dept. of Library and Information Science and U.S. National Science Foundation. GSE&IS, UCLA, Los Angeles, CA, February 15-17, 1996. (35 attendees)

Keynote speaker. Conference on Semantic Information Systems. Getty Information Institute, Los Angeles, CA, October 10-11, 1996. (30 attendees)

Congress on Professional Education. Special invited conference, American Library Association, Washington, DC, April 30–May 1, 1999. (150 attendees)

Bicentennial Conference on Bibliographic Control for the New Millennium: Confronting the Challenges of Networked Resources and the Web. Library of Congress Cataloging Directorate, Washington, DC, November 15-17, 2000. (150 attendees)

Speeches and Workshop Presentations (not otherwise published)

"Educating Librarians and Users for a New Model of Library Service," American Library Association Annual Conference, San Francisco, CA, July 1, 1975.

Principal speaker, two-day workshop entitled "Information Services: Current Research, Future Design," sponsored by the Baltimore Regional Planning Council and the Maryland State Department of Education Division of Library Development and Services. Baltimore, MD, November 18-19, 1976.

"User Studies—Where Do They Lead?" University of Washington School of Librarianship Annual Alumni Day Institute, Seattle, WA, May 20, 1977.

"Information Search Tactics," at an institute "New Techniques in the Teaching of Online Searching," funded by U.S. Office of Education. Univ. of Washington School of Librarianship, Seattle, WA, March 24, 1978.

"Information Search Tactics." Associated Librarians of University of Washington Forum, Seattle, WA, June 5, 1978.

"Information Search Tactics," at a three-day seminar entitled "Indexing in Perspective," sponsored by the National Federation of Abstracting and Indexing Services. Berkeley, CA, May 11, 1979.

"Information Search Tactics." American Society for Information Science, Pacific-Northwest Chapter Annual Meeting, Portland, OR, July 20, 1979.

"Idea Tactics in the Online Environment." American Society for Information Science, Pacific-Northwest Chapter Annual Meeting, Portland, OR, June 27, 1980.

"Search Strategy." Los Angeles Chapter of American Society for Information Science, Los Angeles, CA, Fall, 1981.

"New Ideas in Search Strategy." GSLIS Alumni, UCLA, Los Angeles, CA, Spring, 1982.

"Some Design Principles for Subject Access in Online Catalogs." California Cataloging Colloquium, Los Angeles, CA, March 9, 1984.

"Information World: Information Futures," noon presentation to students at Graduate School of Management, UCLA, Los Angeles, CA, April 9, 1985.

"Rigorous Systematic Bibliography." Lecture to class in bibliography, Royal School of Librarianship, Copenhagen, Denmark, September 24, 1985.

"An Exploratory Paradigm for Online Information Retrieval." Guest Lecture, Royal School of Librarianship, Copenhagen, Denmark, September 24, 1985.

Moderator, Contributed Paper Session "Databases: Organization and Access," American Society for Information Science Annual Meeting, Las Vegas, NV, October 22, 1985.

"Update on Current Work." California Cataloging Colloquium, Los Angeles, CA, February 16, 1986.

"A Design for Subject Access in Online Catalogs." Librarians' Association of the University of California (LAUC), U.C. San Diego Library, March 13, 1986.

"Subject Access in Online Catalogs: A Design Model." Colloquium, Graduate School of Library and Information Science, UCLA, Los Angeles, CA, April 10, 1986.

"What Studies of Information Seeking Suggest for Bibliographic Instruction." California Library Association Annual Conference, California Clearinghouse on Library Instruction, Long Beach, CA, November 15, 1986.

"Things I Wish They'd Told Me (Before I Started Working)." Stanford Professional Women, Athenaeum, California Institute of Technology, Pasadena, CA, March 26, 1987.

"Rethinking Authority Control for Online Catalog Subject Access." American Library Association Annual Conference, New Orleans, LA, July 11, 1988.

"Turning an Idea into a Research Project." Librarians' Association of UCLA (LAUC-LA), Powell Library, UCLA, Los Angeles, CA. April 26, 1989.

"Update on Subject Search Interface Work." California Cataloging Colloquium, San Diego, CA, May 29, 1989.

"Improving Search Term Match Rates through the OPAC User Interface. "American Society for Information Science Midyear Conference, San Diego, CA, May 23, 1989.

"New Directions in the Study of Scholarly Communication." International Communication Association, San Francisco, CA, May 28, 1989.

"Browsing and Search Interface Design." Colloquium, Graduate School of Library and Information Science, UCLA, Los Angeles, CA, January 18, 1990.

Speeches and Workshop Presentations (not otherwise published, (cont'd.)

"Information Seeking of Young People." At workshop titled "Discovery: Information Age Kids and the Research Process." UCLA Extension, Los Angeles, CA,"Does It Make Sense to Use LCSH in Online Catalogs?" Iowa Library Association, Des Moines, IA, October 11, 1990.

Colloquium Moderator, "Music and Censorship: Implications for a Democratic Society." Sponsored by the Friends of the UCLA Library and the Graduate School of Library and Information Science First Amendment Information Resources Center, UCLA, Los Angeles, CA, May 23, 1991.

"Interaction between the Design of the Online Catalog and the Design and Rules of LCSH." California Library Association, Technical Services Chapter, Cerritos, CA, September 26, 1991.

"Some Essential Considerations in the Design of Information Retrieval Systems and Interfaces." New Jersey American Society for Information Science Distinguished Lectureship, Rutgers University, New Brunswick, NJ, October 10, 1991.

"Larger Targets: Improving Subject Retrieval in Megadatabases." At Research Workshop: "User Interfaces for Online Public Access Catalogs," Library of Congress, Washington, DC, November 1, 1991.

"Online Searching in the Humanities: A Joint Project with the Getty." GSLIS Faculty Showcase, Graduate School of Library and Information Science, UCLA, Los Angeles, CA, April 11, 1992.

"Features in Subject Tools that Promote Good Online Retrieval." American Library Association, San Francisco, CA, June 28, 1992.

"Desirable Features in Thesauri for Online End Users in the Humanities." American Society for Information Science, Annual Conference, Pittsburgh, PA, October 27, 1992.

Discussant, "Standards for Information Retrieval Indexes." American Society for Information Science Annual Conference, Pittsburgh, PA, October 28, 1992.

"Humanities Researchers Really are Different: Results of the Getty Online Searching Project." Duke University, Durham, NC, October 30, 1992.

"What do Humanities Online Searchers Need from a Thesaurus?–Results of the Getty Online Searching Project." 21st Annual Conference of Art Libraries Society of North America, San Francisco, CA, February 2, 1993.

"Humanities Researchers Really are Different: Results of the Getty Online Searching Project." Library Seminar, Doheny Memorial Library, University of Southern California, Los Angeles, CA, April 8, 1993.

"Data Organization for Subsequent Retrieval." Catalog Interoperability NASA Science Internet Workshop, San Diego, CA, April 27, 1993.

"Use of Online Searching by Humanities Scholars and End Users." School of Library and Information Science, University of North Carolina, Chapel Hill, NC, September 15, 1993.

"GSLIS and the Professional Schools Restructuring Initiative." Law School, UCLA, Los Angeles, CA, September 20, 1993.

"Future of Library Education in the Los Angeles Area." Association of Jewish Libraries, Los Angeles, CA, September 20, 1993.

"Library Education in Southern California." California Library Association, Corona Public Library, Corona, CA, October 15, 1993.

"Expanded Entry Vocabulary for the *Library of Congress Subject Headings*." U.S. Library of Congress, Washington, DC, October 22, 1993.

"Facet Analysis for Online Retrieval in the Humanities." American Society for Information Science Annual Conference, Columbus, OH, October 25, 1993.

"Humanities Researchers' Use of Online Searching in Their Scholarship." American Society for Information Science Annual Conference, Columbus, OH, October 26, 1993.

"What's New in Library Education and Recruitment." Metronet Libraries. UCLA, Los Angeles, CA, March 30, 1994.

"Chair's Report." Graduate School of Library and Information Science Annual Alumni Luncheon, UCLA, Los Angeles, CA, April 23, 1994.

Graduation Speaker, School of Library and Information Studies, University of California, Berkeley, CA, May 15, 1994.

"Overcoming Communication Barriers among Information System Designers, Librarians, and End Users." American Library Association Annual Conference, Miami Beach, FL, June 25, 1994.

"The Role of Library and Information Science Programs in Research Universities." American Library Association Annual Conference, Miami Beach, FL, June 27, 1994.

"Shared Interests of Our Merged Schools." Orientation Talk, Graduate School of Education and Information Studies, UCLA, Los Angeles, CA, September 28, 1994.

"Future of Academic Indexing." American Society of Indexers, Costa Mesa, CA, October 8, 1994.

"Improving the Role of Classification in the Future." Thirty-Sixth Allerton Institute: New Roles for Classification in Libraries and Information Networks. Monticello, IL, October 25, 1994.

"Campus Visibility in Times of Change." Association for Library and Information Science Education, Council of Deans, Philadelphia, PA, February 3, 1995.

"Chair's Report." UCLA LIS Alumni Association, Annual Luncheon, UCLA, Los Angeles, CA, April 1, 1995.

Reactor, "Unlimiting the Library." Librarians Association of the University of California, Santa Barbara, Santa Barbara, CA, May 5, 1995.

"Reflections on My Life in Information Science." Annual Awards Dinner of the Los Angeles Chapter of the American Society for Information Science, October 25, 1995.

"Learning about Your Users' Information Needs." Guam Library Association, Guam Island, November 4, 1995.

"Training Librarians to Teach Electronic Resources." Southern California Electronic Library Consortium Annual Conference. Center for Scholarly Technology, University of Southern California, Los Angeles, CA, May 3, 1996.

"Human, Database, and Domain Factors in Content Indexing and Access to Digital Libraries and the Internet." Graduate School of Library and Information Science, University of Illinois, Champaign-Urbana, IL, October 25, 1996.

"Organizing Information in Digital Libraries." Libraries, People and Change: A Research Forum on Digital Libraries, Thirty-Eighth Allerton Institute, Monticello, IL, October 28, 1996.

"Information and Technology Literacy." Colloquium Series: Scholarship in a New Media Environment, Office of Instructional Development, UCLA, Los Angeles, CA, February 21, 1997.

"Shaping a Research Idea into a Workable Project." Association for Library and Information Science Education Annual Conference. New Orleans, LA, January 7, 1998.

"Information Science: The Invisible Substrate." Dept. of Information Studies Colloquium, UCLA, Los Angeles, CA, January 22, 1998.

"The Unique Knowledge Domain That Is Classification for Retrieving Information." American Society for Information Science Annual Conference. Pittsburgh, PA, October 26, 1998.

Reactor, "Major Pressures in Higher Education Today." Plenary Session, Association for Library and Information Science Education Annual Conference. Philadelphia, PA, January 27, 1999.

"Educating for User-Centered Design of Information Systems and Services." Association for Library and Information Science Education Annual Conference. Philadelphia, PA, January 28, 1999.

"Information Curriculum for the 21st Century." Congress on Professional Education, American Library Association, Washington, DC, May 1, 1999.

Reactor, "Information Use in the Professions." American Society for Information Science Midyear Meeting. Pasadena, CA, May 25, 1999.

"Designing Information Systems for the Particular Needs of Humanities Scholars." Museums, Libraries, and Archives: Summer Institute for Knowledge Sharing, Getty Information Institute and UCLA, Los Angeles, CA, August 5, 1999.

All-Nordic special graduate course on Cognition, Learning, and Information Seeking, Department of Information Studies, Åbo Akademi University, Turku, Finland, August 31–September 3, 1999. Three lectures:

"Supporting and Thwarting Information Seeking." August 31, 1999.

"Necessary (and Surprising) Learning for Information Seeking." September 1, 1999.

"Embedding Understanding of the User in Information System Design." September 1, 1999.

"The Invisible Substrate of Information Science." Royal School of Library and Information Science, Copenhagen, Denmark, September 10, 1999.

Keynote address. American Society for Information Science Pacific Northwest Chapter Fall 1999 meeting. Redmond, WA, September 17, 1999.

"The Role of the Ph.D. in a Professional Field." Srygley Lecture (endowed). School of Information Studies, Florida State University, Tallahassee, FL, October 16, 1999. http://www.gseis.ucla.edu/faculty/bates/.

"The Influence of Discipline/Domain on Information Seeking Behavior." Panel member. American Society for Information Science Annual Conference. Washington, DC, November 2, 1999.

"New Interface Technologies, Old Search Interaction Models: What Next?" Panel member, American Society for Information Science Annual Conference. Washington, DC, November 3, 1999.

"The Invisible Substrate of Information Science." ALA Student Chapter, San Jose State University, San Jose, CA, March 7, 2000.

"Information Studies at UCLA: The State of the Art" Panel. Librarians' Association of the University of California, UCLA Chapter, UCLA, Los Angeles, CA, March 10, 2000.

"Librarians' Image: Shaping Our Own Future," California Academic and Research Libraries conference, Long Beach, CA, October 6, 2000.

"The Biological and Social Consequences of Information Seeking," Series: Explorations of Emerging Issues. Dominican University Graduate School of Library and Information Science, and North Suburban Library System, Wheeling, IL, October 11, 2000.

"The Biological and Social Consequences of Information Seeking," Lazerow Lecture (endowed), School of Library and Information Science, University of Kentucky, Lexington, KY, October 13, 2000.

"The Relation of Information System Design to Information Literacy," New Technologies, New Literacies: The Pacific Bell/UCLA Summit. UCLA, Los Angeles, CA, October 21, 2000.

"Tenure: 'The Rules,'" Association of Library and Information Science Education 2001 Conference. Washington, DC, January 10, 2001.

"Natural and Represented Information." Information Studies Departmental Colloquium. UCLA, Los Angeles, CA, October 11, 2001.

"The State of User Studies." American Society for Information Science and Technology. 2001 Conference. Washington, DC, November 5, 2001.

"United States/Canadian Practices in Library and Information Science Education." Portuguese Culture Ministry, Lisbon, Portugal, September 9, 2002.

Keynote address. Information Seeking in Context Conference. Universidade Lusiada, Lisbon, Portugal, September 11, 2002.

"On Documents and Antelopes." School of Information, University of Michigan, Ann Arbor, MI, September 18, 2002.

"Toward an Integrated Model of Information Seeking and Searching." Michigan Chapter, Association for Computing Machinery, Special Interest Group on Human-Computer Interaction. Ann Arbor, MI, September 18, 2002.

"Conceptualizing Users and Uses." American Society for Information Science and Technology Annual Conference. Philadelphia, PA, November 18, 2002.

"Conceptions of Information as Evidence." American Society for Information Science and Technology Annual Conference. Philadelphia, PA, November 20, 2002.

"Science Users and Information Literacy." Science and Engineering Librarians (SEAL), California State University, Pomona, CA, April 25, 2003.

NOTE: *Most speeches given after 2003 were not recorded.*

"The nature of the information professions." University of Wisconsin-Milwaukee School of Information Studies, September 11, 2012.

"The nature of the information professions." University of Toronto iSchool Colloquium Series. September 27, 2012.

DEPARTMENTAL AND UNIVERSITY COMMITTEE SERVICE

Within College of Library and Information Services, University of Maryland

Doctoral Committee (committee that supervised doctoral program), member, 1972–1973.

Curriculum Committee, member, 1972–1973.

Doctoral Committee, Chair, 1973–1974.

Curriculums Programs, and Research Committee, Chair, 1973–1974.

Faculty Search Committee, member, 1974.

College Collegium (governing body of College), Chair, 1975–1976.

Ad Hoc Committee on Extension Courses, member, Spring 1976.

Within Division of Human and Community Resources, University of Maryland

(*The University of Maryland was divided into five divisions at that time, each headed by a provost.*)

Service Study Committee, member, Fall 1972.

Ad Hoc Affirmative Action Committee, member, Fall 1975.

Representative, as Chairperson of College Collegium, to "Faculty Assembly Chairpersons" meetings on regular basis with Provost of Division, 1975–1976.

University-wide, University of Maryland

Adjunct Committee on Professional Ethics and Faculty Conduct, member, 1972–1973.

Adjunct Committee on Research, member, 1973–1974.

Graduate Faculty, Associate Member, 1973–1975. Regular Member, 1975–1976.

To the Maryland Community

Representative for the College of Library and Information Services to the Technical Committee on Library Services of the Baltimore Regional Planning Council, 1975–1976.

Within School of Librarianship, University of Washington

Task Force on Team Teaching, Chair, 1976–1977.
Curriculum Task Force, member, 1976–1977.
Space Needs Task Force, member, May–June 1978.
Awards, Scholarships, and Loans Committee, Chair, 1979–1981.

University-wide, University of Washington

Graduate Faculty, member, 1977–1981.
University Faculty Senate, Senator, 1978–1980 term.

Within Graduate School of Library and Information Science, UCLA

Chair, Ph.D. Committee (Supervises Ph.D. program—admissions, examinations, policy, etc.), 1982–1984.
Library Committee, Chair, 1985–1986.
Comprehensive Examination Committee, Chair, Spring 1986; Winter 1987; Fall 1987; Fall 1988–Spring 1990; Fall 1993–Spring 1994.
United Way Campaign Chair, 1988–1990.
Awards Committee, member, 1989–1990.
Search Committee for Recruitment and Placement Officer for GSLIS, member, 1990.
Search Committee for three Faculty positions, Chair, 1990.
Core Curriculum Committee, Chair Pro Tem, Fall 1990.
Admissions Committee, Member, 1990–1991, Chair, 1991–1992.
Various sub-committees on curriculum revision, member, 1991–1992.
Preparation of internal departmental report on national standing of GSLIS faculty, Fall 1992.
Preparation of several drafts of GSLIS Self Review document for Graduate Council Review, Spring 1993.
Interests and Welfare Committee, Chair, 1992–1993.
Awards Committee, member, 1992–1993.
Associate Dean, 1992–1993.
Search Committee for Faculty position, Chair, 1993–1994.
Ph.D. Program Committee, member, 1993–1994.

Within Graduate School of Education and Information Studies

GSE&IS Executive Committee, member, 1994–95.
Task Force on Education/LIS Programs, member, 1994–95.
Revenue-Centered Management Task Force, member, 1995.
Faculty Executive Council, Member-at-Large, 2000–01.

Within Department of Information Studies

Chair of Department, 1993–1995.
Ph.D. Program Committee, Chair, 1994–95, member, 1995–96.
Search Committee for two Faculty positions, Chair, 1994–95.
Professional Programs Committee, member, 1995–96, Chair, 1997–2001.
Admissions, Awards, and Recruitment Committee, member, 1996–97.
Undergraduate Education Ad Hoc Committee, Chair, 1996–97.

Continuing Education Ad Hoc Committee, member, 1996–97.

Search Committee for Faculty position, Chair, 1997–98.

All-curriculum Ad Hoc Committee, member, 1997–98.

Professional Programs Committee, member and Chair, 1998–2001.

Information Studies Department Diversity Council Curriculum Committee, Chair, 1999–2000.

Doctoral Program Committee, member, 2001–02.

Instructional Services Committee, Chair, 2002–03.

Faculty Secretary, 1999–2003.

University-wide, UCLA

Academic Senate Committee on Committees, member, 1987–90.

Faculty Executive Chair, Graduate School of Library and Information Science, 1993–94.

UCLA Legislative Assembly Representative, 1993–94.

Computer Science Department Chancellor's Review Committee, member, 2000–01.

—Revised April 14, 2016

APPENDIX B

Content list of Volumes I, II, and III

INFORMATION AND THE INFORMATION PROFESSIONS: SELECTED WORKS OF MARCIA J. BATES, VOL. I

ISBN 978-0-9817584-1-1

All entries are by Bates unless otherwise noted.

INFORMATION SEARCHING THEORY AND PRACTICE: SELECTED WORKS OF MARCIA J. BATES, VOL. II

ISBN 978-0-9817584-2-8

All entries are by Bates unless otherwise noted.

INFORMATION USERS AND INFORMATION SYSTEM DESIGN: SELECTED WORKS OF MARCIA J. BATES, VOL. III

ISBN 978-0-9817584-3-5

All entries are by Bates unless otherwise noted.

Index

library science/information science degrees,
doctoral studies, *continued*
    deciding whether to get a doctorate,
      285–86
    nature of, 287–88
    program quality, 290–91
    to supplement the MLS, 286–87
    where to apply, 289
library science student, becoming a, 312–18
literature domains, *Journal of ASIS* articles on,
  178–79
literature reviews, 236–38
Llewellyn C. Puppybreath III, 108, 332
logons, 85, 86
Losee, Robert, 75–76
Luhn, H. P., 153

Maack, Mary Niles, 199, 200, 252, 265
Machlup, Fritz, 250–52
MacKay, Donald, 85–86
Madden, A. D., 5, 73–74
management of information organizations
  dimension of professional practice, 301–2
Manzari, L., 251
Markey, K., 102
Maron, M. E. (Bill), 315, 317, 318, 326
matter and energy. *See* patterns of organization
  of matter and energy
Meadow, Charles, 325
meaning
  assigning/ascribing, 17, 57, 58
  information and, 27–28, 72–73
    "meaning-free" vs. "meaning-full" sense
      of information, 57
mechanics, the world of, 70
medical informatives, 143–44
medicine, recorded information in, 134, 136
memorization, 132. *See also* "external memory"
memory, heritage disciplines, and knowledge,
  137, 137f
Merritt, Leroy, 313
meta-disciplines, 126–28, 127f, 144
  vs. content disciplines, 110, 142–44
  defined, 110, 126
  See also meta-fields
meta-fields, 110, 126, 278–79. *See also*
  meta-disciplines
meta-language, 153, 167
metadata, 241
metatheoretical positions, 6–7, 257, 258
  defined, 6
  See also social science metatheories in
    information studies
metatheories, 262
  categorization, 265
  defined, 25, 257
  historical perspective on, 262
  in library and information science (LIS), 262,
    264–68
  models, theories, and, 256–69
  nature of, 257
  paradigms and, 257
  proliferation of, in social sciences, 262
  sources of, 262–63
metrical information content, 85

metrons, 85, 86
Miller, George, 73, 326
misinformation and misinforming, 79
models
  defined, 257
  See also under metatheories
Montgomery, Christine, 327
Mooers, Calvin, 154
Morgenstern, Oskar, 73, 116
Mote, L. J. B., 99, 101
museum studies, 61, 133, 137, 143, 145, 204–5
museums, 61, 62, 139, 143, 147
museums, storage in, 62

narrative (writing)
  describes a situation, 246–47
  vs. expository writing, 246
natural information, 42, 54
  defined, 42, 43t
Nauta, Doede, 74–75
neural-cultural flow line of information, 44–46,
  62
neural-cultural information, 44, 47, 89–90,
  131–33
  defined, 43t
  types of, 48–50, 133
  See also phenotype
news, collective impact of. *See* information
  whomps
Nilan, M., 321–22
nomothetic, defined, 263
nomothetic approach/worldview, 80, 83
nomothetic-idiographic contrast, 263–64
nouns, indexing terms, 245

object of study, 61
online searching, 98–102
  sources of variation in studies of, 101–2
online systems, 240, 241
organization (in life), levels of, 70
organization of information, 107
orienting strategies, 6, 263
orthogonal disciplines. *See* meta-disciplines
Otlet, Paul, 311

Packer, K. H., 99, 101
Paisley, William, 110, 267, 320, 321
paradigms, 29, 108, 232, 262. *See also under*
  information science(s)
Parker, Edwin, 7, 153
patterns (of organization), 28, 40–42
  definitions and senses of "pattern," 8, 39–41
  of everything, information as, 22, 41–42
  first- vs. second-order, 8, 39–40
patterns of organization of matter and energy,
  information as, xvii, 7–8, 16, 18, 23, 27, 28,
  30, 88
  underlying assumptions, 9
  whose pattern it is, 9–10, 15–20
    the evolutionary context, 10–15 (*see also*
      evolutionary psychology)
Ph.D.
  role in professional fields, 276–77, 283, 284
  See also library science/information science
    degrees

phenotype, 45, 47, 54, 62, 89
  defined, 43t
phenotype, *continued*
  embodied, 44, 45, 47, 62, 63
  extended, 45
philosophical-analytic approach (metatheory), 266
philosophical analysis, 234
philosophical approaches to information, 5–6, 13
physical approach to information transfer (metatheory), 267–68
physical processes, role of information in, 71
physics, 70, 93
Plans and the Structure of Behavior (Miller), 117, 326
Poole, H., 259–61
Popper, Karl, 68, 83–85
  and the problem of Popper's worlds, 83–85
portraying vs. being, 112
postmodernism, 93, 121
Powell, Ronald, 243
Pratt, Allan D., 5, 76
prediction. *See* description, prediction, and explanation
Price, Derek J. de Solla, xx, 250
Prisoner's Dilemma, 116
processes, 110
  information and, 75
Production and Distribution of Knowledge in the United States, The (Machlup), 250–51
"Productivity Impacts of Libraries and Information Services" (Koenig & Manzari), 251
professions and professional knowledge, 277. *See also* information professions; Ph.D.
proto-theory, 258. *See also* models
psychoanalysis, 229–31
psycholinguistics, 117
psychology, 6. *See also* creative process in the information sciences; evolutionary psychology; social science metatheories in information studies
Public Knowledge: An Essay Concerning the Social Dimension of Science (Ziman), 84

quantitative, information as, 77
question and answering, the 55 percent rule in, 243–45

Ranganathan, S. R., 139, 142, 197–98
Rayward, W. B., 5
reading deeply, not just widely, 235
reality, internal and external, 86–87
reasoning, heuristics of, 234
recorded information, 84, 134
  defined, 43t, 51, 90, 133
  disciplines interested in, 134, 190
  embedded information and, 46, 47t, 52, 53, 60–63, 90, 133–34, 137, 139
  exosomatic flow line and, 46, 47t, 62–63, 90, 133
  exosomatic information and, 133
  expressed information and, 52, 53
  historical perspective on, 134, 148
  information science and, 62, 109–13, 118–20, 137

librarianship and, 120, 133
  in medicine, 134, 136 (*see also* medical informatives)
  vs. memorization, 132
  vs. more evanescent or ephemeral forms, 120
  moving through exosomatic flow line, 46, 47t
  overview and nature of, 51–52
  statistical properties of, 119
  types of, 62, 134
  universe of, 110, 113, 118, 119, 134, 280, 282
  See also exosomatic information
"Records Continuum Model" (McKemmish et al.), 197
records management, 143, 197
reference librarians and reference desks at libraries, 243–45
registration (cataloging), 139
relevance, 93, 234
  Journal of ASIS articles on, 160–61
Relevance: Communication and Cognition (Sperber & Wilson), 234
representation, 85, 107, 113
  defined, 44
represented information, 42, 44, 47, 89
  defined, 42, 43t, 47, 89
  See also representing (information)
representing (information), 112–13
  vs. being, 112–13
  See also represented information
research
  contextualization of, in literature reviews, 236
  nomothetic vs. idiographic approaches to, 263–64
  See also library and information science (LIS) research
research disciplines, spectrum of traditional, 125–26, 126f, 130, 130f
research metatheories, 6
research practices in sciences vs. humanities, 281
research traditions, drawing on a variety of, 232–34
residue lineage/information flow channel, 46, 47t, 53–54, 89, 90, 132f, 133, 134
retrieval. *See* information retrieval
Rosenberg, Victor, 153, 323–24

Salton, Gerard, xiii, 108, 154, 305, 317, 328
satisficing, 259–61
scatter fields, 99
  "high scatter," 99, 100
scholarly communication, *Journal of ASIS* articles on, 175–76
science, 84, 108, 262
  contrasted with humanities, 281, 282
  history of, 70, 71
  society and, 70
  See also *specific scientists*
Science Citation Index (SCI), 305–7, 309
science-oriented fields, 129. *See also* information science(s)
scientific communication, *Journal of ASIS* articles on, 175–76
search strategy, 101

search strategy and evaluation, *Journal of ASIS* articles on, 169–71
search tactics, information, 232
searching, interface design specific to, 241–43
selective dissemination of information (SDI), 99, 101
selective information content, 85–86
semiosis, 56–58
    defined, 56
semiotic definitions of information, 73–76
semiotics, 55–56, 59, 93
    defined, 55
sensemaking, 29, 259–60
service dimension of professional practice, 300–301
services and functions carried out by information professionals, 139
Shannon, Claude E., 5, 15, 23, 67–69, 71–73, 117, 326
Shoffner, Ralph, 313–15
signs and information, 55–56
situation, narrative writing describing a, 246–47
situation and event coding for indexing, 247, 247t
situations, 93, 246
Skinner, B. F., 326
social informetrics, 267
social processes, role of information in, 71
social science metatheories in information studies, 6–7, 24–27
social sciences and information science, 119, 121
socially embedded information, 81
socio-cognitive approach (metatheory), 267
sociological constructivist theory, 265
sociotechnical methodological stance of information science, 120, 278
Soergel, D., 98, 99
Spencer-Brown, G., 117
Sperber, Dan, 234
Starr, Kevin, 292
static structures, 70
Stearns, Charles, 153
Stevens, Mary E., 318
Stonier, T., 21
storage efficiency, 12–13
structural information content, 85
structures, recognition of underlying, 116–18
subject expertise, 113–15
subjective experience, 48. *See also* experienced information
Swank, Ray, 324, 327–28
syntactic structures, Chomsky's theory of, 117
systems theory, 116

Talja, Sanna, 25–27, 29, 30, 265, 266
Taube, Mortimer, 153

term-entry system/approach, 153–54
theory
    definitions, 256–58
    See also metatheories
thermostats, *See* homeostasis
Thompson, Frederick, 79–80
thoughts, 93
tool making dimension of professional practice, 298, 299
topic vs. event indexing, 246, 247t
trace information, 47, 53, 55, 90
    defined, 43t, 90
Turoff, Murray, 153
Two Kinds of Power: An Essay on Bibliographical Control (Wilson), 236–37, 324–25

understanding, 39
University of California at Berkeley (U.C. Berkeley) School of Library and Information Studies in 1960s, 311–12, 327–28
    becoming a master of library science (M.L.S.) student, 312–14
    going for the doctorate, 315–18
    and the social context of the times, 318–20
    women at, 312, 318–19, 323, 333
user-centered design approach (metatheory), 268
users, librarians' knowledge about, 297
utterances, 93

Valian, Virginia, 334
value seeking vs. information seeking, 239–40, 260
values of information science, 122
variable fields
    defined, 110
    vs. level fields, 110
verb indexing terms, 247–48
visitor studies, 139
von Neumann, John, 116

Wanger, J., 100
Watson, James D., 325
White, Howard, 234
Wiener, Norbert, 21, 67–68, 70–73, 116–17, 326
Wilson, Deirdre, 234
Wilson, Patrick, 236–37, 324–25, 328
women
    equality vs. inequality of, 318–19, 332–34
    in information science, xiii, xx, 310, 318, 323, 327, 332–36
    at U.C. Berkeley in 1960s, 312, 318–19, 323, 333
worlds, Popper's three, 83–85

Ziman, John, 84
Zipf, G. K., 260–61, 267

CPSIA information can be obtained
at www.ICGtesting.com
Printed in the USA
BVHW062236221221
624720BV00012B/241